Parasitology Research Monographs

Volume 15

Series Editor

Heinz Mehlhorn, Department of Parasitology, Heinrich Heine University, Düsseldorf, Germany

This book series "Parasitology Research Monographs" presents carefully refereed volumes on selected parasitological topics. Parasites have an increasing impact on animal and human health in the present times of globalization and global warming. Parasites may be agents of diseases and- often at the same time- vectors of other agents of disease such as viruses, bacteria, fungi, protozoa and/or worms. The growth in knowledge of parasitic physiology, cell structure, biotechnological and genetic approaches, ecology, therapeutic capabilities, vaccination, immunology, diagnosis, transmission pathways and many other aspects of parasitology is increasing dramatically, even in the face of the breakthroughs that have already been made. Reflecting these most recent achievements and the importance of parasites as a threat to human and animal health, the series' broad scope concentrates on particularly hot topics that were recently covered by review articles in the journal "Parasitology Research" or in other journals. These reviews offer compact but intense insights into the ongoing research and into the methods and technologies used to control parasites. The volumes in the series build on these topics, and the volume editors are well-known experts in their respective fields. Each volume offers 10 to 20 comprehensive reviews covering all relevant aspects of the topic in focus.

More information about this series at http://www.springer.com/series/8816

Kun Yang • Heinz Mehlhorn
Editors

Sino-African Cooperation for Schistosomiasis Control in Zanzibar

A Blueprint for Combating other Parasitic Diseases

Springer

Editors
Kun Yang
Jiangsu Institute of Parasitic Diseases
Wuxi, Jiangsu, China

Heinz Mehlhorn
Department of Parasitology
Heinrich Heine University
Düsseldorf, Germany

ISSN 2192-3671 ISSN 2192-368X (electronic)
Parasitology Research Monographs
ISBN 978-3-030-72164-0 ISBN 978-3-030-72165-7 (eBook)
https://doi.org/10.1007/978-3-030-72165-7

This Springer imprint is published by the registered company Springer Nature Switzerland AG.
The registered company address is: Gewerbestrasse 11, 6330 Cham, Switzerland

Preface

Schistosomiasis in China has been known since at least 2150 years killing millions of people. In 1955, the Chinese Chairman Mao Zedong called for the "elimination of schistosomiasis" and introduced a campaign for the prevention and treatment of schistosomiasis in China, where at least 12 million people and 1.2 million buffaloes had been diagnosed to be infected in an area of 14.8 billion square meters that had been proven to be contaminated with local vector snails. In total, at least 100 million people and at least the same amount of animals had been endangered in China at that time. The huge efforts of this project and the great success of protecting millions of Chinese inhabitants and their animals have been documented in 2019 in a book titled "Schistosomiasis Control in China" published by Zhongdao Wu, Yiwen Liu, and Heinz Mehlhorn in the series Parasitology Research Monographs.

Based on the knowledge obtained during these years of hard work in China, the Chinese government together with the WHO started projects to help people worldwide in the fight against schistosomiasis.

The present volume reports on the very hard, but very successful work that was done to reduce up to elimination the large numbers of severe human schistosomiasis on the islands of Zanzibar. This was only possible by delegations of large amounts of Chinese specialists and high financial support by China and the WHO. The present book offers insights into the control methods, which can be used as a blueprint in other regions of the world.

References

1. Mehlhorn H, Klimpel S (eds.) (2019) Parasite and disease spread by major rivers on earth. Parasitol Res Monographs, Vol. 12, Springer International.

2. Wu Z, Liu Y, Mehlhorn H (eds.) (2019) Schistosomiasis control in China. Parasitol Res Monographs, Vol. 11, Springer International.

3. Wu Z (2019) Epidemic status of schistosomiasis in Guangdong province. In: Wu Z, Liu Y, Mehlhorn H (eds.) (2019) Schistosomiasis control in China. Parasitol Res Monographs, Vol. 11, Springer International.

Düsseldorf, Germany Heinz Mehlhorn

Preface

Schistosomiasis is a global public health problem, it is endemic in 78 countries and regions in Asia, South America, the Middle East, and Africa, and 85% of people infected with schistosomiasis are located in sub-Saharan Africa. Based on China's successful experience and products of schistosomiasis control and elimination, the activities of participating in global public health governance and promoting the transfer of China's schistosomiasis control expertise and strategies to Africa may become a model for China–Africa cooperation in the field of public health. On May 21, 2014, China, the World Health Organization (WHO), and Zanzibar signed a memorandum of understanding (MOU) on cooperation of schistosomiasis control in Zanzibar. In 2017, the China–Zanzibar schistosomiasis control project was officially launched. In the past three years, 30 Chinese experts and African staffs jointly carried out on-site work to get a preliminary understanding of the local schistosomiasis epidemic, comprehensive control measures were implemented, and a local technical team was created. The human infection rate of schistosomiasis in the project stie dropped from 8.92% to 0.64%, achieving the goal of schistosomiasis elimination.

In order to further promote the effectiveness of the project, we organized the project staff to compile the "Sino-African Cooperation for Schistosomiasis Control in Zanzibar: A Blueprint for Combating Other Parasitic Diseases," which covers the global status of schistosomiasis control and elimination and the experiences and achievements of China–Zanzibar schistosomiasis project. This book can be used as a reference for local schistosomiasis control professionals in Africa, China staffs engaged in China-aid project of public health, and teachers and students of medical school. The publication of this book will make a positive contribution to promoting the elimination of schistosomiasis and will play an important role in the major strategy of my country's "Belt and Road" health cooperation.

The publication of this book has received strong support from the Ministry of Commerce of the People's Republic of China, National Health Commission, China International Development Cooperation Agency, the Chinese Embassy in Tanzania, the Chinese Consulate General in Zanzibar, the Chinese Center for Disease Control and Prevention, the World Health Organization, and Ministry of Health of Zanzibar. The publication of this book has been partially funded by the Jiangsu International Science and Technology Cooperation Project (BZ2020003).

Wuxi, China Kun Yang

Contents

Chapter 1
Schistosoma haematobium (Bladder Fluke): Life Cycle and Morphology

Heinz Mehlhorn

Abstract This species belongs to the unique group of trematodes, where male and females live as adults lifelong (up to 20 years) closely united together, whereby the male worm keeps the female hidden inside a furrow formed by the evolving lateral sides of its flattened body. This construction is also called "canalis gynaecophorus". It guarantees the survival and reproduction in a host even when only low numbers of this worm species are present in a host. The present book deals with the report of the successful eradication program in Zanzibar which was supported and directed by the Chinese government and executed by Chinese scientists.

Keywords Schistosomiasis · *Schistosoma haematobium* · Bladder fluke disease · Disease control · Prevention

1.1 Characteristics

1.1.1 Name of the parasite: Schistosoma haematobium

Greek: *schisis* = splitting; *soma* = body; *haima* = blood; *bios* = life. The name refers to the fact that the male and the female are closely connected, since the male bears the female in a fold (Fig. 1.2).

1.1.2 Geographic Distribution/Epidemiology

The so-called bladder fluke *Schistosoma haematobium* endangers the health of animals and humans in many regions of Africa (Fig. 1.1). It is found in North-Western to North-Eastern Africa, in the delta of the Nile up to its source in the

H. Mehlhorn (✉)
Parasitology, Heinrich Heine University, Düsseldorf, Germany
e-mail: mehlhorn@uni-duesseldorf.de

K. Yang, H. Mehlhorn (eds.), *Sino-African Cooperation for Schistosomiasis Control in Zanzibar*, Parasitology Research Monographs 15,
https://doi.org/10.1007/978-3-030-72165-7_1

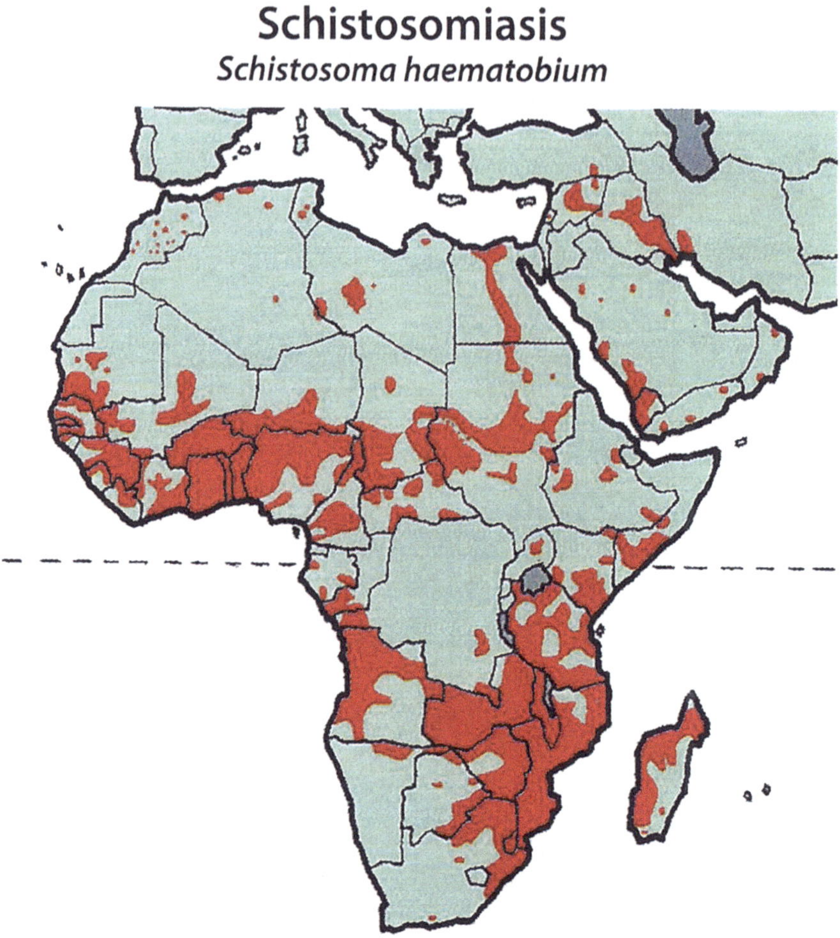

Fig. 1.1 Diagrammatic representation of the map of Africa and neighbouring countries and the island of Madagascar showing regions in red, where *Schistosoma haematobium* worms are parasitizing in humans (from WHO)

floodplains of Eastern and Western Africa, on the Western side of Madagascar, on the Arabian Peninsula, in Iraq as well as in some small spots in India.

1.1.3 Biology/Morphology

The adult ♀ and ♂ worms of *S. haematobium* (Trematoda, Platyhelminthes) stay together lifelong in a permanent copula. The 1.5 cm long, dorso-ventrally flattened male carries the in cross-sections circular appearing female (up to 2.6 cm long) in its ventral fold (canalis gynaecophorus). The couples stay in the veins of the urogenital system and bladder of their human hosts (Fig. 1.2). Every day each female lays up to

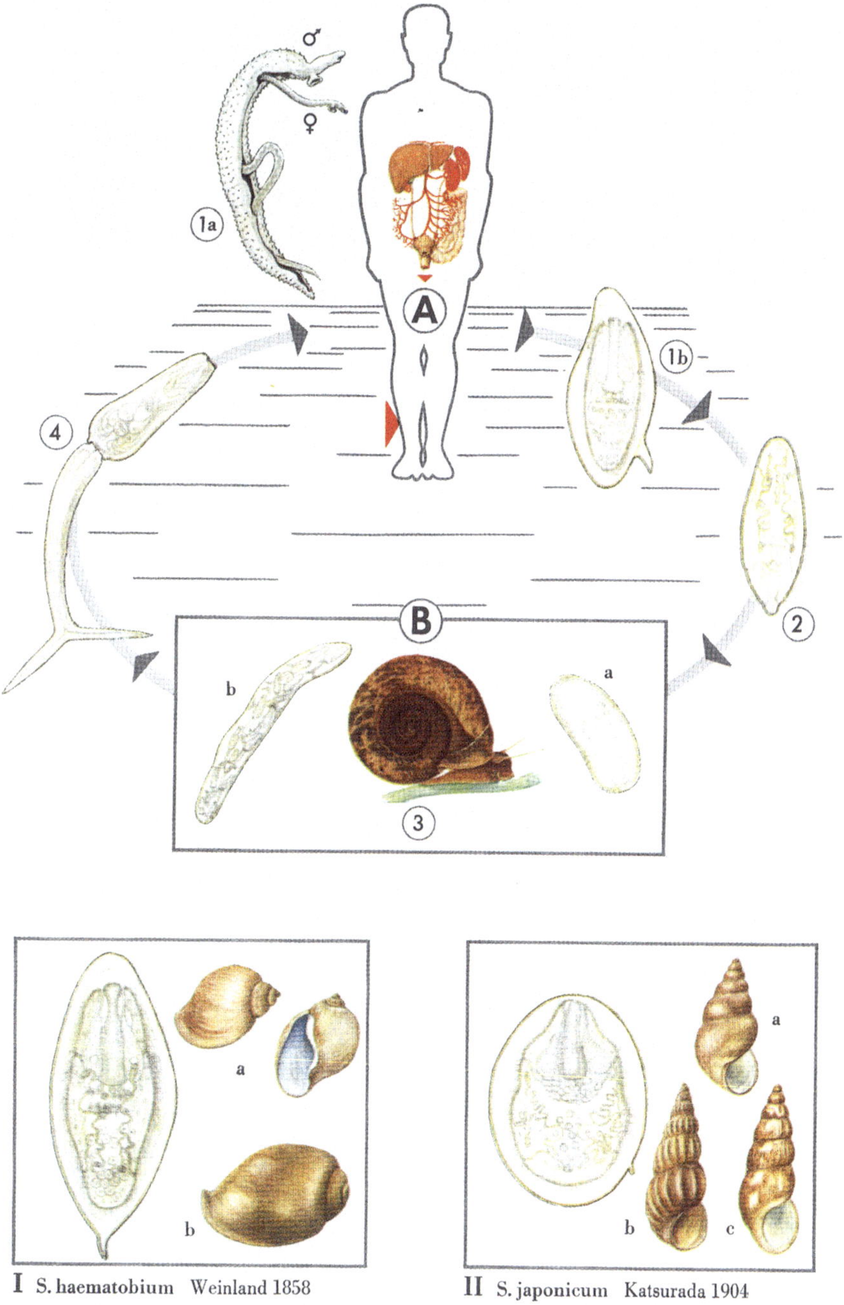

Fig. 1.2 Life cycle of *Schistosoma mansoni* and typical eggs and intermediate hosts of *S. haematobium* and *S. japonicum* (According to Piekarski 1987). **A** Final host: man; site of the worms: mesenteric vessels. **1a** Sexually mature pair of flukes of *S. mansoni*. **b** Mature egg of *S. mansoni* (lateral spine). **2** Miracidium. **B** Intermediate hosts: aquatic snails (Planorbidae, e.g., *Planorbis boissyi*, *Australorbis glabratus*). **3a** Sporocyst of the first order (mother sporocyst); **b**

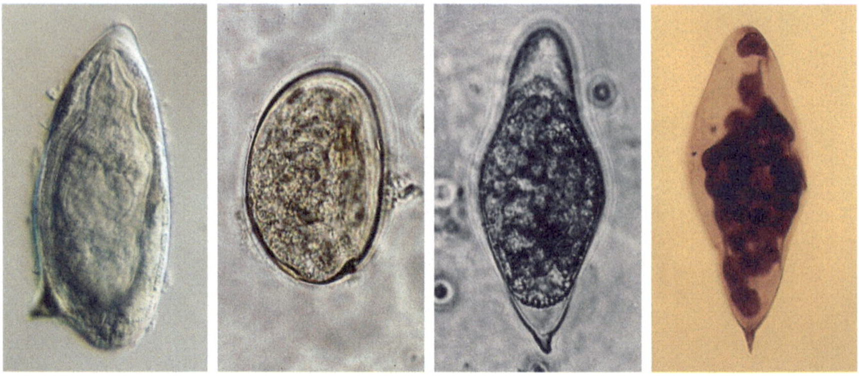

Fig. 1.3 Light micrographs of the eggs of four important *Schistosoma* species (from left: *S. mansoni, S. japonicum, S. intercalatum,* and *S. haematobium*)

300 eggs, which are quite big (110–170µm × 70µm) and possess a small characteristic terminal spike (Fig. 1.3). These eggs penetrate the bladder wall being supported in this process by proteases excreted by the contained miracidium larva. As soon as the egg reaches the mucosa of the bladder, it can be observed endoscopically inside a whitish pseudotuberculum of 1–2 mm in size. When the egg reaches freshwater being excreted within human urine, the miracidium (Fig. 1.4) hatches from the egg, starts swimming, searches and penetrates actively into water snails (Fig. 1.2). In the hepatopancreas of the snail (=intermediate host) a proliferation process via sporocysts takes place. The daughter sporocyst generation finally produces many larvae: the so-called bifurcated cercariae (Fig. 1.5). These stages leave the snail, swim around and have to find their hosts in the water within 24–48 hours. In the case that humans are working on paddy fields or bathing in the contaminated water, they may become infected (Fig. 1.2). While penetrating the skin of the human host, the cercariae throw off their "tail". After a rather short staying in the skin (as so-called **schistosomulum**-stage) and a following phase in the lung while adapting to the immune system of the host, the parasites finally take their way via blood vessels to the urogenital system. There the females reach sexual maturity within 5–6 weeks while staying in constant copulation. This couple is able to fix itself inside the human blood vessels by help of the two big suckers at the mouth and belly of the male bearing the female inside its so-called canalis gynaecophorus (Fig. 1.6).

Fig. 1.2 (continued) Sporocyst of the second order (daughter sporocyst). **4** Free cercaria ("forked-tail cercaria"). **I** *S. haematobium* (Bilharz 1852): egg with miracidium (terminal spine); shells of the intermediate hosts of the species (**a**) *Bulinus truncatus* (North Africa) and (**b**) *Bulinus globosus* (West Africa). **II** *S. japonicum* (Katsurada 1904): the egg containing a miracidium has a very small lateral spine. **A-C** Shells of intermediate hosts of the genera (**a**) *Schistosomorpha*, (**b**) *Oncomelania*, and (**c**) *Katayama*. Reprinted from Mehlhorn (2016a) with Permission from Springer Nature

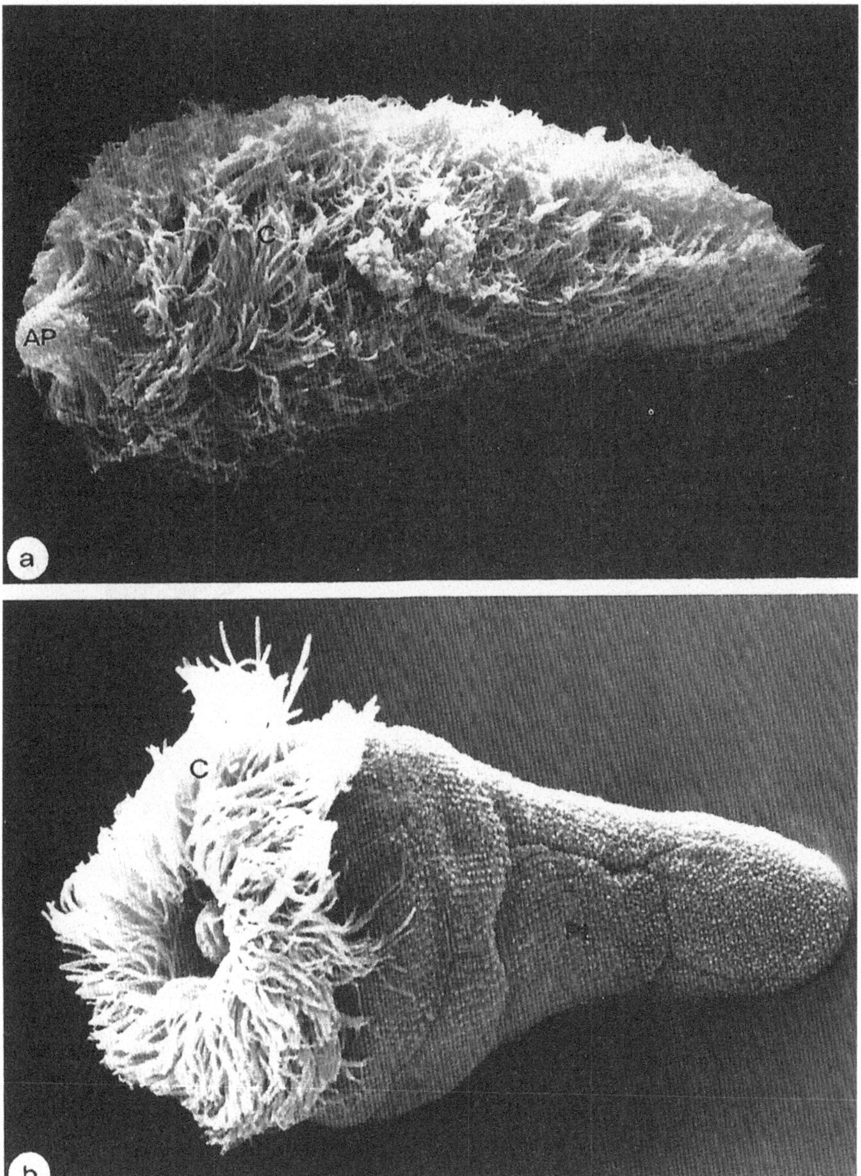

Fig. 1.4 Scanning electron micrographs of the typical miracidium larva with all cilia (**a**) and after removal of some cilia in order to show cell borders (**b**). Reprinted from Mehlhorn (2016a) with Permission from Springer Nature

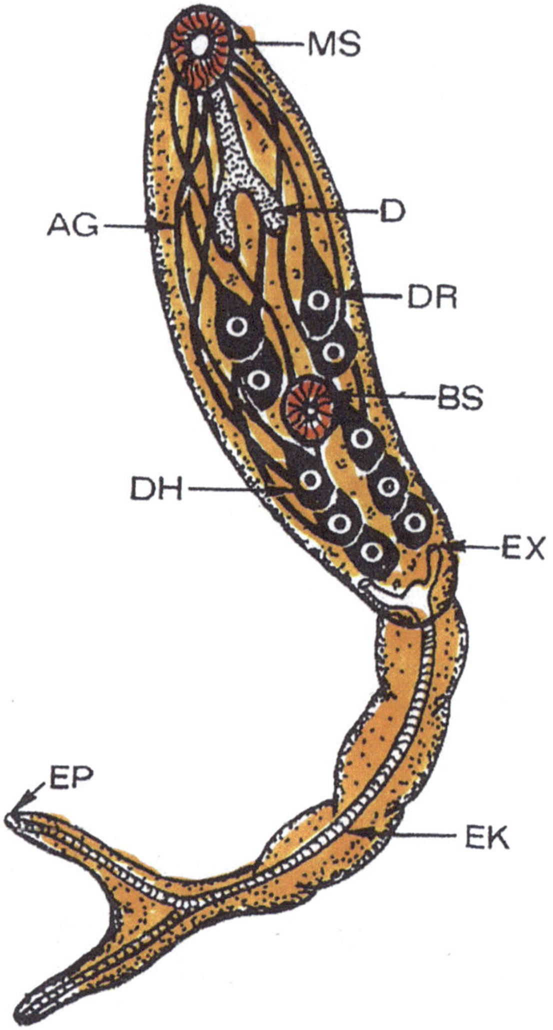

Fig. 1.5 Diagrammatic representation of a *Schistosoma* cercaria. **AG** Excretory ductule; **BS** Ventral sucker; **D** Intestine; **DH** Hind glands; **DR** Intestinal gland; **EK** Excretory channel; **EP** Excretory pores; **EX** Excretory channel; **MS** Mouth sucker

1.1.4 Symptoms of the Disease (Schistosomiasis of the Bladder, Bladder-Bilharziasis)

The severity of the symptoms of disease depends on the number of couples of parasites, which may live for about 25 years according to some records published in the literature. In general the following symptoms of disease are shown in infected humans:

(a) **Pruritus** at the entry point of the cercariae at the surface of the skin. Repeated infections may lead to a **cercarial dermatitis**.

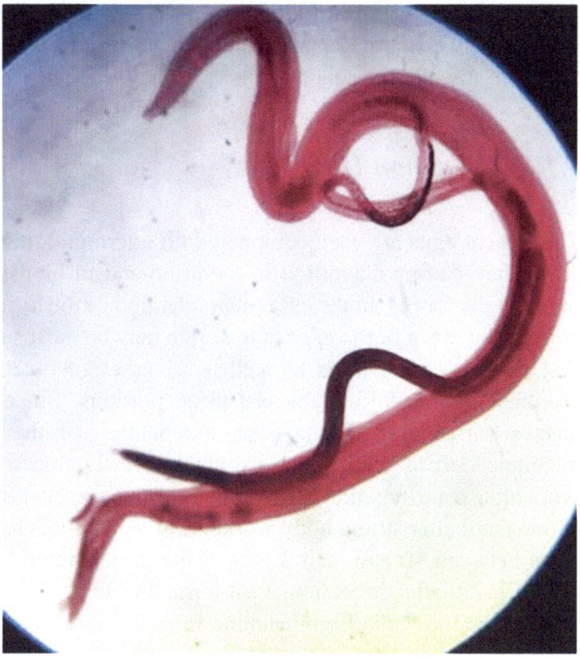

Fig. 1.6 Light micrograph of a stained couple of *Schistosoma* worms. Without staining the worms appear whitish. In this preparation the female is about to leave the ventral fold of the larger male. Inside hosts they always stay together

(b) In the case of a first infection of a human host, an **unspecific fever** occurs after 4–7 weeks as a general reaction of the host's sensibilization.
(c) Eggs (Fig. 1.2) are found in the urine starting in general about 2–3 months after infection. Blood traces (= **haematuria**) are observed especially in the last portion of the urine. Then patients complain a burning feeling in the urethra and urinary tissues.
(d) The further chronic course of the disease does not show blood and eggs all the time. Therefore, it is recommended to repeat the investigation of the urine, if the test was negative primarily in order to verify the absence of excreted eggs.
(e) During bladder inspection with a cytoscope egg granulomas may be observed (pseudotubercles) protruding at the inner side of the bladder wall. They appear as whitish nodules of 1–2 mm in diameter.
(f) *S. haematobium* infections initiate or increase the following complications:

- Secondary bacterial infections of the bladder may occur due to incomplete emptying. This causes granuloma-borne fibrosis and calcification. In this case the contraction ability of the bladder will be considerably reduced.
- Ulcers and also carcinoma of the bladder wall may occur due to permanent tissue irritations and secondary infections.
- Painful alterations of the genitalia (so-called elephantiasis by chronical oedema) have also been observed.

- A slight swelling of the liver can be found as well due to the formation and growth of granulomas, which—if appearing in large numbers—may induce life-threatening cancer.

1.1.5 Common Diagnosis

The typical eggs are each equipped with a terminal thorn (Table 1.1; Figs. 1.2 and 1.3). They can be diagnosed in the urine and in biopsies of the bladder, but only rarely in the faeces. In the case of an infection of the host with enormous amounts of worms, the proof of the eggs in the urine may be most successful in the spontaneous first portion of the urine as well as in cases of quantitative examination of the sediment obtained from the last urine portions. Since the excretion of the eggs always varies over the daytime, a repetition of the test procedure is strongly recommended as well as the eventual investigation of the sediment of the total urine after centrifugation. The number of the excreted eggs has its peak in daytime at noon and after strong body workout. In general it is recommended to examine the urine between 10 a.m. and 2 p.m. If the urine cannot be investigated immediately within 1–2 h after collection, 1 ml formalin (37%) /100 ml urine has to be added to prevent the miracidia from hatching from the egg.

Table 1.1 Morphological criteria of *Schistosoma* species of humans

Characteristics	*S. haematobium*	*S. japonicum*	*S. mansoni*	*S. intercalatum*
Male, outer surface[a]	Small tubercles provided with many hooks	No tubercles of the tegument	Large tubercles (bosses)	Tubercles with central hook-free spots
Female, ovary	Midbody region	Posterior half of the body	Anterior half of body	Past mid of body
Number of eggs in uterus	20–100	50–300	1–10 (often one)	10–20
Eggs shape	With terminal thorn	With tiny lateral thorn	With large lateral thorn	With terminal thorn
Egg size (μ)	110–170 × 40–70	70–100 × 55–65	115–180 × 45–70	140–230 × 50–80
Site of parasitizing	Veins of the urogenital system	Veins of the intestine	Veins of the intestine	Veins of the intestine
Intermediate hosts/snails[b]	*Bulinus* species	*Oncomelania* species	*Biomphalaria* species	*Bulinus* species
Reservoir hosts	Monkeys, hamsters	Mice, dogs, cats, pigs, cattle	Rats, mice, monkeys	Goats, sheep, rodents

[a]The ovoid eggs of *S. mekongi* have no thorn and measure only 50–60 × 30–50 μm, thus being considerably smaller than those of *S. japonicum*. In Thailand and Malaysia, *S. malayensis* has been described as own species, which has a similar life cycle like *S. japonicum*. *Truncula* snail species are intermediate hosts for *S. mekongi*, while *Robertsiella* species are intermediate host for *S. malayensis*

[b]Only very common species are listed here

If a low-graded infection is suspected, the investigation of larger amounts of urine after collection over 24 h should be done as follows: sedimentation of the urine for 1 h in a large glass vessel. In the case of the occurrence of huge amounts of sediments, it is necessary to wash them with cold 1% NaCl solution several times, followed by decantation, centrifugation of the remnant fluid (max. 2000 g for 2 min) before microscopical investigation of the sediment. Smaller amounts of urine (e.g. the last 10–50 ml portion of a urine from noon) can be filtered. Filtering is done by help of a polycarbonate filter (nylon or paper can be used as well). The filtered material is transferred onto a glass slide (after staining by Lugol's solution) and can be examined microscopically for eggs at a low magnification (40–100×). In the case of studying unfixed material, the vitality of the eggs can be decided by observation of the motions of the cilia of the miracidia by help of the so-called miracidium hatching test or by supravital coloration with 0.5% trypan blue. The proof of the eggs in biopsates can be documented with a squeezing preparation (see also intestinal schistosomiasis).

1.1.6 Immunodiagnosis

ELISA, PCR techniques should be used in cases of a suspected infection without successful proof of eggs. Eosinophilia occurs most frequently during the period of migration and growing up of many schistosomula. In probable cases of lung or brain schistosomiasis, the use of techniques like radiology, sonography and computer tomography is recommended. This is also needed in cases of undefined fibrosis.

The **disease** due to infections by *Schistosoma* species is today no longer called bilharziasis but **schistosomiasis** and has now obtained the rank of the second most socioeconomically devastating disease following closely behind human malaria. The different species of human schistosomiasis have the highest death rates among the top ten of the most important infectious diseases of humans, following closely behind HIV (AIDS) infections and tuberculosis, since the parasites are still today present in 71 countries having infected at least 200 million humans and also many millions of animal hosts, which may serve as reservoirs. This broad propagation makes it extremely difficult to eradicate this disease especially in warm and humid rural regions, especially in politically instable countries. Since several papers have shown that humans might also get infected by those *Schistosoma* species, which are mainly found in animals, the number of human infections may constantly increase considerably.

1.1.7 Pathway of Infection

The infection of final hosts occurs by percutaneous penetration of the anterior portion of the cercaria, which represents the later worm, while the long, bifurcated tail stays outside of the final host's skin (Fig. 1.5).

1.1.8 Prophylaxis

Avoidance of unprotected contacts with potentially cercariae-containing water from lakes, rivers, etc. in endemic regions. It is recommended to enter water of lakes or rivers only when wearing rubber gloves and boots.

1.1.9 Incubation Period

(a) General reactions after skin penetration of cercariae: 6 h until 2 days;
(b) First host reactions on the presence of pre-adult and adult worms:

S. mansoni 2–3 weeks.
S. japonicum 2–3 weeks
S. intercalatum 4–7 weeks
S. haematobium 2–20 months

1.1.10 Prepatent Period

The data of this period describe the needed time until eggs occur first in faeces and in urine of human hosts:
S. mansoni 4–7 weeks
S. japonicum 4–5 weeks
S. intercalatum 6–8 weeks
S. haematobium 8–12 weeks

1.1.11 Patency

5–25 years (depending on the species), if hosts had not been treated.

1.1.12 Therapy

The drug of choice (according to WHO) is **praziquantel**, which acts successfully against all human *Schistosoma* species. The recommended dosage is 3 × 20 mg/kg bodyweight per day on 3 days. **Oxamniquine** (1 × 15 mg/kg bodyweight) acts significantly only against *S. mansoni*. In cases of the so-called **Katayama syndrome**, it is needed to give additionally corticosteroids. In cases of hypertension symptoms combined with varioles along the oesophagus should be ameliorated by the help of endoscopic sclerosis.

1.1.13 Genome of Schistosoma *Species*

- Nuclear genome size: ≥270 Mb
- Chromosomes: eight pairs (=seven pairs autosomes, one pair sex chromosomes)
- Male ZZ
- Female ZW (= heterogametic)
- Gene copies: ~14,000
- Repetitive content: 40–60%
- Mitochondrial genome: 14,500 nucleotides

Further Reading

Bamgbola OF (2014) Urinary schistosomiasis. Pediatr Nephrol 29:2113–2120

Bilharz T (1852) Ein Beitrag zur Helminthographis humana (a contribution to the distribution of human helminths). Z Wiss Zool 4:53–76

Cioli D, Pica-Mattoccia L, Basso A, Guidi A (2014) Schistosomiasis control: praziquantel forever? Mol Biochem Parasitol 195:23–29

Enigk K (1986) History of helminthology in German speaking countries. G. Fischer, Stuttgart

Figueiredo JC et al (2015) Prostate adenocarcinoma associated with prostatic infection due to *Schistosoma haematobium*. Parasitol Res 114:351–358

Gönnert R, Andrews P (1977) Praziquantel, a new broad-spectrum antischistosomal agent. Parasitol Res 52:129150

Gower CM, Vince L, Webster JP (2017) Should we be treating animal schistosomiasis in Africa? The need for a one health economic evaluation of schistosomiasis control in people and their livestock. Trans R Soc Trop Med Hyg 111:244–247

Gray DJ et al (2014) A multicomponent integrated approach for the elimination of schistosomiasis: a review. Parasit Vect 8:156–166

Grys PC (2016) Schistosomiasis control in China: diagnostics and control strategies leading to success (challenges in public health). Peter Lang GmbH, Internationaler Verlag der Wissenschaften, Frankfurt/Main

Katsurada F (1904) Determination of the cause of a new parasite disease seen in Yamanashi and other prefectures. Tokyo Iji Shinshi 1371:13–32

Lai YS et al (2015) Spatial distribution of schistosomiasis and treatment needs in sub-Saharan Africa: a systematic review and geostatistical analysis. Lancet Infect Dis 15:927–940

Leger E, Webster JP (2017) Hybridizations within the genus *Schistosoma*: implications for evolution, epidemiology and control. Parasitology 144:65–80

Liu R, Dong HF, Guo Y, Zhao QP, Jiang MS (2011) Efficacy of praziquantel and artemisinin derivatives for the treatment and prevention of human schistosomiasis: a systematic review and meta-analysis. Parasit Vect 4:201

Mazingo HD et al (2013) Epidemiology and interactions of human immunodeficiency virus-1 and *Schistosoma* in sub-Saharan Africa. Infect Dis Poverty 2:2

McCreesh N, Booth M (2013) Challenges in predicting the effects of climate change in *Schistosoma mansoni* and *S. haematobium* transmission potential. Trends Parasitol 29:548–557

Mehlhorn H (2016a) *Schistosoma* species. In: Mehlhorn H (ed) Encyclopedia of parasitology, 4th edn. Springer, Berlin, Heidelberg. https://doi.org/10.1007/978-3-662-43978-4_2822

Mehlhorn H (2016b) Human parasites: diagnosis, treatment, prevention. Springer International Publishing, Cham

Mehlhorn H (2016c) Animal parasites: diagnosis, treatment, prevention. Springer International Publishing, Cham

Mehlhorn H, Becker B, Andrews P, Frenkel JK (1981) In-vivo and in-vitro experiments on the effects of praziquantel on *Schistosoma mansoni*. A light and electron microscope study. Drug Res 31:544–554

Mehlhorn H, Frenkel JK, Andrews P, Thomas H (1982) Light and electron microscopical studies on *Schistosoma mansoni* granulomas following treatment with praziquantel. Trop Med Parasitol 33:229–239

Mehlhorn H, Frenkel JK, Bunnag G, Rim A (1983) Ultrastructural investigations on the effects of praziquantel on human trematodes from Asia: *Clonorchis sinensis, Metagonimus yokogawai, Opisthorchis viverinii, Paragonimus westermani* and *Schistosoma haematobium*. Drug Res 33:91–98

Othman AA, Soliman RH (2015) Schistosomiasis in Egypt: a never-ending story? Acta Trop 148:179–190

Panic G, Keiser J (2018) Acting beyond 2020: better characterization of praziquantel and promising antischistosomal leads. Curr Opin Pharmacol 42:27–33

Piekarski G (1987) Medical parasitology in plates. Springer, Heidelberg

Price A et al (2015) Are health education interventions effective for the control and prevention of urogenital schistosomiasis in sub-Saharan Africa? A systematical review. Trans Roy Soc Trop Med Hyg 109:239–244

Shimada M, Kirinoki M, Shimizu K, Kato-Hayashi N, Chigusa Y, Kitikoon V, Pongsasakulchoti P, Matsuda H (2010) Characteristics of granuloma formation and liver fibrosis in schistosomiasis: a morphological comparison between *Schistosoma mekongi* and *S. haematobium*. Parasitology 137:1781–1789

Steinmann P, Keiser J, Bos R, Tanner M, Utzinger J (2006) Schistosomiasis and water resources development: systematic review, meta-analysis and estimate of people risk. Lancet Infect Dis 6:411–425

Tchuem Tchuenté LA, Rollinson D, Stothard JR, Molyneux D (2017) Moving from control to elimination of schistosomiasis in sub-Saharan Africa: time to change and adapt strategies. Infect Dis Poverty 6:42

Wang W, Wang L, Liang YS (2012) Susceptibility or resistance of praziquantel in human schistosomiasis. Parasitol Res 111:1871–1877

Wu Z (2016) Schistosomes. In: Mehlhorn H (ed) Encyclopedia of parasitology, 4th edn. Springer, New York

Wu Z, Lin Y, Mehlhorn H (eds) (2019) Schistosomiasis control in China. The successful example of the Jianpxi Province. Parasitology Research Monographs, vol 11. Springer Nature, Switzerland

Xu J, Xu JF, Li SZ, Zhang LJ, Wang Q, Zhu HH, Zhou XN (2015) Integrated control programs for schistosomiasis and other helminth infections in P.R.China. Acta Trop 141:332–341

Zheng H, Zhang LJ, Zhu R, Xu J, Li SZ, Guo JG, Xiao N, Zhou XN (2012) Schistosomiasis situation in People's Republic of China in 2011. Zhongguo Xue Xi Chong Bing Fang Zhi Za Zhi 24:621–626

Zwang I, Olliaro PL (2014) Clinical efficacy and tolerability of praziquantel for intestinal and urinary schistosomiasis. PLoS Negl Trop Dis 8(11):e3286

Chapter 2
Praziquantel: A Broad-Spectrum Drug Acting against Trematodes and Cestodes Parasitizing Humans and Animals

Heinz Mehlhorn

Abstract The chemical compound praziquantel has been developed by Bayer company in Leverkusen, Germany, is now also produced in China and sold worldwide as treatment of human and animal diseases due to infections with tapeworms and trematodes. It was used successfully by Chinese scientists for elimination of human and animal schistosomiasis in China and Zanzibar as well as in many other countries with endemics or returning tourists.

Keywords Praziquantel · Schistosomiasis control · Tapeworm and trematode control

2.1 Historical Background

Schistosomiasis (also known as bilharziosis honoring its discoverer—the German physician Theodor Bilharz (1825–1862)) belongs today to the most important Neglected Tropical diseases (NTD) affording even today the death of millions of humans and also of huge numbers of animals, that are grown up to deliver food for millions of persons in warm countries around the globe (Mehlhorn et al. 2014). Thus it was necessary to control spreading of this disease, especially in those countries with huge populations, which paid every year a high human death toll besides enormous losses among strongly needed farm animals (see Figs. 2.1, 2.2, 2.3, and 2.4).

Tapeworms—especially the larval stages (hydatids) are known since long. Already Aristophanes (444–380 BC) and Hippocrates (460–370 BC) described these stages in humans and animals. However, detailed research started only in the nineteenth century especially by German scientists like Goeze (1717–1786 AD), von Siebold as well as Küchenmeister (1821–1890) and Leuckart (1822–1898 AD), who

H. Mehlhorn (✉)
Heinrich Heine University, Düsseldorf, Germany
e-mail: mehlhorn@uni-duesseldorf.de

K. Yang, H. Mehlhorn (eds.), *Sino-African Cooperation for Schistosomiasis Control in Zanzibar*, Parasitology Research Monographs 15,
https://doi.org/10.1007/978-3-030-72165-7_2

Fig. 2.1 Praziquantel formula

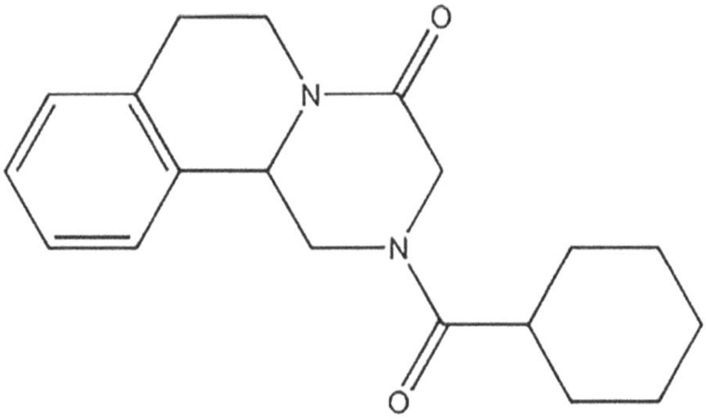

both described in the years 1855 and 1861 the full life cycles of the species *Taenia solium* (in pigs and humans) and *T. saginata*. The adults of both species live in the intestines of humans, while the larvae (cysticerci) of both species parasitize in muscles or other tissues of the intermediate hosts pigs, respectively, cattle. In contrast to *T. saginata*, where the cysticercus stage parasitizes only in cattle, the eggs of *T. solium* may induce in humans a cysticercus stage, too, thus becoming a potential intermediate host, where such stages will be found in muscles as well as in the brain. Especially in South America many human cases are reported (see Figs. 2.5, 2.6, 2.7, and 2.8).

2.2 Recent Control Measurements

2.2.1 History

The scientists of the German company E. Merck at Darmstadt synthesized a large number of pyrazino isoquinoline compounds when looking for potential tranquilizers. In an agreement between the companies Merck in Darmstadt and Bayer at Wuppertal and Leverkusen it was agreed that Bayer was allowed to start around 1970 a study dealing with a large series of these isoquinoline compounds for their activities as antihelminthic drugs. This intense search was done by the group of Prof. Gönnert (Wuppertal) and finally led to the finding of a very active compound called EMBAY 8440 which finally was named **praziquantel**. This compound showed very high activities against practically all very important species of trematodes and cestodes parasitizing farm animals and humans. These first breakthrough results had been obtained in animal tests and were published by Professor Gönnert's group at Bayer (Thomas and Gönnert 1977; Gönnert and Andrews 1977). First tests on human volunteers followed in the year 1978 by Leopold et al. and by Katz et al. (1979) treating humans who had been infected with *Schistosoma mansoni*. Later followed tests by Davis et al. (1979) treating people suffering from *S. haematobium* infections and tests by Ishizaki et al. (1979) treating persons being infected with stages of *Schistosoma japonicum*.

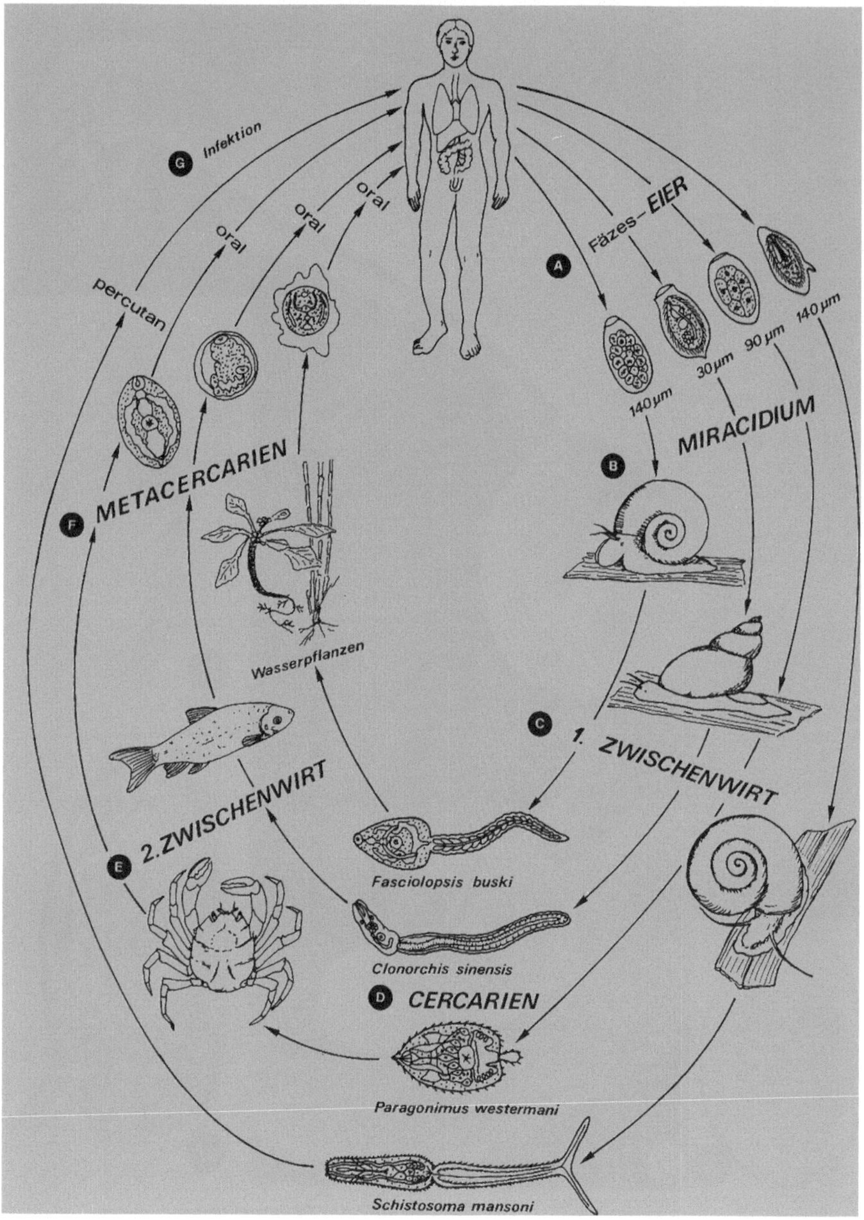

Fig. 2.2 Diagrammatic representation of the life cycle stages of four important trematode species infecting humans. Eier = eggs; Faezes = feces; Zwischenwirt = intermediate host; Cercarien = cercariae; Infektion = infection. Reprinted from Mehlhorn (2016b) with permission from Springer Nature

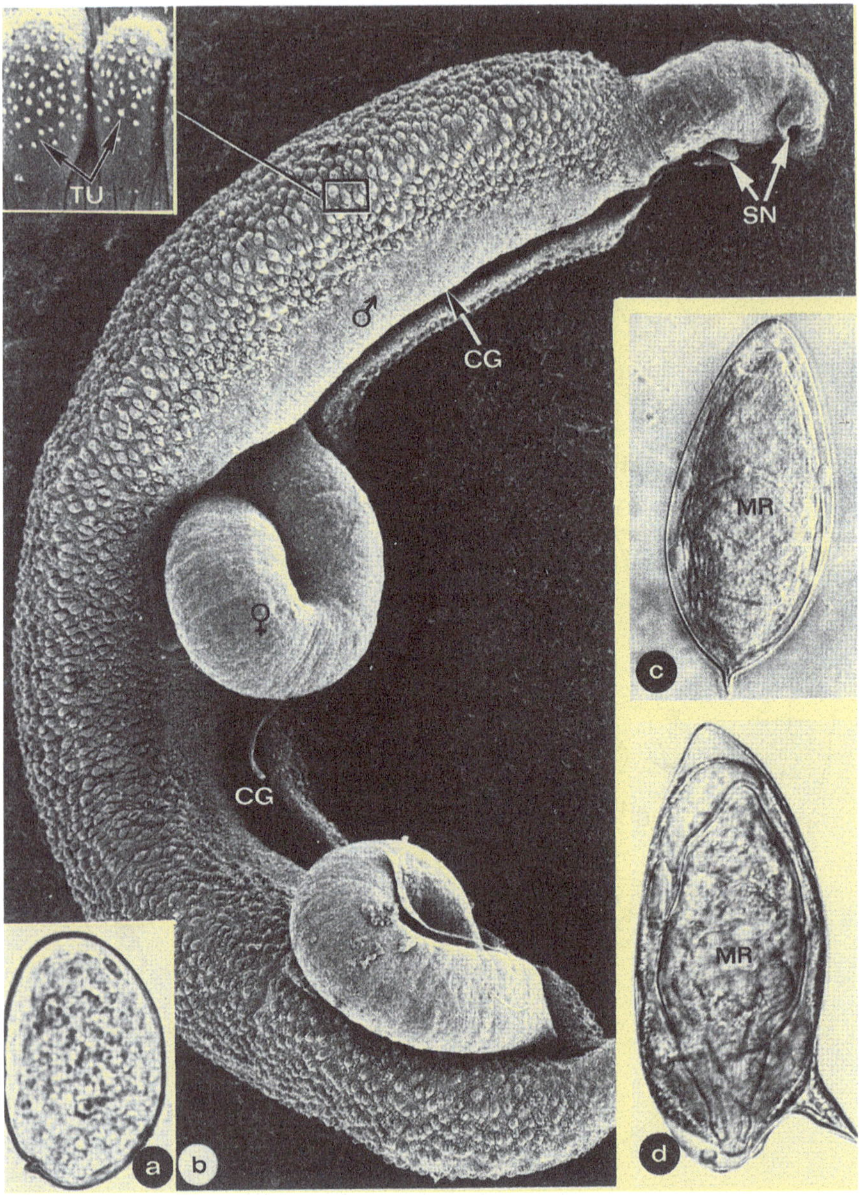

Fig. 2.3 Scanning electron micrograph of a couple of *Schistosoma mansoni*. The female is permanently transported in the longitudinal body folding of the male. **(a)** Light micrograph of an egg of *Schistosoma japonicum*. **(b)** Scanning electron micrograph of an adult couple. **(c)** Light micrograph of an egg of *S. haematobium*. **(d)** Light micrograph of an egg of *S. mansoni*, note the typical large lateral spike

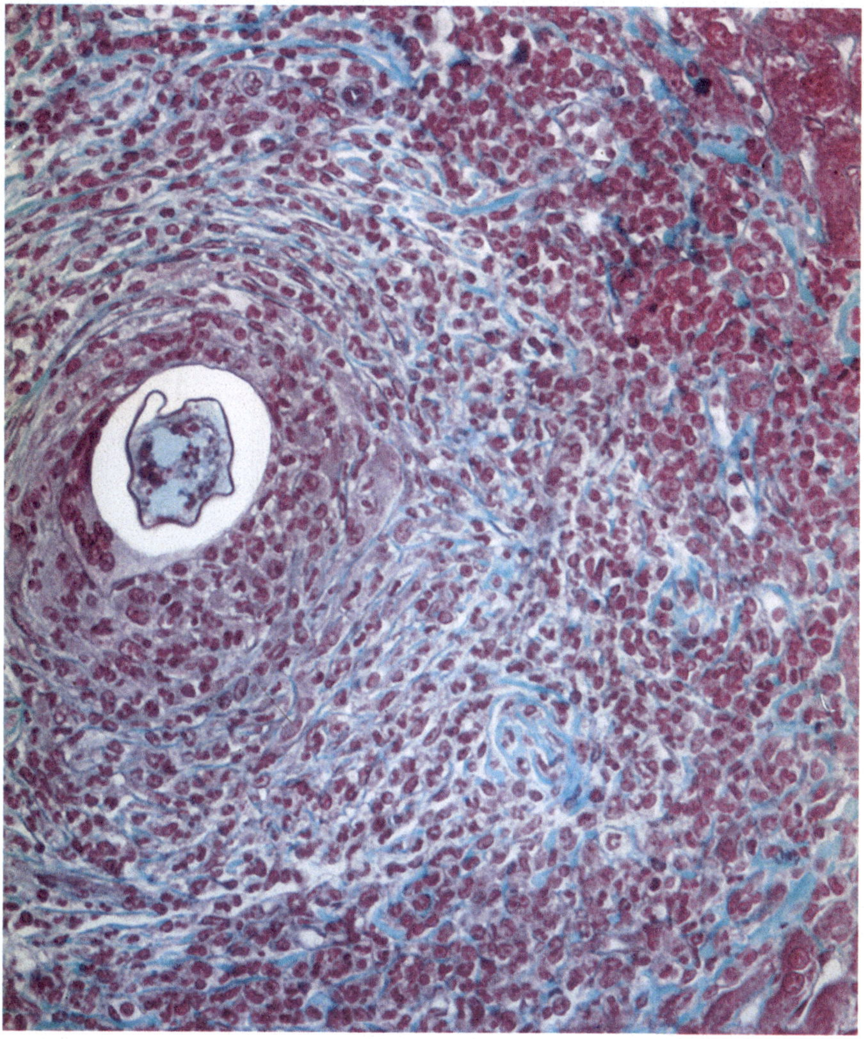

Fig. 2.4 Light micrograph of a section through a human liver containing a shrunken *Schistosoma* egg, which induced the formation of a granulome. Masses of such granulomas may induce liver tumors

In the year 1980 a praziquantel containing preparation became marketed as **Biltricide®** and was later worldwide distributed as **Cesol®** and **Cysticide®** as well as under a big bunch of other names (see below). It kills trematodes as well cestodes. In the year 2007 the German Merck KGaA spent for 10 years 25 million tablets for free and since 2012 always 250 million tablets per year, which helped to decrease or even eliminate schistosomiasis in many countries.

Praziquantel (Tables 2.1 and 2.2) is also used to deworm ornamental fish (Tremazol® produced by Alpha-Biocare GmbH, Neuss, Germany and distributed

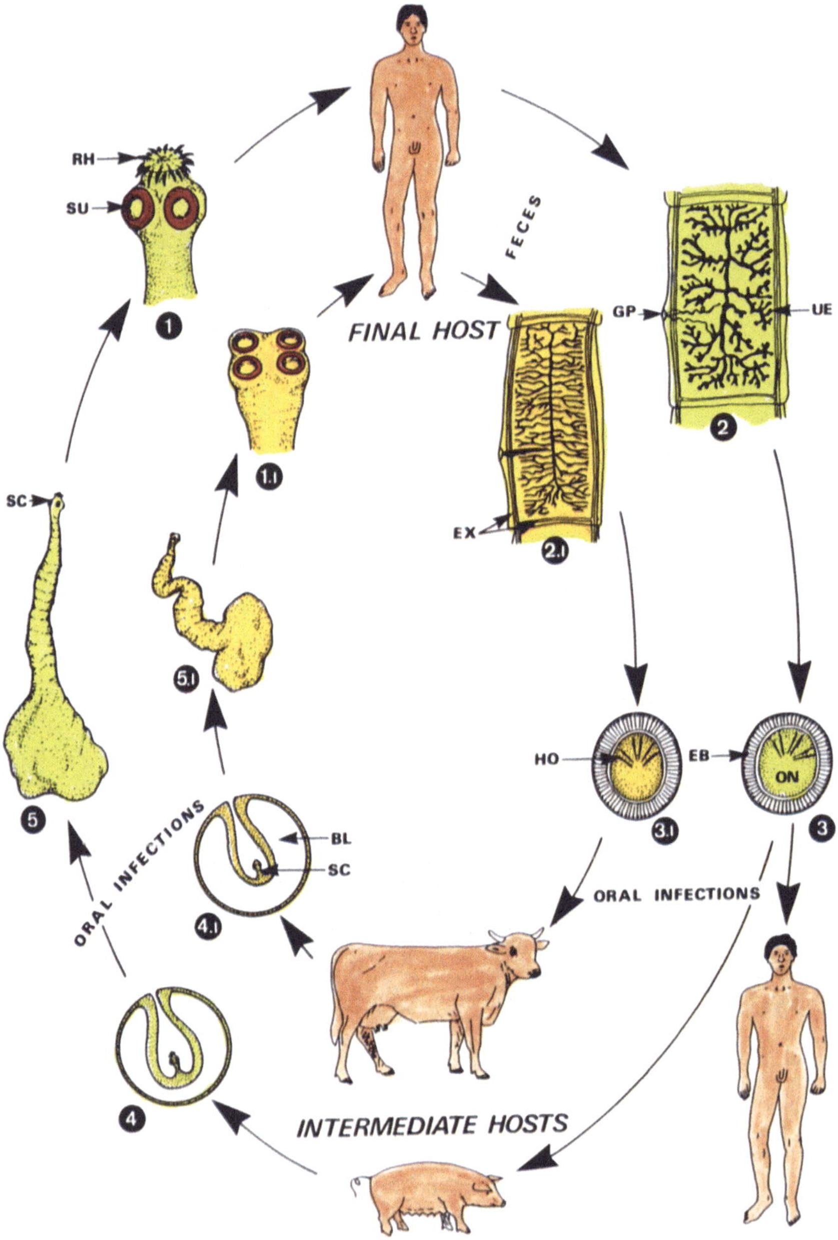

Fig. 2.5 Life cycles of ***Taenia solium*** (**1–5**) and ***T. saginata*** (**1.1–5.1**). (**1–2.1**) Adult worms live exclusively in the intestine of man and reach a length of 4–6 m (*T. solium*) or 6–10 m (*T. saginata*) often with about 2000 proglottids. The scolex of *T. solium* is endowed with an armed rostellum (**1**). The terminal, whitish appearing proglottids (10–20 × 5–7 mm) are characterized (when excreted in the feces of the host) by a typically branched uterus filled with up to 100,000 eggs. On each day 6–7 of these proglottids detach and may either pass out with the feces or actively migrate out of the anus (**3, 3.1**). As soon as an excreted proglottid begins to dry up, a rupture occurs along the midventral

by Sera, Heinsberg, Germany) (Fig. 2.1; Tables 2.1 and 2.2). If this fluid formulation is added to aquaria containing infected ornamental fish, all worms on gills and skin or in the intestine are killed.

The group of the author of the present chapter showed in several in vitro and in vivo experiments, why praziquantel is still today highly effective under both conditions against trematodes and cestodes although it is already more than 40 years in use (Mehlhorn 2016a, b). Only very few reports exist on findings or indications of slight local resistances, which, e.g., had been described, e.g., in the case of the treatment of cases of *Dipylidium caninum* (Jesudoss Chelladurai et al. 2018).

2.2.2 Chemical Composition of Praziquantel

Figure 2.1 shows the diagrammatic representation of the structure of the compound. Its commercial preparation is a **racemate**, which consists of two equal parts: a levo (R-) and a dektro (S+) isomer, wherefrom only the enantiomer has an antischistosomal activity, while both isomers have essentially the same toxicity on parasites as was shown by Liu et al. (1986). Wu et al. (2019), however, showed that patients treated only with 20 mg/kg praziquantel reached the same high cure rate, but suffered from significantly lower side effects than persons treated with 40 mg/kg of the racemic preparation.

Analysis of the chemistry showed that praziquantel is a more or less whitish, crystalline powder, which initiates a bitter taste and melts at 136–140 °C leading to complete decomposition. On the other hand, it is more or less insoluble in water, a bit soluble in ethanol but totally soluble in solvents like chloroform. In chemical terms praziquantel is described as 2-(cyclohexylcarbonyl)-1,2,3,6,7,11b hexahydro-4H-pyrazinol-[2,1-a] isoquinoline-4-one (Fig. 2.1).

Fig. 2.5 (continued) and terminal regions and allows eggs to escape. The spherical eggs (40–45μm) are indistinguishable between species. Originally they have a hyaline outer membrane (eggshell) which is usually lost during the time the eggs are voided within the feces. Thus, the eggs are bordered by a thick, striated embryophore surrounding the oncosphaera (ON). (**4, 4.1**) When ingested by the intermediate host, the **oncosphaera** hatches in the duodenum, penetrates the mucosa, enters a venule and is carried throughout the body. A bladder worm (cysticercus) of about 7–9 × 5 mm is formed, reaching infectivity in about 2 months (*C. cellulosae* in *T. solium*; *C. bovis*, *C. inermis* in *T. saginata*). When humans ingest eggs of *T. solium* within contaminated food or a terminal proglottid is destroyed inside the intestine, cysticerci may readily develop in many organs including brain and eyes. These infections lead to severe dysfunctions depending on the parasitized organ (**cysticercosis**). (**5**) Persons become infected when a bladder worm is eaten along with raw or insufficient cooked meat of intermediate hosts. The evaginating scolex becomes attached to the mucosa of the small intestine and matures within about 5–10 weeks. **BL** bladder of cysticercus, **EB** embryophore, **EX** excretory vessels, **GP** genital pore, HO hooks of oncosphaera, **ON** oncosphaera, **RH** rostellar hooks, **SC** scolex, **SU** sucker, **UE** uterus filled with eggs. Reprinted from Mehlhorn (2016b) with permission from Springer Nature

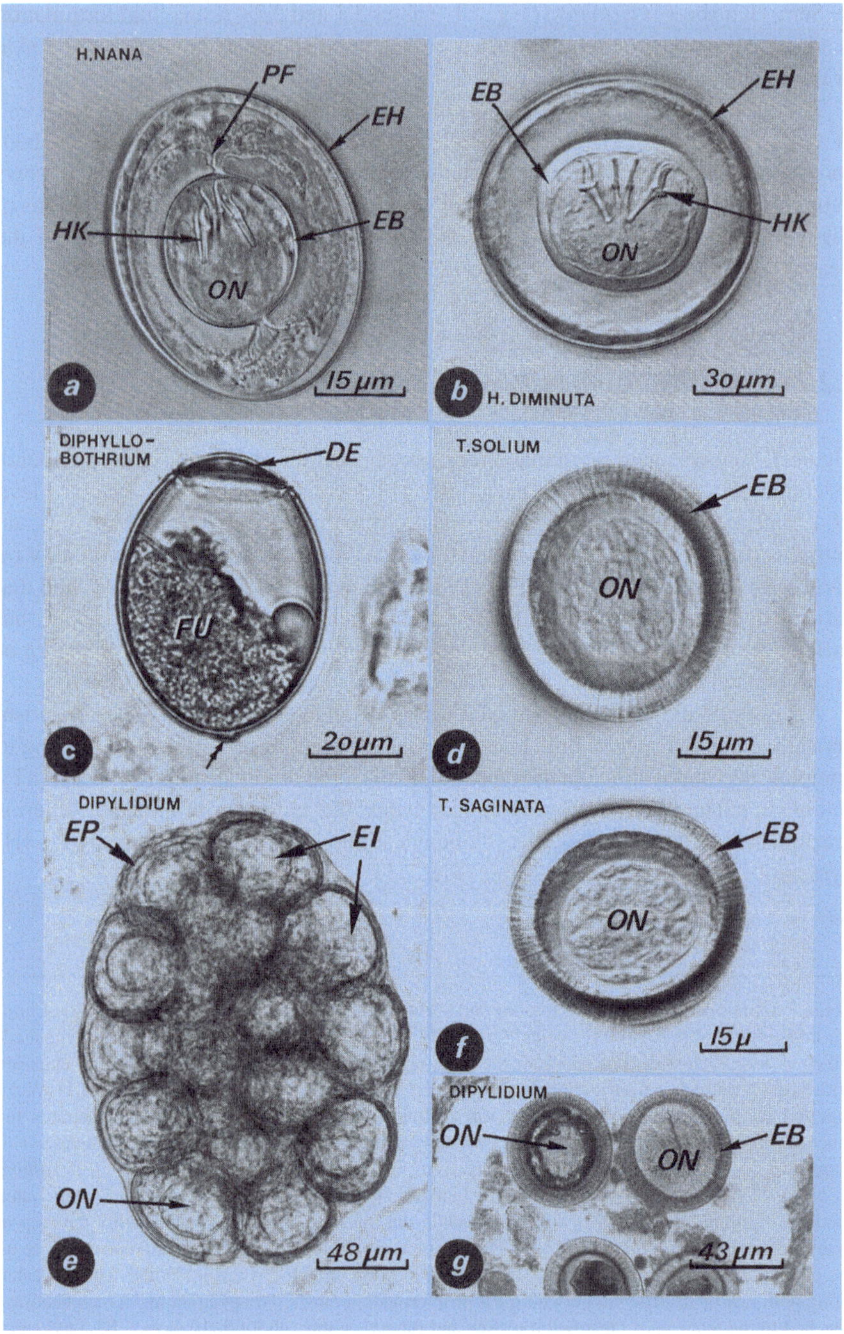

Fig. 2.6 Light micrographs of eggs of important tapeworm species. **DE** cover, operculum, **EB** embryophore, **EH** egg shell, **EI** egg, **EP** package of eggs, **FU** developmental stage during cell division, **HK** hooks, **ON** oncosphaera larva, **PF** polar filament

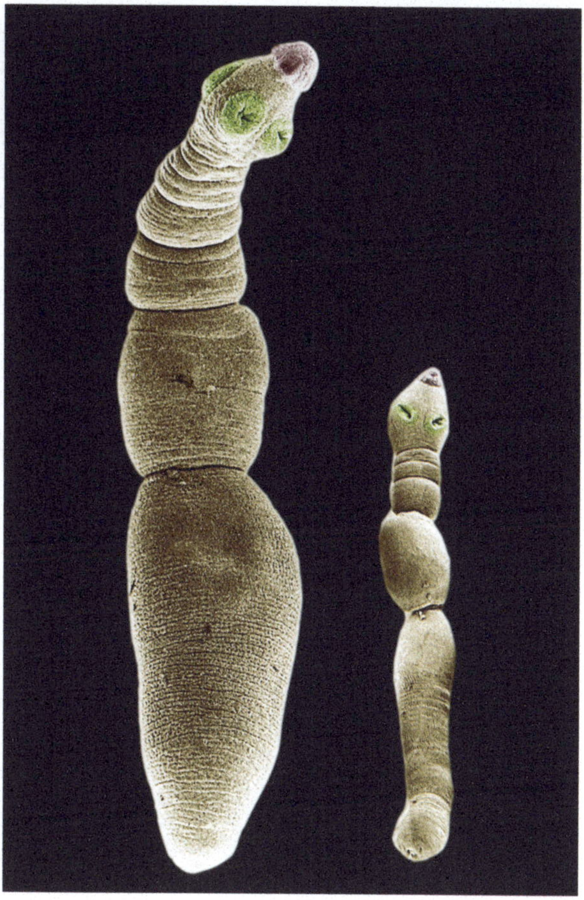

Fig. 2.7 Scanning electron micrograph of the adult worms of *Echinococcus multilocularis* (right) and *E. granulosus*, both of which produce only few proglottids

2.2.3 First Efficacy Tests

The WHO organized clinical trials very early to proof efficacy on *S. mansoni* (Katz et al. 1979), on *S. haematobium* (Davis et al. 1979) and on *S. japonicum* (Ishizaki et al. 1979). The results obtained were very promising, so that praziquantel became worldwide rather quickly the **drug of choice** for the treatment of *Schistosoma* infections of humans and animals. In the year 1983 a Korean company developed a new, cheaper synthesis of praziquantel followed by production in several countries and thus it became worldwide distributed. It is still today the drug of choice against the members of this group of worms (Wu 2019; Wu et al. 2012, 2019; Mehlhorn et al. 2014; Mehlhorn 2016a, b, c).

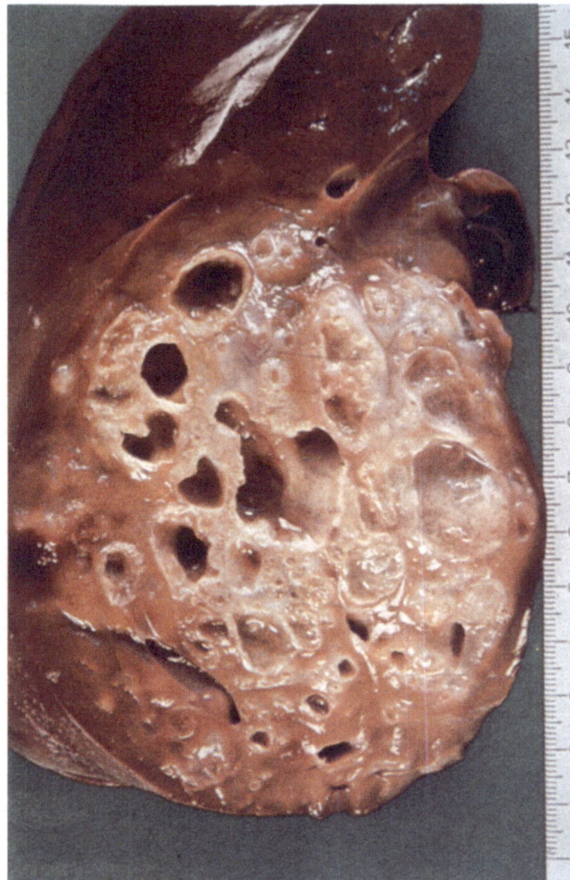

Fig. 2.8 Section through the liver of a human deceased due to the liver destruction by the fox tapeworm *E. multilocularis* cysts. Note the large alveolar caverns

2.2.4 Dose and Administration

The WHO recommended a treatment dose range between 40 and 60 mg/kg body weight, whereby the 40 mg dose is recommended for treatment of *S. mansoni* and *S. haematobium* infections. The 60 mg dose should be split into two administrations especially in cases with the Asian species *S. japonicum* and *S. mekongi* (WHO 2016; Thomas and Timson 2018).

2.2.5 Use during Pregnancy

Although praziquantel has never been officially clinically tested intensively among larger groups of pregnant or lactating women, it is presumed to be safe based on many animal and human studies. Thus today praziquantel is grouped into the "**Pregnancy category B**" (WHO 2002).

Table 2.1 Praziquantel (data sheet)

General name	Praziquantel
Chemical name	(RS)-2-(Cyclohexylcarbonyl)-2,3,4,6,7,11b-hexahydro-1H-pyrazino[2,1-a]isochinolin-4-on (IUPAC)
Sum formula	$C_{19}H_{24}N_2O_2$
Description	Whitish, polymorphous, crystalline, hygroscopic powder
CAS number	55268-74-1
ATC code	P02BA01
Drug class	Anthelmintics
Molar mass	31,241 $g \times mol^{-1}$
Melting points	136 °C (racemic compound) 110 °C (enantiomers)
Dissolving capacity	Difficult in water (400 $mg \times l^{-1}$) Easy in ethanol and dichloromethane
Toxicity data	>200 $mg \times kg^{-1}$ (LD50, dog, orally) 2840 $mg \times kg^{-1}$ (LD50, rat, orally) 24,540 $mg \times kg^{-1}$ (LD50, mouse, orally)
Product trade names	Biltricide, Cesol, Cysticide
International product or trade names (selection)	Anipracit, Aniprazol, Bihelminth, Caniquantel, Cestocur, Droncit, Drontal, Equimax, Milbemax, Prazinex, Profender, Tremazol, Vermis-Ex, VetBancid etc.

2.2.6 Side Effects

The tested side effects on humans after drug use had been rather smooth but have found to be present in 30–60% of the tested persons. These in general mild symptoms comprised (as single alternatives) headache, anorexia, nausea, vomiting, diarrhea, slight fever, dizziness, etc. and in extremely rare cases skin rash (Cioli and Pica-Mattoccia 2003). But compared to the potential killing effects of large numbers of these worms, these effects seem to be justified to reach survival.

2.2.7 Toxicity Studies

Toxicity of praziquantel was found to be very low in animal studies even when it was used in long term studies (Cioli and Pica-Mattoccia 2003) and toxic effects had not observed in humans when being used in the recommended dosages.

2.2.8 Mode of Action

Praziquantel has significant **killing effects** on most species of **cestodes** (Tables 2.3 and 2.4) and **trematodes** (Table 2.5) parasitizing in humans and farm animals such as sheep, cattle, horses, dogs, etc. Only a few parasites like the so-called liver fluke

Table 2.2 Primary and secondary effects induced in trematodes as function of time after drug exposure. Similar effects can be seen also in cestodes

Helminth species	Drug concentration	Observed effect	Time after drug exposure of observable effect
Primary effects			
All schistosomes	0.32–0.6μM	Surface blebbing, vacuolization	Seconds
		Spastic paralysis	Seconds
		Increase in muscle tension	Seconds
S. mansoni	1μM	Ca^{2+} influx	Seconds
Effects associated to primary membrane effects			
S. mansoni		Depolarization or alterations of ion fluxes (Na^+, K^+); uptake of glucose and adenosine, lactate excretion, glycogen breakdown, alterations of membrane associated enzymes: ATPase	Seconds until minutes
Secondary effects			
S. mansoni		Phagocytic cells associated with parasite	4 h
S. mansoni		Invasion of phagocytes	17 h
S. mansoni		Lysis of parasite's tissues	Few days

Fasciola hepatica are not susceptible to praziquantel—even not when used in very high doses. This is apparently due to their very thick surface coat and the thick non-cellular tegument as outer layer. Both together (surface coat and tegument) apparently block the entrance of the active compounds of this drug into the fluke's body. During in vitro tests two main effects have been documented on movie pictures. Both occur immediately after such trematodes have been placed into a medium containing praziquantel: the surface of the worms becomes immediately ruptured and the body shrinks considerably after a phase of huge contraction. In tapeworms especially the "neck zone" below the suckers is hit by these effects, which apparently have the consequence that the worm cannot remain anchored at the intestinal wall of the host, if the applied dose is high enough in vivo (see Tables 2.3 and 2.4).

2.2.9 *Tolerability*

The **tolerability** of praziquantel using it against trematodes is apparently rather high as was shown in relevant tests. Studies in mice, rabbits and dogs showed that ingestion up to five times of the recommended dose was tolerated without adverse effects. However, a tenfold overdose caused transitory vomiting and signs of depression. On the other hand, there was no evidence for **embryotoxicity** or **teratogenicity** in reproduction studies in rats, rabbits, cats and dogs. Neither such effects had been described in human cases. The following Tables 2.1, 2.2, and 2.3

Table 2.3 Morphological features of important *Schistosoma* species of humans

Features	*S. mansoni*	*S. japonicum/S. mekongi*	*S. intercalatum*	*S. haematobium*
Males, surface	Large tubercles	No tubercles	Tubercles with hook-free corona	Small tubercles with many hooks
Females, ovary	Anterior half of the body	Posterior half of the body	Behind the center of the body	Close to the center of the body
Number of eggs in uterus	1–10 (mostly a single one)	50–300	10–20	20–100
Shape of eggs	With big lateral thorn	With small lateral thorn	With thorn at posterior end	With thorn at posterior end
Excretion site of eggs	Feces	Feces	Feces	Urine
Size of worms (mm)	M: 6–12 × 1 F: 7–16 × 0.3	M: 12–20 × 0.5 F: 12–28 × 0.3	M: 11–15 × 0.4 F: 10–14 × 0.2	M: 10–15 × 1 F: 20–26 × 0.3
Location of adult worms in human host	Intestinal mesenteric blood vessels	Intestinal mesenteric blood vessels	Intestinal mesenteric blood vessels	Veins of urogenital tract
Intermediate host/snail	*Biomphalaria* sp.	*Oncomelania* sp.	*Bulinus* sp.	*Bulinus* sp.
Reservoir host	Rats, mice, monkeys	Mice, dogs, cats, pigs, cattle	Goats, sheep, rodents	Monkeys, hamsters
Drug of choice	Praziquantel 1 × 60 mg/kg body weight in three doses	Praziquantel 1 × 60 mg/kg body weight in three doses	Praziquantel 1 × 60 mg/kg body weight in three doses	Praziquantel 1 × 60 mg/kg body weight in three doses
Incubation period	2–3 weeks	2–3 weeks	4–7 weeks	2–10 weeks
Prepatent period	4–7 weeks	4–5 weeks	6–8 weeks	8–12 weeks
Patency	5–20 years	5–20 years	5–20 years	5–20 years

contributed by Prof. Dr. Achim Harder (former coworker of Bayer company, Germany) give some examples of activities of praziquantel on some important cestodes (tapeworms) and trematodes (flukes).

2.2.10 *Efficacy of Praziquantel and Other Drugs on Tapeworms*

This compound (see Tables 2.1 and 2.2) is effective against **trematodes** (flukes, Tables 2.3 and 2.4) as well as against **cestodes** (tapeworms, Table 2.5).

Results summarized and reprinted from Mehlhorn (2008) and Mehlhorn (2016c) with permission from Springer Nature

Table 2.4 Trematocidal activity of praziquantel

Parasite	Treatment	Infection/comments
1. *Schistosoma mansoni* **Adults** live in the venous system of the intestine, eggs (embryonated, large, oval, with lateral spine, pass into feces; the latter must be deposited in fresh water so that miracidia can hatch and reach appropriate snails), endemic in Africa, Middle East, and parts of South America; intestine and liver are primarily affected; **eggs** (soluble antigens in tissues) induce severe inflammatory reactions related to intensity of infection and thus host responses; damage caused in acute phase is followed by irreversible fibrosis of liver and adjacent tissues.	Drug of choice: **praziquantel** (PZQ) (e.g., *Biltricide Bayer, others) (40 mg/kg/d in two doses × 1 d: adult/pediatric) alternative: oxamniquine (15 mg/kg once, pediatric 20 mg/kg/d in two doses × 1 d adult/pediatric: in East Africa, the dose should be increased to 30 mg/kg, and in Egypt and South Africa to 30 mg/kg/d × 2 d; some experts recommended 40–60 mg/kg over 2–3 d in all Africa [Shekhar KC (1991) Drugs 42: 379].	PZQ is (was) a major therapeutic breakthrough in control of schistosomiasis (cure rates 85%–100%); side effects are common, but mild: headache, diarrhea, rash, fever; single dose treatment results in a very high cure rate; levo-PZQ (150 mg/kg b. w.); administered to mice infected with *S. mansoni* caused damage to the tegument of adults including severe swellings, vacuolization, fusion of the tegumental ridges, and loss or shortening of the spines on the tubercles, collapse, and peeling (dextro PZQ was proven to be inactive), oxamniquine (*Vansil, Pfizer) (regional differences in efficacy.
2. *S. mekongi* adults live in the host's mesenteric venules, **eggs** resemble closely *S. japonicum* eggs, pass into feces; this species belongs to "minor" species group found in Southeast Asia.	Drug of choice: **Praziquantel** (PZQ) (60 mg/kg in three doses × 1 d: Adult/pediatric).	**Endemic** along the Mekong river, including Laos, Cambodia, Thailand. **Diagnostic problems**: Fecal debris adheres to shell of *S. japonicum* and *S. mekongi* eggs, thus eggs might be overlooked in fecal preparations; spine of the egg is inapparent and difficult to see.
3. *S. intercalatum* adults live in mesenteric venules, **eggs** resemble closely *S. haematobium* eggs but are larger and are passed within feces.	Drug of choice: **Praziquantel** (PZQ) (60 mg/kg in three doses × 1 d: Adult/pediatric).	*S. intercalatum* is a "minor" species of man in west and Central Africa: *S. bovis*, *S. mattheei*, or *S. nasalis* are primarily parasites in other mammals (e.g., equines, ruminants), but may also infrequently infect humans.
4. *S. haematobium* adults live in the venous plexus of the urinary tract (mainly bladder), **eggs** are embryonated, large oval, with terminal spine pass into urine; the latter must be deposited in fresh water so that miracidia can hatch and thus may reach appropriate snails; endemic in 54 countries of	Drug of choice: **Praziquantel** (PZQ) (e.g., *Biltricide Bayer, others) (40 mg/kg/d in two doses × 1 d: Adult/pediatric), considered as safe in children over 4 years of age who tolerate it better than do adults (**contraindications**: Ocular cysticercosis cf. general information on schistosomiasis).	Single dose treatment results in a cure rate equal to or greater than 85%; in persons not cured, the egg burden is markedly decreased; side effects caused by PZQ are common but mild such as nausea, abdominal discomfort, dizziness, headache, and diarrhea; rash, pruritus, urticaria, fever; myalgia, and

(continued)

Table 2.4 (continued)

Parasite	Treatment	Infection/comments
Africa and in eastern countries along the Mediterranean Sea.		eosinophilia are noted occasionally and are related to parasite burden.
5. Fascioliasis ***Fasciola hepatica***, ***F. gigantica*** adults (2.5–5 cm long, 0.6–1.4 cm wide), *F. gigantica* up to 7.5 cm long: Both flukes live in bile ducts, liver tissues, and in aberrant sites, e.g., lung and/or subcutaneous tissues, **eggs**: unembryonated, large, broadly ellipsoidal, operculum indistinct; shape of egg (both species) resembles closely to that of *F. buski* and are passed into feces; cosmopolitan distribution, closely related to flukes of herbivores and other mammals, rarely man; if no eggs are found (aberrant sites) induce serological tests.	Drug of choice: **Triclabendazole** (a benzimidazole) (TCBZ) (*Egaten, Novartis) (10 mg/kg once or twice: Adult/pediatric) [Richter J et al. (2002). Curr Treat Option Infect Dis 4:313] (cf. characteristics of TCBZ in animals Tables 2.1 and 2.2 ↑) alternative: Bithionol (a bisphenol) (*Bitin, Tanabe, Japan) (30–50 mg/kg on alternate days × 10–15 doses: Adult/pediatric) (praziquantel proved ineffective against *Fasciola*); TCBZ: There may be availability problems.	Aquatic snails (*Lymnaea* spp.) serve as intermediate hosts; infection is acquired by ingestions of encysted cercariae (metacercariae) attached to wet grass and herbs (e.g., watercress); adults of *Fasciola hepatica* (global distribution, most common in sheep and cattle, wild ruminants, but also dog, cat, swine, horse, kangaroo, man) and *F. gigantica* (throughout Asia, Middle East, Africa, the Americas, and Hawaii), most common in cattle borrow tunnels through the liver parenchyma and feed on hepatocytes and blood.
6. Dicroeliasis ***Dicrocoelium dendriticum:*** Adults live in fine branches of bile ducts, gallbladder, **eggs** are embryonated, ovoid, small, have an indistinct operculum, brown shell, are passed into feces; species is a cosmopolitan in herbivores, rabbits, pigs, dogs, deer, occasionally in humans, too).	Praziquantel, or albendazole (may be effective at dose regimens recommended for other small liver flukes, cf. *Clonorchis sinensis* and *O. viverrini*).	Infections in man are rare but globally widespread, cf. *F. hepatica*, not as pathogen as *F. hepatica*; intermediate (IM) hosts are land snails (first IM) and ants (second IM); infection is acquired by ingestions of ants; in advanced cases extensive cirrhosis of liver, clinical signs may be anemia, edema, and emaciation.
7. Clonorchiasis: ***Clonorchis sinensis*** (Chinese or oriental liver fluke): Adults live in bile ducts, sometimes pancreatic duct and duodenum; **eggs** are embryonated, ovoid, small and have a seated operculum, are passed into feces, occur in Japan, Korea, Vietnam, and China.	Drug of choice: **Praziquantel** (PZQ) (75 mg/kg in three doses × 1d: Adult/pediatric) or albendazole (*Albenza or *Eskazole GlaxoSmithKline) (10 mg/kg × 7 d: Adult/ pediatric).	Fish-eating mammals (e.g., weasel, mink, dog, cat, pig, rats) serve as reservoir hosts; adult worms may live in host for up to 25 years; cercariae emerge from snail, come in contact with fish and encyst under scales; man becomes infected by ingestion of raw or undercooked freshwater or brackish water fish meat.
8. Opisthorchiasis: ***Opisthorchis felineus*** (syn. *O. tenuicollis*); ***O. viverrini***, (southeast Asian liver fluke),	Drug of choice: **Praziquantel** (PZQ) (e.g., *Biltricide Bayer, others) (75 mg/kg in three doses × 1 d: Adult/pediatric);	Distribution and reservoir hosts of closely related liver flukes: *O. felineus* (Russian Federation, Eastern Europe;

(continued)

Table 2.4 (continued)

Parasite	Treatment	Infection/comments
Metorchis conjunctus (north American liver fluke); adults live in gall bladder, bile ducts of liver. **Eggs** (embryonated, small, seated (or small: *Metorchis*); difficult to distinguish from those of *C. sinensis*), are always passed via feces.	mebendazole has been reported to be effective.	cats, civets, dogs, pigs, rats, other mammals, and man), *O. viverrini* (northern Thailand, and Laos; dog, cat, fox, pig, endemic in man), *M. conjunctus* (areas of northern America; dog, cat fox, mink, raccoon, other wildlife, man); *M. albidus* (Europe).
9. Paragonimiasis: Asia: ***Paragonimus westermani***, *P. heterotremus*, *P. skrjabini*. Africa: *P. uterobilateralis*, *P. africanus*. Canada: *P. kellicotti*. Peru, Ecuador: *P. mexicanus*. Larvae form cysts, capsules in lung parenchyma or in aberrant sites as brain, spinal cord, peritoneum, liver, spleen, kidneys, testes/ovary, muscles, intestinal wall, mesenteric lymph nodes. **Eggs** (unembryonated, prominent operculum, different sizes: *P. westermani* much larger than others, dark shell) pass up from lung into sputum (**eggs**: Either dislodged by coughing or swallowed and pass into feces); **eggs** can be confused with smaller cestode eggs like those of *Diphyllobothrium latum*.	Drug of choice: **Praziquantel** (PZQ) (e.g., *Biltricide Bayer, others) (75 mg/kg in three doses × 2 d: Adult/pediatric); PZQ is better tolerated than bithionol; PZQ is contraindicated in ocular disease; alternative: **Bithionol** (*Bitin Tanabe, Japan) (30–50 mg/kg on alternate days × 10–15 doses: Adult/pediatric) (*availability problems). **Surgical:** Excision of extra-pulmonary lesions, shunt in case of hydrocephalus.	Human lung flukes may infect an estimated 21 million people worldwide (~ten million in **China**: Asiatic species), *Paragonimus* infections being endemic in Central China, **Philippines**, **Thailand**, **Korea**, **Laos**, but found also in **Taiwan**, **Japan**, **Malaysia**, **Indonesia**, and **India**; other species cause infections in Asia and the Pacific, Africa, **Canada**, **central** and **South America** (scattered reports); lung flukes are common in crustacean- eating wild carnivores (e.g., otter, fox, mink, mongoose, dog, cat, wildcat, raccoon, tiger, leopard, panther, wolf, and omnivores like bush rat, rat, pig, monkeys, and other mammals including man; transmission: Humans acquired infection by ingestion of raw undercooked crabs or crayfish.
10. Heterophyiasis: ***Heterophyes heterophyes***: Adults are attached to the wall of the small intestine, eggs are small, have an embryonated, inconspicuous operculum; egg resembles that of *C. sinensis*, are passed into the feces.	Drug of choice: **Praziquantel** (PZQ) (75 mg/kg in three doses × 1 d: Adult/pediatric).	Small intestinal fluke, which is uncommon but widely distributed (Middle East, Turkey, eastern and southeastern Asia). It occurs in dog, cat, fox, and man; many species of fish (brackish or freshwater fish) act as second intermediate host; only heavily infected individuals may show nonspecific diarrhea, abdominal pain, and eosinophilia.
11. Metagonimiasis: ***Metagonimus yokogawai***: Adults are attached to the wall	Drug of choice: **Praziquantel** (PZQ) (75 mg/kg in three doses × 1 d: Adult/pediatric).	Small intestinal fluke; most common heterophyid fluke in the Far East (also found in

(continued)

Table 2.4 (continued)

Parasite	Treatment	Infection/comments
of the small intestine, eggs (embryonated, small, egg resembles those of *C. sinensis* and *Heterophyes* but it has an obvious operculum).		Mediterranean basin); it may occur in dog, cat, pig, and man; several species of fresh-water fish act as second intermediate host; only heavily infected individuals may develop nonspecific diarrhea and vague abdominal complaints.
12. Echinostomiasis: ***Echinostoma ilocanum***, *E. lindoense*, *E. hortense* adults (attached to wall of small intestine), eggs (usually large, oval, unembryonated).	Drug of choice: **Praziquantel** (PZQ) (e.g., *Biltricide Bayer, others) (75 mg/kg in three doses × 1 d: Adult/pediatric).	Are primarily parasites of birds and rodents; *E. ilocanum* may be common in humans (Korea, Philippines, Indonesia) whereas *E. hortense* is principally a parasite of rodents; same snail or neighboring snails (some echinostomatids: Fish, clams, and tadpoles) also serve as second intermediate host; mild infections are asymptomatic but heavy infection can be accompanied with diarrhea, and intestinal colic (similar to fasciolopsiasis).
13. Gastrodiscoidiasis: ***Gastrodiscoides hominis*** adults (attached to the wall of colon and cecum), eggs (unembryonated, large, ovoid, egg resemble closely to that of *F. hepatica* or *F. buski*).	Drug of choice: **Praziquantel** (PZQ) (75 mg/kg in three doses × 1 d: Adult/pediatric).	Occurs in India Southeast Asia and parts of the former USSR; pigs (natural host), monkeys and man, field rats serve as hosts; incorrect egg diagnosis may occur; man acquired infection by eating uncooked aquatic plants; only a massive infection may produce mucous diarrhea.
14. Fasciolopsiasis: ***Fasciolopsis buski*** adults (~7.5 cm long, 2 cm wide, attached to wall of small intestine), eggs (unembryonated, large, broadly ellipsoidal, operculum indistinct; egg resembles closely to that of *F. hepatica*).	Drug of choice: **Praziquantel** (PZQ) (75 mg/kg in three doses × 1 d: Adult/pediatric)	Large intestinal fluke that occurs in Far East (India, China, Taiwan, Thailand, Indonesia, and other parts of Asia); fresh water snails serve as intermediated hosts; mature cercariae emerge from snail, attach to aquatic plants (water caltrop, water bamboo, water chestnut, lotus on the roots, and other aquatic vegetables) and encyst to become metacercariae; man becomes infected by ingestion of uncooked vegetation being contaminated with

(continued)

Table 2.4 (continued)

Parasite	Treatment	Infection/comments
		metacercariae. In severe infections occur thousands of worms; flukes may also attach to the ileum or colon; intestinal flukes cause inflammation, ulceration, and mucous secretion at the site of attachment; symptoms may be eosinophilia, diarrhea, and edema, severe infections may also cause intestinal obstruction or malabsorption leading to hypoalbuminemia, ascites, and obstruction of common bile duct; pigs are an important **reservoir host**.
15. *Nanophyetus salmincola* adults (small or large intestine), eggs (unembryonated, indistinct operculum, much smaller than those of *P. westermani*); pass into feces; this occurs in Eastern Siberia, Northwestern of USA.	Drug of choice: **Praziquantel** (PZQ) (e.g., *Biltricide Bayer, others) (60 mg/kg in three doses × 1 d: Adult/pediatric).	Fluke infects various fish-eating mammals (dog, cat, fox, otter, mink, lynx, and some piscivorous birds, and man); cercariae emerge from snail, come in contact with fish (family Salmonidae) and encyst there under scales; man becomes infected by ingestion of raw or undercooked fish or via contaminated utensils used to prepare fish or vegetables for cooking. This fluke penetrates deeply into the mucosa of the duodenum or attaches to the mucosa of other parts of the small and large intestine thereby causing superficial or hemorrhagic **enteritis**. Adult *N. salmincola* may harbor rickettsial organisms causing an often fatal disease in dogs or other Canidae (so-called "**salmon poisoning**," and "**Elokomin fluke fever**", which may cause high morbidity); *N. salmincola* and lung flukes like *Paragonimus* spp. belong to the same family (Troglotrematidae).

Results summarized and reprinted from Mehlhorn (2008) and Mehlhorn (2016c) with permission from Springer Nature

Table 2.5 Cestocidal drugs

Parasite disease Distribution, pathology	**Stage affected** (location), morphology	**Chemical class** and other information (1*, 2*, etc. refer to linked chemicals and comments)	**Nonproprietary name** adult dosage (pediatric dosage) (oral route), d = days, comments	**Miscellaneous comments**
Diphyllobothriasis: First intermediate host are copepods, second intermediate host are various freshwater fishes due to lack of specificity of the plerocercoid; the sparganum stage lacks a bladder; see sparganosis); humans become infected by ingestion of raw or undercooked intermediate hosts (fish) containing **plerocercoids.** Adult tapeworms may live for 25 years or longer in humans and may cause (rarely) **vitamin B12** deficiency (due to absorption of B12 by the worm) and **megaloblastic anemia**; *D. latum* eggs (possess a small knob at the end, opposite to the operculum) may be confused with those of *Paragonimus westermani* (cf. Trematodocidal drugs/Table 2.3)				
Diphyllobothrium latum (fish tapeworm), common where cold, clear lakes are abundant; occurs in North Europe, North America and Japan in fish-eating mammals and in man.	Adult worms are 5–15 m long and stay in the upper half of the intestine. Eggs are unembryonated, appear ovoid, operculate, passed via feces into water.	**1* pyrazino-isoquinolines** contraindicated in ocular cysticercosis. **2* halogenated salicylanilides** (availability of problems).	**1* praziquantel** (drug of choice) (5–10 mg/kg once). Alternative: **2* niclosamide** (2 g once or 50 mg/kg once).	**1***safe and well tolerated, occasionally skin rashes **2***safe drug, also in pregnant or debilitated patients; occasionally abdominal pain, and pruritus.
Dipylidium caninum (dog tapeworm)	Humans become infected by occasional swallowing fleas containing infective larval stages (cysticercoids).			
Distribution worldwide; rare in humans; pathogenic effects are unknown; infection is due to close contact with infected dogs and their fleas.	**Adults** (approx. 50 cm long) live in the small intestine, **eggs** are thin-shelled, small, spherical; contain six-hooked oncosphere larvae are passed in feces (typical **egg packets:** Each containing up to 15 eggs).	**1* pyrazino-isoquinolines** **2* salicylanilides** (availability problems).	1* **Praziquantel** (drug of choice) (5–10 mg/kg once). Alternative: **2* niclosamide** (2 g once or 50 mg/kg once).	Definite hosts (dogs, cats) should be regularly treated to avoid infection in humans (especially those of children); eradication of fleas (cleaning of pet's carpets, rugs, and other sites) is needed.
Hymenolepiasis: Humans become usually infected by ingestion of embryonated eggs (via contaminated food and water); human infection also is possible by ingestion of arthropods (e.g., beetles) containing the infective larval stage (cysticercoid).				

(continued)

Table 2.5 (continued)

Hymenolepis nana (dwarf tapeworm): occurs worldwide; infects simian primates, rodents, man; oncosphere can hatch also from embryonated egg within intestinal tract causing autoinfection; this makes eradication difficult; amounts of cysticercoids may damage intestinal mucosa and produce diarrhea.	**Adult** (approx. 2–4 cm long), (in lumen of small intestine; oncosphere develops in lamina propria of a villus to preadult cysticercoid) eggs (small, thin- shelled, spherical to subspherical, 6-hooked larva surrounded by a membrane with 2 polar "knobs" from which 4–8 filaments arise, pass in feces).	**1* pyrazino-isoquinolines 2* salicylanilides 3* aminoglycoside** (<1% of dose is absorbed from gut, excreted unchanged in urine) **4*// 5-nitrothiazole benzamide** (first described in 1984 as a human cestodocidal drug).	1* **praziquantel** (drug of choice) (25 mg/kg once) alternative: **2* niclosamide** (2 g × 1d; 1 g × 6d) **3* paromomycin** minimal side effects **4* nitazoxanide** (500 mg × 3d) (tablet = 500 mg) (investigational drug, check registration).	**1*** e.g., **Biltricide** is active against both juvenile and adult stages e.g., **2*Yomesan** acts only against adults and mature cysticercoids, therefore treatment for 5–7 days **3*** (40 mg/kg/d for 7d) may cause total elimination of worm burden.
H. diminuta (rat tapeworm) occurs worldwide; common parasite of rats, may occasionally infect men.	**Adult** (20–60 cm long) (small intestine); **eggs** (large, thick-shelled, spherical, six-hooked oncosphere are passed in feces).	**Pyrazino-isoquinolines salicylanilides** (availability problems).	**Praziquantel** (drug of choice) (25 mg/kg once) alternative: **Niclosamide** (2 g × 1 d; 1 g × 5–7 d).	Clinical manifestations are inconspicuous (unknown).
Taeniasis: Humans become infected by ingestion of raw or undercooked muscle meat (beef or pork) containing infective cysticercus; **diagnostic problems**: Eggs of human and animal taeniid species (*Taenia* and *Echinococcus* spp.) are all indistinguishable from each other; *T. saginata* and *T. solium* diagnostic is usually done by examination of gravid proglottids injected with India ink or stained by permanent stains to visualize the characteristic number of lateral uterine branches (*T. saginata*: 15–30, and *T. solium*: 7–13); some pollen grains found in feces may closely resemble eggs of *Taenia* spp.; taeniid eggs contain a characteristic 6-hooked embryo (oncosphere); caution should be used in handling unidentified gravid (mature) taeniid proglottids since *T. solium* and *Echinococcus* spp. eggs are infective to humans; they can induce larval tapeworm infections of brain, liver, lungs, and other organs (cf. larval tapeworm infections)				
Taenia saginata (beef tapeworm), worldwide distribution. It lives in the human intestine, but the adult worm causes no distinct lesions. Cattle act as intermediate	Adults (up to 8 m long) live in the upper half of small intestine, eggs are embryonated, have a spherical, yellow-brown, thick shell being striated radially, they	**1* pyrazino-isoquinolines 2* salicylanilides** (availability problems) **3* benzimidazole carbamates.**	**1* Praziquantel** (drug of choice) (5–10 mg/kg once). Alternative: **2* niclosamide** (2 g once) **3* mebendazole**, Vermox (200 mg BID × 3 d).	In most cases there are no clinical signs; on an individual level infection can be prevented by thorough cooking of meat or by freezing it at −18 °C for 7 d.

hosts containing the cysticercus larva in meat.	measure 30–44μm Ø, are passed in feces.			
T. solium (pork tapeworm), worldwide distribution. The adult worms live in the intestine of humans, pigs are intermediate hosts = cysts are found in muscles.	**Adults** are up to 6 m long and live in the small intestine, eggs are passed in feces. Eggs of *T. solium* may also be infective to humans and can cause cysticercosis e.g. in the brain, which leads to death.	**1* pyrazino-isoquinolines** (1* has a moderate effect on larval stages) **2* salicylanilides 3* benzimidazole carbamates**	**1* Praziquantel** (drug of choice) (5–10 mg/kg once). Alternative: **2* niclosamide** (2 g once) **3* mebendazole** (200 mg BID × 3 d)	**2*** has no ovicidal action on larvated egg; larva liberated from a shed gravid segment in upper intestine can cause cysticercosis; do not use a purgative; an enhanced peristalsis increases risk of cysticercosis.
Larval cestode diseases: Humans may be infected by ingestion of an oncosphere-containing (embryonated) egg or by ingestion of a larval stage belonging to the genus *Spirometra* (cf. sparganosis).				
1. Cysticercosis, neurocysticercosis: Contaminated food containing *T. solium* eggs may infect humans during meal; **autoinfection** is possible; cysticerci with a fibrous tissue capsule may undergo calcification and release antigens that can cause inflammatory reactions.				
Taenia solium (larva type: a single scolex invaginated into a bladder), **cerebral cysticercosis** may occur where pig rearing is done; (prevention of infection cf. "Taeniasis").	**Cysticercus cellulosae** larva may reside in almost any tissue, often brain: larva tends to grow slowly to a large cyst that causes space-occupying lesions and/or hydrocephalus (larva in ventricle blocks CSF circulation).	**1* benzimidazole carbamates 1***Eskazole, Albenza, others **2* pyrazino-isoquinolines 2*** Biltricide experimental drugs: Mebendazole flubendazole.	Treatment of choice (see below) alternative: **1* albendazole** (ABZ) (400 mg BID × 8–30 d, can be repeated as necessary) **2* praziquantel** (PZQ) (50–100 mg/kg/d in three doses for × 30 d).	Treatment of choice (see below) alternative: **1* albendazole** (ABZ) (15 mg/kg/d, <max.800 mg> in 2 doses × 8–30d; can be repeated as necessary) **2* praziquantel** (PZQ) (50–100 mg/kg/d in three doses for × 30d).
2. Coenuriasis (infrequent): Adult worms occur in Canidae like dog, fox, coyote, and others (**definitive hosts**); herbivores serve as **intermediate hosts** as cattle, horse, sheep, wild herbivores, rarely man; latter becomes infected by ingestion of contaminated food containing embryonated eggs; human *Coenurus cerebralis*-disease mainly occurs in Africa; space-occupying larva usually invades brain.				

(continued)

Table 2.5 (continued)

Taenia multiceps (= *Multiceps multiceps*) (larva type: multiple invaginated scoleces into a bladder), infrequent occurrence.	**Coenurus cerebralis**, larval cyst contains several hundred proto-scoleces and is found in central nervous system, usually brain.	Highly effective prevention is not possible. Reservoir hosts include numerous species of wild carnivores.	Drug medication is unknown against space-occupying lesions due to larva.	The only reliable means is the excision of the accessible cyst.
3. Cystic hydatid disease (hydatidosis or cystic echinococcosis = CE): *Echinococcus granulosus* larva-type is a unilocular hydatid cyst infection of intermediate hosts (usually sheep, other herbivores as cattle, horse, and occasionally man) due to close contacts with dogs and other canids (definitive hosts), which may be infected with large numbers of adult worms (3–6 mm long, with a single gravid proglottid that is longer than wide and that contains typical *Taenia*-like eggs); feces of dogs contain embryonated eggs infective to potential intermediate hosts (occasionally humans). The larval stage in human tissues is characterized by multiple daughter bladders or "brood capsules" with multiple invaginated protoscoleces budding from their walls (inner layer of germinal epithelium of cystic cavity). The cyst content (materials) consists of degenerated scolices and debris in a milky fluid which is referred as hydatid sand (calcareous corpuscles).				
Echinococcus granulosus: humans are accidental intermediate hosts and are usually infected by handling an infected dog.	**Hydatid cyst**, unilocular (solitary or multiple or oval masses on imaging, frequently seen in liver and lungs, but also in other tissues).	**1* benzimidazole carbamates mebendazole** (40–50 mg/kg/d in three divided doses is less efficacious than albendazole in long-term regimen).	**1* albendazole** (drug of choice), (400 mg bid × 1-–6 months) **2*praziquantel** preoperatively use.	**1***(15 mg/kg/d, <max. 800 mg> × 1–6 months); adverse effects include nausea, hepatoxicity, neutropenia, rarely alopecia (long-term treatment).
4. Alveolar disease (or alveolar echinococcosis = AE): *Echinococcus multilocularis* larva-type: Multilocular or alveolar multilocular cyst; large numbers of adult worms (3–5 mm long, morphology cf. *E. granulosus*, and eggs of taeniid type) primarily occurring in foxes (definitive host), occasionally in dogs, cats, and wolves. Several microtine rodents serve as intermediate host (**sylvatic cycle**). Major source of the infrequent human infections are fruits and vegetables contaminated with larvated eggs from fox feces. In infected humans, "metastasis" (branches) of the laminated membrane by the alveolar cyst resemble lesions of a neoplasm. Lesions in the liver are usually membranous. In contrast to the hydatid of *E. granulosus*, neither protoscoleces nor hydatid sand (calcareous corpuscles) are identifiably in human cysts.				
Echinococcus multilocularis (infrequent); ***E. vogeli*** (infrequent) causing **polycystic** echinococcosis.	**Growth of alveolar (multilocular) cyst** found peripherally and invasive; metastases occur frequently (liver, adjacent tissues).	**Experimental drugs:** flubendazole, fenbendazole, oxfendazole, amphotericin B nitazoxanide.	**Treatment** if early diagnosed, surgical excision of lesions is a reliable means of treatment.	No therapy is fully effective against the tumor-like growth of the alveolar type of cyst.

Clinical manifestations produced by the alveolar larva of *E. multilocularis* are related to the extent of tumor-like lesions; **AE** is characterized by a chronic course lasting for months or years; clinical symptoms are variably and usually only follow a long asymptomatic period (5–15 years); in liver, alveolar hydatid presents a mass-producing inflammatory process; metastasis in lungs may be seen as multiple small solid foci and pathological findings may be characterized by marked foreign-body reaction in bones; treatment: **AE** is **difficult to treat**; current treatment approach is preferable surgery, or secondarily benzimidazole chemotherapy; radical surgical resection of complete alveolar lesion(s), e.g., in right or left liver lobes, is the only potential curative treatment [cf. details Craig P (2003), Curr Opin Dis 16: 473–444]. **Albendazole** (ABZ: 10–15 mg/kg/d: Long-term treatment: 6.5 years) is the drug of choice for human AE in patients with nonresectable AE; ABZ may be a useful adjunct therapy to surgery; overall its use orally resulted in improved 10-year survival rates up to 80–83% compared to 6–25% for untreated historical controls [a WHO group recommended long-term treatment: 3–24 months: ABZ (10 mg/kg), **mebendazole** (40–50 mg/kg); high doses are necessary to affect larva WHO Group Bull WHO, 74:231, 1996]. **Nitazoxanide** (NTZ)-treated in vitro cultured metacestodes of *E. multilocularis* were nonviable when implanted into susceptible mice; as NTZ is much better absorbed than ABZ following oral administration it provides an attractive alternative for medical treatment of human AE. Status of echinococcosis: The disease of humans due to the *Echinococcus* species belong also to the group of "Neglected tropical diseases" (NTD) due to regionally (locally) reduced hygienic safety in many countries (Table 2.6). Thus people traveling to such coutries should be warned and informed by their family physicians.				
5. Sparganosis: Larva type: Solid-bodied larvae lacking a bladder adult worm (pseudophyllidean tapeworm) occur in cats, dogs, wild canids, or felids and are of little significance to definitive hosts. First intermediate host is a copepod, second one any vertebrate due to lack of specificity of the plerocercoid. Humans become infected (**1**) by ingestion of a copepod (crustaceans) containing procercoids (first larval stage), (**2**) by ingestion of raw or undercooked flesh or organs of any vertebrate (amphibians, reptiles, mammals, particularly feral pigs raised for human consumption) containing plerocercoids (= sparganum, second larval stage, easily mistaken for nerves), or (**3**) by local application of flesh (poultice to wounds or to the eye) containing plerocercoids sparganum.				
Spirometra spp. (infrequent), life cycle is similar to that of *D. latum* (fish tapeworm of humans).	**Procercoid** (migrates in subcutaneous and muscular tissue), **plerocercoid** (migrates in connective tissue of muscles, abdomen, hind legs; peritoneum, pleura).	There is no reliable chemotherapy (praziquantel may show some larvicidal effects).	Treatment covers surgical prevention, but is difficult because of entrenched eating habits and other customs.	Larval worms cause during subcutaneous migration painful edema, urticaria, inflammations, fibrosis. Spargana grow into irregular nodules of subcutaneous tissues.

Table 2.6 Neglected tropical diseases (NTD)—data from German Ministry of Health

No.	Disease	Agent of disease (AD)	Vaccination	Drugs	Therapy	Diagnostics	Operational activities
1	African trypanosomiasis	P	0	V	V	V	E
2	Buruli ulcer	B	0	T, V	V, S	V	E
3	Chagas disease American trypanosomiasis	P	E, T	V, S	E	V	E
4	Dengue fever Chikungunya fever	V	T, V, S	E, T	V, S	V, S	V
5	Dracontiasis *Dracunculus medinensis*	P	0	E	0	T	0
6	**Echinococcosis**	**P**	**E, T**	**T, V**	**E, T, V**	**V**	**E, V**
7	Treponematosis	B	E	V	V	V, S	E, V
8	Geohelminthosis	P	E	V	T	T	E, V
9	Leishmaniasis	P	E, T	E, T	E, T	0	E
10	Leprosis	B	T	S	S	E	V
11	Lymphatic filariasis	P	0	T	T	T	E
12	Mycetoma	B	0	V	V	V	V
13	**Onchocerciasis**	**P**	**T**	**T, V, S**	**T, V**	**E, V**	**V**
14	**Schistosomiasis**	**P**	**E, T**	**S**	**V, S**	**V**	**V**
15	Snake bites	P	ß	V	V	E	V
16	Scabies	P	E	S	V, S	V	E
17	**Cysticercosis**	**P**	**V**	**S**	**S**	**V**	**V**
18	Rabies	V	V, S	E	E	V	E, V
19	Trachoma	B	E	0	S	V	V
20	**Trematodes**	**P**	**0**	**V, S**	**S**	**V**	**V**

Evaluation: Done by German Ministry of Health, Berlin
Abbreviations of agents of diseases: B = bacteria, P = parasite, V = virus
Note: Abbreviations (E, T, V, S, 0) have different meanings in groups 1–5
1. Vaccinations

E: Vaccinations not available
T: Vaccine candidates available but must be tested
V: Vaccines available, but not very effective
S: Vaccines for special groups (e.g. children) must be developed or ameliorated
0: Further research has no priority
2. Drugs
E: Drugs are not available but must be developed
T: Drug candidate available, needs to be tested in clinical trials
V: Drugs are registered, but not very effective
S: Drugs for special groups (e.g. children) must be developed or ameliorated
0: Further research has no priority
3. Therapy
E: Therapy options not available or poor—must be developed
T: Therapy options available—need clinical trials
V: Therapy options available—need ameliorations
S: Therapy options for special groups (e.g. children) must be developed
0: Further research has no priority
4. Diagnostics
E: Diagnostic products are not available and must be urgently developed
T: Diagnostic products are available, but clinical trials have to be done
V: Diagnostic products not highly effective
S: Diagnostic products for special groups (e.g. children) have to be developed
0: Further research not needed/has no priority
5. Operational activities
E: Efficient control measurements are available, but functions must be ameliorated
V: Implementation possibilities are low and must be ameliorated
0: Further research has no priority

Further Reading

Abdel-Ghaffar F, Sobhy HM, Metwaly KS, AbdEl-Monem S, Mehlhorn H (2019) Nile River, impact on human health in Egypt from pharaohs until now. In: Mehlhorn H, Klimpel S (eds.) Parasite and disease spread by major rivers on earth. Parasitology Research Monographs, vol 12. Springer International Switzerland, Cham, pp. 237–278

Almeida GT, Lage RC, Anderson L, Venancio TM, Nakaya HI, Miyasato PA, Rofatto HK, Zerlotini A, Nakano E, Oliveira G, Verjovski-Almeida S (2015) Synergy of omeprazole and praziquantel in vitro treatment against *Schistosoma mansoni* adult worms. PLoS Negl Trop Dis e0004086:9

Andrews P (1981) A summary of the efficacy of praziquantel against schistosomes in animal experiments and notes on its mode of action. Drug Res 31:538–541

Bagchus WM, Bezuidenhout D, Harrison-Moench E, Kourany-Lefoll E, Wolna P, Yalkinoglu O (2019) Relative bioavailability of orally dispersible tablet formulations of levo- and racemic praziquantel: two phase I studies. Clin Transl Sci 12:66–76

Becker B, Mehlhorn H, Andrews P, Thomas H (1980a) Scanning and transmission electron microscope studies on the efficacy of praziquantel on *Hymenolepis nana* (Cestoda) in vitro. Z Parasitenkd (Parasitol Res) 61:121–133

Becker B, Mehlhorn H, Andrews P, Thomas H, Eckert J (1980b) Light and electron microscopic studies on the effect of praziquantel on *Schistosoma mansoni*, *Dicrocoelium dendriticum* and *Fasciola hepatica* (Trematoda) in vitro. Z Parasitenkd (Parasitol Res) 63:113–128

Becker B, Mehlhorn H, Andrews P, Thomas H (1981) Ultrastructural investigations on the effect of praziquantel on the tegument of five species of cestodes. Z Parasitenkd (Parasitol Res) 64:257–269

Bergquist R, Utzinger J, Keiser J (2017) Controlling schistosomiasis with praziquantel: how much longer without a viable alternative? Infect Dis Poverty 6:74

Calvopiña M, Romero D, Castañeda B, Hashiguchi Y, Sugiyama H (2014) Current status of *Paragonimus* and paragonimiasis in Ecuador. Mem Inst Oswaldo Cruz 109:849–855

Cioli D, Pica-Mattoccia L (2003) Praziquantel Parasitol Res 90(Supp 1):S3–S9

Cioli D, Pica-Mattoccia L, Basso A, Guidi A (2014) Schistosomiasis control: praziquantel forever? Mol Biochem Parasitol 195:23–29

Colley DG, Bustinduy AL, Secor WE, King CH (2014) Human schistosomiasis. Lancet 383:2253–2264

Coulibaly JT, Panic G, Silué KD, Kovač J, Hattendorf J, Keiser J (2017) Efficacy and safety of praziquantel in preschool-aged and school-aged children infected with *Schistosoma mansoni*: a randomised controlled, parallel-group, dose-ranging, phase 2 trial. Lancet Glob Health 5:e688–e698

Cupit PM, Cunningham C (2015) What is the mechanism of action of praziquantel and how might resistance strike? Future Med Chem 7:701–705

da Silva VBR, Campos BRKL, de Oliveira JF, Decout JL, do Carmo Alves de Lima M (2017) Medicinal chemistry of antischistosomal drugs: Praziquantel and oxamniquine. Bioorg Med Chem 25:3259–3277

Davis A, Biles JE, Ulrich AM (1979) Initial experiences with praziquantel in the treatment of human infections due to *Schistosoma haematobium*. Bull World Health Organ 57:773–779

Doenhoff M, Kimani G, Cioli D (2000) Praziquantel and the control of schistosomiasis. Parasitol Today 16:364–366

Dömling A, Khoury K (2010) Praziquantel and schistosomiasis. ChemMedChem 5:1420–1434

Eckert J, Thompson RC, Mehlhorn H (1983) Proliferation and metastases formation of larval *Echinococcus multilocularis*. I. Animal model, macroscopical and histological findings. Z Parasitenkd (Parasitol Res) 69:737–748

Fenwick A (2015) Praziquantel: do we need another antischistosoma treatment? Future Med Chem 7:677–680

Fischer PU, Weil GJ (2015) North American paragonimiasis: epidemiology and diagnostic strategies. Expert Rev Anti-Infect Ther 13:779–786

Friedman JF, Olveda RM, Mirochnick MH, Bustinduy AL, Elliott AM (2018) Praziquantel for the treatment of schistosomiasis during human pregnancy. Bull World Health Organ 96:59–65

Gazzinelli-Guimaraes PH, Dhanani N, King CH, Campbell CH, Aurelio HO, Ferro J, Nala R, Fenwick A, Phillips AE (2018) Accuracy of the WHO praziquantel dose pole for large-scale community treatment of urogenital schistosomiasis in northern Mozambique: is it time for an update? PLoS Negl Trop Dis 12:e0006957

Gönnert R, Andrews P (1977) Praziquantel, a new board-spectrum antischistosomal agent. Z Parasitenkd (Parasitol Res) 52:129–150

Gryseels B, Mbaye A, De Vlas SJ, Stelma FF, Guissé F, Van Lieshout L, Faye D, Diop M, Ly A, Tchuem-Tchuenté LA, Engels D, Polman K (2001) Are poor responses to praziquantel for the treatment of *Schistosoma mansoni* infections in Senegal due to resistance? An overview of the evidence. Tropical Med Int Health 6:864–873

Ishizaki T, Kamo E, Boehme K (1979) Double-blind studies of tolerance to praziquantel in Japanese patients with *Schistosoma japonicum* infections. Bull World Health Organ 57:787–791

Jesudoss Chelladurai J, Kifleyohannes T, Scott J, Brewer MT (2018 Nov) (2018) Praziquantel resistance in the zoonotic cestode *Dipylidium caninum*. Am J Trop Med Hyg 99(5):1201–1205

Jiang B, Zhou XN, Zhang HB, Tao Y, Huo LL, Liu N (2017) Slow-release praziquantel for dogs: presentation of a new formulation for echinococcosis control. Infect Dis Poverty 6:140

Katz N, Rocha RS, Chaves A (1979) Preliminary trials with praziquantel in human infections due to *Schistosoma mansoni*. Bull World Health Organ 57:781–785

Kaewpitoon N, Kootanavanichpong N, Kompor P, Chavenkun W, Kujapun J, Norkaew J, Ponphimai S, Matrakool L, Tongtawee T, Panpimanmas S, Rujirakul R, Padchasuwan N, Pholsripradit P, Eksanti T, Phatisena T, Loyd RA, Kaewpitoon SJ (2015) Review and current status of *Opisthorchis viverrini* infection at the community level in Thailand. Asian Pac J Cancer Prev 16:6825–6830

Leopold G, Ungethüm W, Groll E, Diekmann HW, Nowak H, Wegner DH (1978) Clinical pharmacology in normal volunteers of praziquantel, a new drug against schistosomes and cestodes. An example of a complex study covering both tolerance and pharmacokinetics. Eur J Clin Pharmacol 14:281–291

Liu YH, Qian MX, Wang XG, Jia J, Wang QN, Jiang YF, Wang RQ, Yan SH, Chen BY, Li JS et al (1986) Comparative efficacy of praziquantel and its optic isomers in experimental therapy of schistosomiasis japonica in rabbits. Chin Med J 99:935–940

Mehlhorn H (2008) Encyclopedia of parasitology, 3rd edn. Springer, Berlin, Heidelberg

Mehlhorn H (2016a) Animal parasites. Diagnosis, treatment. In: prevention. Springer International Publishing, Cham

Mehlhorn H (2016b) Worms of humans. In: Human parasites. Diagnosis, treatment, prevention. Springer International Publishing, Cham. https://doi.org/10.1007/978-3-319-32802-7_4

Mehlhorn H (2016c) Encyclopedia of parasitology, 4th edn. Springer-Verlag, Berlin Heidelberg

Mehlhorn H, Becker B, Andrews P, Thomas H, Frenkel JK (1981) In vivo and in vitro experiments on the effects of praziquantel on *Schistosoma mansoni*. A light and electron microscopic study. Drug Res 31:544–554

Mehlhorn H, Klimpel S (eds.) (2019) Parasite and disease spread by major rivers on earth. Parasitol res monographs, vol 12, Springer International Switzerland, Cham

Mehlhorn H, Kojima S, Rim HJ, Ruenwongsa P, Andrews P, Thomas H, Bunnag B (1983) Ultrastructural investigations on the effects of praziquantel on human trematodes from Asia: *Clonorchis sinensis*, *Metagonimus yokogawai*, *Opisthorchis viverrini*, *Paragonimus westermani* and *Schistosoma japonicum*. Drug Res 33:91–98

Mehlhorn H, Schmahl G, Frese M, Mevissen I, Harder A, Krieger K (2005) Effects of a combinations of emodepside and praziquantel on parasites of reptiles and rodents. Parasitol Res 97 (Suppl 1):S65–S69

Mehlhorn H, Wu Z, Ye B (eds) (2014) Treatment of human schistosomiasis in traditional Chinese medicine. Parasitol res monographs, vol 6. Springer, Berlin, New York

Meister I, Kovac J, Duthaler U, Odermatt P, Huwyler J, Vanobberghen F, Sayasone S, Keiser J (2016) Pharmacokinetic study of praziquantel enantiomers and its main metabolite R-trans-4-OH-PZQ in plasma, blood and dried blood spots in *Opisthorchis viverrini*-infected patients. PLoS Negl Trop Dis e0004700:10

Montero R, Ostrosky P (1997) Genotoxic activity of praziquantel. Mutat Res 387:123–139

Mwangi IN, Sanchez MC, Mkoji GM, Agola LE, Runo SM, Cupit PM, Cunningham C (2014) Praziquantel sensitivity of Kenyan *Schistosoma mansoni* isolates and the generation of a laboratory strain with reduced susceptibility to the drug. Int J Parasitol Drugs Drug Resist 4:296–300

Olliaro P, Delgado-Romero P, Keiser J (2014) The little we know about the pharmacokinetics and pharmacodynamics of praziquantel (racemate and R-enantiomer). J Antimicrob Chemother 69:863–870

Olveda RM, Acosta LP, Tallo V, Baltazar PI, Lesiguez JL, Estanislao GG, Ayaso EB, Monterde DB, Ida A, Watson N, McDonald EA, Wu HW, Kurtis JD, Friedman JF (2016) Efficacy and safety of praziquantel for the treatment of human schistosomiasis during pregnancy: a phase 2, randomised, double-blind, placebo-controlled trial. Lancet Infect Dis 16:199–208

Panic G, Coulibaly JT, Harvey N, Keiser J, Swann J (2018) Characterizing the biochemical response to *Schistosoma mansoni* infection and treatment with praziquantel in preschool and school aged children. J Proteome Res 17:2028–2033

Panic G, Keiser J (2018) Acting beyond 2020: better characterization of praziquantel and promising antischistosomal leads. Curr Opin Pharmacol 42:27–33

Peters U, Kaiser G, Trapp R, Rosin H, Mehlhorn H (1984) Multiple parasitosis in an eosinophilic patient (*Schistosoma mansoni, Ancylostoma duodenale*). Forum Clin 35:427–430

Portokalidou S, Gkentzi D, Stamouli V, Varvarigou A, Marangos M, Spiliopoulou I, Dimitriou G (2019) *Dipylidium caninum* infection in children: clinical presentation and therapeutic challenges. Pediatr Infect Dis J 38:e157–e159

Qian MB, Utzinger J, Keiser J, Zhou XN (2016) Clonorchiasis Lancet 387:800–810

Schmahl G, Mehlhorn H (1985) Treatment of fish parasites. 1. Praziquantel effective against Monogenea (*Dactylogyrus vastator, Dactylogyrus extensus, Diplozoon paradoxum*). Z Parasitenkd (Parasitol Res) 71:727–737

Seubert J, Pohlke R, Loebich F (1977) Synthesis and properties of praziquantel, a novel broad spectrum anthelmintic with excellent activity against schistosomes and cestodes. Experientia 33:1036–1037

Shimada M, Kirinoki M, Shimizu K, Kato-Hayashi N, Chigusa Y, Kitikoon V, Pongsasakulchoti P, Matsuda H (2010) Characteristics of granuloma formation and liver fibrosis in murine schistosomiasis mekongi: a morphological comparison between *Schistosoma mekongi* and *S. japonicum* infection. Parasitology 137:1781–1789

Staudt U, Schmahl G, Blaschke G, Mehlhorn H (1992) Light and scanning electron microscopy studies on the effects of the enantiomers of praziquantel and its main metabolite on *Schistosoma mansoni* in vitro. Parasitol Res 78:392–397

Sun Q, Mao R, Wang D, Hu C, Zheng Y, Sun D (2016) The cytotoxicity study of praziquantel enantiomers. Drug Des Devel Ther 10:2061–1068

Taraschewski H, Mehlhorn H, Bunnag D, Andrews P, Thomas H (1986) Effects of praziquantel on human intestinal flukes (*Fasciolopsis buski* and *Heterophyes heterophyes*). Zentralbl Bakteriol Mikrobiol Hyg A 262:542–550

Thomas CM, Timson DJ (2018) The mechanism of action of praziquantel: six hypotheses. Curr Top Med Chem 18:1575–1584

Thomas H, Andrews P, Mehlhorn H (1982) New results on the effect of praziquantel in experimental cysticercosis. Am J Trop Med Hyg 31:803–810

Thomas H, Gönnert R (1977) The efficacy of praziquantel against cestodes in animals. Z Parasitenkd (Parasitol Res) 52:117–127

Thomas H, Gönnert R (1978) The efficacy of praziquantel against cestodes in cats, dogs and sheep. Res Vet Sci 24:20–25

Vale N, Gouveia MJ, Rinaldi G, Brindley PJ, Gärtner F, Correia da Costa JM (2017) Praziquantel for schistosomiasis: single-drug metabolism revisited, mode of action and resistance. Antimicrob Agents Chemother 61:e02582–e02516

Wahn V, Mehlhorn H (1984) Four kinds of parasites in an 8-year-old boy. Therapeutic effects of praziquantel against *Fasciola hepatica*. Dtsch Med Wochenschr 109:1486–1488

Wan F, Jiang M, Zhan A (eds) (2017) Biological invasions and its management in China, vol 2. Springer Nature, Singapore

Webbe G, James C (1977) A comparison of the susceptibility to praziquantel of *Schistosoma haematobium, S japonicum, S mansoni, S intercalatum* and *S mattheei* in hamsters. Z Parasitenkd (Parasitol Res) 52:169–177

WHO (2016) WHO urges increased access to praziqantel as schistosomiasis treatment coverage improves. In: Geneva: Head Quarter Information

Wu W, Qian X, Huang Y, Hong Q (2012) A review of the control of clonorchiasis sinensis and *Taenia solium* taeniasis/cysticercosis in China. Parasitol Res 111:1879–1884

Wu Z (2019) Epidemic status of schistosomiasis in Guangdong Province. In: Mehlhorn H, Klimpel S (eds.) Parasite and disease spread by major rivers on earth. Parasitology research monographs, vol 12. Springer International Switzerland, Cham, pp. 279–284

Wu Z, Liu Y, Mehlhorn H (eds.) (2019) Schistosomiasis control in China. The successful example of Jiangxi province. Parasitology research monographs, vol 11. Springer International Switzerland, Cham

Yoshida A, Doanh PN, Maruyama H (2019) *Paragonimus* and paragonimiasis in Asia: An update. Acta Trop 105074:199

Young ND, Chan KG, Korhonen PK, Min Chong T, Ee R, Mohandas N, Koehler AV, Lim YL, Hofmann A, Jex AR, Qian B, Chilton NB, Gobert GN, McManus DP, Tan P, Webster BL, Rollinson D, Gasser RB (2015) Exploring molecular variation in *Schistosoma japonicum* in China. Sci Rep 5:17345

Zwang J, Olliaro P (2017) Efficacy and safety of praziquantel 40 mg/kg in preschool-aged and school-aged children: a meta-analysis. Parasit Vectors 10:47

Chapter 3
The Global Status and Control of Human Schistosomiasis: An Overview

Wei Wang and Kun Yang

Abstract Schistosomiasis is an ancient but highly debilitating disease. Currently, this infectious disease is identified as the second most devastating parasitic disease after malaria in terms of its morbidity and mortality and significant medical and veterinary importance in tropical and subtropical regions across the world. In this chapter, we describe the epidemiology, diagnosis, management, and control of human schistosomiasis. In addition, we propose the potential challenges during the progress towards elimination of schistosomiasis in this wormy world.

Keywords Schistosomiasis · Epidemiology · Diagnosis · Treatment · Prevention · Elimination

3.1 Introduction

Schistosomiasis is a highly debilitating infectious disease of poverty that leads to chronic ill health (McManus et al. 2018). This parasitic disease, which was firstly documented by a German physician Theodor Bilharz in Egypt in 1851, was initially named bilharzia, and currently, the most widely accepted term for this disease is schistosomiasis, although bilharzia remains to be used sometimes (Olveda et al. 2014a). As a global neglected tropical zoonosis, schistosomiasis is identified as the

W. Wang (✉)
Key Laboratory of National Health Commission on Parasitic Disease Prevention and Control, Jiangsu Provincial Key Laboratory of Parasites and Vector Control Technology, Jiangsu Institute of Parasitic Diseases, Wuxi, Jiangsu Province, China
e-mail: wangwei@jipd.com

K. Yang
Key Laboratory of National Health Commission on Parasitic Disease Prevention and Control, Jiangsu Provincial Key Laboratory of Parasites and Vector Control Technology, Jiangsu Institute of Parasitic Diseases, Wuxi, Jiangsu Province, China

School of Public Health, Nanjing Medical University, Nanjing, Jiangsu Province, China

K. Yang, H. Mehlhorn (eds.), *Sino-African Cooperation for Schistosomiasis Control in Zanzibar*, Parasitology Research Monographs 15,
https://doi.org/10.1007/978-3-030-72165-7_3

second most devastating parasitic disease after malaria in terms of its morbidity and mortality and significant medical and veterinary importance in tropical and subtropical regions across the world (LoVerde 2019). In this chapter, we describe the epidemiology, diagnosis, management, and control of human schistosomiasis. In addition, we also analyze the potential challenges during the progress towards elimination of schistosomiasis in this wormy world.

3.2 Epidemiology

Among the genus *Schistosoma*, there are currently six main species of the parasite that infects humans, including *S. japonicum*, *S. mansoni*, *S. haematobium*, *S. intercalatum*, *S. mekongi*, and *S. guineensis* (Colley et al. 2014). Each species of *Schistosoma* has its specific intermediate host snails (Table 3.1), and the geographical distribution of the schistosomiasis is determined by the location of the intermediate host snails (Sanogo et al. 2018). Among these six species of human schistosomes, *S. japonicum*, *S. mansoni,* and *S. haematobium* are the three most common and dangerous parasites. *S. haematobium* is the most common species and is prevalent in more than 50 countries across the world (Toor et al. 2020), and *S. mansoni* has been reported in sub-Saharan Africa, Brazil, the Caribbean islands, Puerto Rico, Suriname, and Venezuela (Toor et al. 2020), while *S. japonicum* is endemic in the People's Republic of China, Indonesia, and the Philippines (Zhou et al. 2010). Unlike *S. mansoni* and *S. haematobium*, *S. japonicum* is a zoonotic species with more than 40 mammalians serving as its reservoir hosts (Van Dorssen et al. 2017).

Currently, schistosomiasis is estimated to affect around 240 million in 78 tropical and subtropical countries throughout the world (Fig. 3.1), with up to 780 million at risk of infections (McManus et al. 2018). In 2017, this parasitic disease was responsible for 1.43 million disability-adjusted life years (DALYs) (GBD 2017 DALYs and HALE Collaborators, 2018) and 1.09 million years lived with disability (YLDs) (GBD 2017 Disease and Injury Incidence and Prevalence Collaborators

Table 3.1 Intermediate hosts and geographical distribution of the species of *Schistosoma*

Parasite species	Intermediate host	Geographical distribution
S. japonicum	*Oncomelania* spp.	China, Indonesia, the Philippines
S. mansoni	*Biomphalaria* spp.	Africa, the Middle East, the Caribbean, Brazil, Venezuela and Suriname
S. haematobium	*Bulinus* spp.	Africa, the Middle East, Corsica (France)
S. intercalatum	*Bulinus* spp.	West Africa and Central Africa
S. mekongi	*Neotricula* spp.	Several districts of Cambodia and the Lao People's Democratic Republic
S. guineensis	*Bulinus* spp.	West Africa and Central Africa

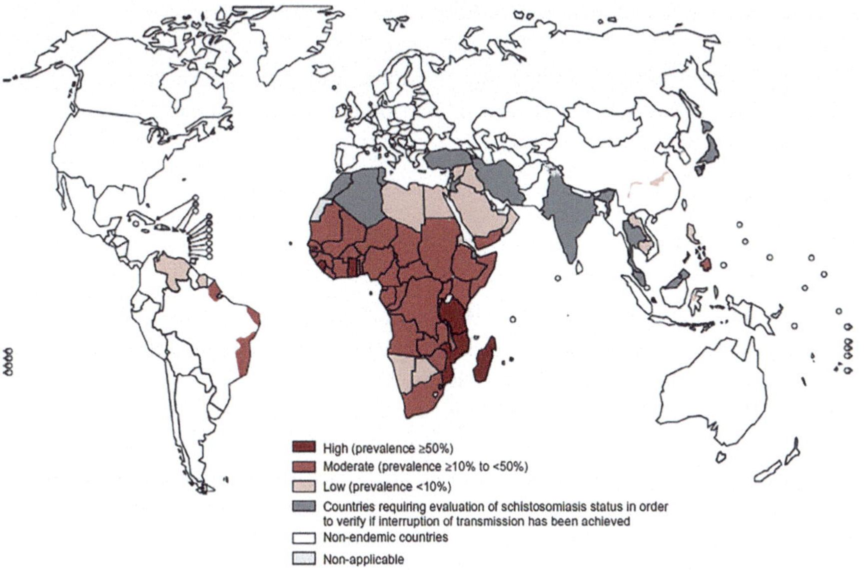

Fig. 3.1 Global distribution of schistosomiasis

2018); however, the burden of disease due to schistosomiasis was considered to be greatly underestimated (King and Galvani 2018).

3.3 Diagnosis

Currently, parasitological diagnosis, by means of detection of parasite eggs or miracidia in stool or urine sample, or rectal biopsy using microscopy, remains the gold standard for the identification of human *Schistosoma* infections (Gray et al. 2011). However, the conventional parasitological approaches are time-consuming and have a high rate of missing diagnosis, notably in areas with a low endemicity (Lin et al. 2008). Antibody- and antigen-based serological assays provide a rapid tool for the large-scale epidemiological survey and screening of potential therapeutic targets for human infections; however, the problems, including high sensitivity, high specificity, and discrimination between action infections and previous infections, remain to be solved (Hinz et al. 2017). The great advances in molecular biology facilitate the use of DNA detection in schistosomiasis, and PCR (Nelwan 2019), real-time PCR (Guegan et al. 2019), loop-mediated isothermal amplification (LAMP) assay (Vicente et al. 2018), recombinase polymerase assay (RPA) (Poulton and Webster 2018) recombinase-aided isothermal amplification (RAA) assay. (Song et al. 2018) have shown promising values in the diagnosis of human schistosomiasis. In addition, imaging tools, such as ultrasound, computed tomography (CT), and magnetic resonance imaging (MRI), are feasible for the accurate assessment of schistosomiasis-induced morbidities (Olveda et al. 2014b). Ultrasound contributes greatly to the diagnosis of intestinal and urinary schistosomiasis, and CT and MRI scanning provides additional diagnosis of ectopic forms of schistosomiasis, notably neuroschistosomiasis and spinal cord schistosomiasis (Olveda et al. 2014b).

3.4 Treatment

Since its introduction in middle 1970s, praziquantel has been replaced other antischistosomal agents and become the drug of choice for the treatment of human schistosomiases because of its high efficacy, low toxicity, easy administration, and low cost (Bergquist et al. 2017). Praziquantel, given at a dose of 40 or 60 mg/kg body weight, achieves a satisfactory cure rate and causes few adverse events (Kabuyaya et al. 2018). However, this agent fails to prevent reinfections and is ineffective against juvenile parasites (Wu et al. 2011), and there is a growing worry about the emergence of praziquantel-resistant parasite strains, since there has been evidence of resistance to praziquantel detected in schistosomes (Wang et al. 2012).

The antimalarial artemether shows a preventive activity against multiple species of *Schistosoma* (Bergquist and Elmorshedy 2018), and dihydroartemisinin, another antimalarial, was reported to be active against both adult and juvenile worms (Zhang et al. 2014); however, the use of artemisinin and derivatives alone or in combination with praziquantel is not recommended for schistosomiasis chemoprophylaxis or therapy in malaria-endemic areas because of the possible selection of artemisinin-resistant *Plasmodium* spp. parasites. Mefloquine, a quinine-derived antimalarial, has shown active against *S. mansoni*, *S. haematobium*, and *S. japonicum* in in vitro assays and animal experiments (Xiao 2013); however, results from a randomized, exploratory, open-label trial showed that the addition of mefloquine or mefloquine-artesunate does not increase the efficacy of praziquantel against chronic *S. haematobium* infection (Keiser et al. 2014). In addition, some traditional Chinese medicines show activity against schistosomiasis and schistosomiasis-induced hepatic fibrosis (Wu and He 2013).

Most cases with schistosomiasis do not require surgical treatment; however, patients with the disease suffering from variceal bleeding require surgery treatment with band ligation, endoscopic sclerotherapy, splenectomy, portal systemic shunts, and combinations of gastro-oesophageal devascularization (Ferraz et al. 2001). In addition, surgical treatment is recommended for neuroschistosomiasis to remove the foci in the brain (Shu et al. 2009; Wu et al. 2015).

3.5 Management

Currently, mass drug administration (MDA) with praziquantel remains the global strategy for schistosomiasis control (Mutapi et al. 2017). However, lessons from the past schistosomiasis control programs have shown that praziquantel MDA is unlikely to eliminate schistosomiasis in the world (Inobaya et al. 2015; Secor 2015), and an integrated strategy, which combines praziquantel administration, snail control, safe water provision, sanitation improvement, and health education, is required (Ross et al. 2017). Such an integrated strategy has shown great successes for schistosomiasis in China (Qian et al. 2018), and schistosomiasis has been eliminated as a public health problem in the country according to the WHO criteria (Xu et al. 2020).

3.6 Prevention

Since there have been no schistosomiasis vaccines until now (Tebeje et al. 2016), prevention of infections requires to avoid the contact with the infested water. Therefore, provision of safe water and improved sanitation facilities may greatly reduce the risk of exposure to infested water and the resultant risk of infections (Campbell et al. 2018). For those with occupational exposure like boatmen and fishermen, anticercarial cream and watertight suits may be useful for prevention of infections (Chun-Li et al. 2016).

3.7 Challenges for the Global Schistosomiasis Elimination Program

Considering great successes achieved in the national schistosomiasis control, an ambitious goal of schistosomiasis elimination was set by the WHO (Rollinson et al. 2013); however, there are still many challenges for the global schistosomiasis elimination program (Deol et al. 2019). (1) The natural hybridization between human and animal schistosome species may post a great impact on host range, praziquantel efficacy, host morbidity and hence ultimately transmission of schistosomiasis in the current era of elimination (Leger and Webster 2017). (2) Introduction of snail hosts to previously non-infesting areas and invasive *Schistosoma* species may cause emerging schistosomiasis, which affects local schistosomiasis elimination programs (Wang et al. 2013). (3) In the context of MDA with praziquantel as the global schistosomiasis elimination strategy, the shortage and drug resistance of praziquantel will be a big blow to the global schistosomiasis elimination program (Tchuem Tchuenté et al. 2017).

3.8 Concluding Remarks

The elimination of schistosomiasis as a public health problem in China, which once bore the highest burden of disease across the world (Chen 1989; Song et al. 2016), greatly encourages the ongoing global schistosomiasis elimination program (Wang and Yang 2020). More importantly, China is transferring its experiences and available tools into schistosomiasis-affected countries (Wang et al. 2019), which may facilitate the adjustment of the global schistosomiasis elimination strategy. Although there are still challenges ahead during the progress moving towards schistosomiasis elimination, geospatial tools, fortunately, provide an effective weapon for precision mapping of schistosomiasis in disease-affected regions, which provide valuable insights into the subsequent precision interventions (Bergquist et al. 2015). Given the importance of praziquantel, screening and development of novel antischistosomal agents, as alternatives of praziquantel, should be given a high priority. In addition, a global sensitive and effective surveillance and response system to tackle the unknown risk is of great urgent and importance for schistosomiasis elimination in the world (Tambo et al. 2014).

Acknowledgements This study was funded by the grants from Jiangsu Provincial Department of Science and Technology (grant No. BZ2020003), Jiangsu Provincial Commission of Health (grant Nos. ZDRCA2016056, QNRC2016621 and H2018097), and Wuxi Municipal Bureau of Science and Technology (grant No. CSE31N1730).

References

Bergquist R, Elmorshedy H (2018) Artemether and praziquantel: origin, mode of action, impact, and suggested application for effective control of human schistosomiasis. Trop Med Infect Dis 3:125

Bergquist R, Utzinger J, Keiser J (2017) Controlling schistosomiasis with praziquantel: how much longer without a viable alternative? Infect Dis Poverty 6:74

Bergquist R, Yang GJ, Knopp S, Utzinger J, Tanner M (2015) Surveillance and response: tools and approaches for the elimination stage of neglected tropical diseases. Acta Trop 141:229–234

Campbell SJ, Biritwum NK, Woods G, Velleman Y, Fleming F, Stothard JR (2018) Tailoring water, sanitation, and hygiene (WASH) targets for soil-transmitted helminthiasis and schistosomiasis control. Trends Parasitol 34:53–63

Chen MG (1989) Progress and problems in schistosomiasis control in China. Trop Med Parasitol 40:174–176

Chun-Li C, Shi-Zhu L, Xiao-Nong Z (2016) Impact of schistosomiasis transmission by catastrophic flood damage and emergency response in China. Zhongguo Xue Xi Chong Bing Fang Zhi Za Zhi 28:618–623

Colley DG, Bustinduy AL, Secor WE, King CH (2014) Human schistosomiasis. Lancet 383:2253–2264

Deol AK, Fleming FM, Calvo-Urbano B, Walker M, Bucumi V, Gnandou I, Tukahebwa EM, Jemu S, Mwingira UJ, Alkohlani A, Traoré M, Ruberanziza E, Touré S, Basáñez MG, French MD, Webster JP (2019) Schistosomiasis—assessing progress toward the 2020 and 2025 global goals. N Engl J Med 381:2519–2528

Ferraz AA, Bacelar TS, Silveira MJ, Coelho AR, Câmara Neto RD, de Araújo Júnior JG, Ferraz EM (2001) Surgical treatment of schistosomal portal hypertension. Int Surg 86:1–8
GBD 2017 DALYs and HALE Collaborators (2018) Global, regional, and national disability-adjusted life-years (DALYs) for 359 diseases and injuries and healthy life expectancy (HALE) for 195 countries and territories, 1990–2017: a systematic analysis for the global burden of disease study 2017. Lancet 392:1859–1922
GBD 2017 Disease and Injury Incidence and Prevalence Collaborators (2018) Global, regional, and national incidence, prevalence, and years lived with disability for 354 diseases and injuries for 195 countries and territories, 1990–2017: a systematic analysis for the global burden of disease study 2017. Lancet 392:1789–1858
Gray DJ, Ross AG, Li YS, McManus DP (2011) Diagnosis and management of schistosomiasis. BMJ d2651:342
Guegan H, Fillaux J, Charpentier E, Robert-Gangneux F, Chauvin P, Guemas E, Boissier J, Valentin A, Cassaing S, Gangneux JP, Berry A, Iriart X (2019) Real-time PCR for diagnosis of imported schistosomiasis. PLoS Negl Trop Dis e0007711:13
Hinz R, Schwarz NG, Hahn A, Frickmann H (2017) Serological approaches for the diagnosis of schistosomiasis—a review. Mol Cell Probes 31:2–21
Inobaya MT, Olveda RM, Tallo V, McManus DP, Williams GM, Harn DA, Li Y, Chau TN, Olveda DU, Ross AG (2015) Schistosomiasis mass drug administration in the Philippines: lessons learnt and the global implications. Microbes Infect 17:6–15
Kabuyaya M, Chimbari MJ, Mukaratirwa S (2018) Efficacy of praziquantel treatment regimens in pre-school and school aged children infected with schistosomiasis in sub-Saharan Africa: a systematic review. Infect Dis Poverty 7:73
Keiser J, Silué KD, Adiossan LK, N'Guessan NA, Monsan N, Utzinger J, N'Goran EK (2014) Praziquantel, mefloquine-praziquantel, and mefloquine-artesunate-praziquantel against *Schistosoma haematobium*: a randomized, exploratory, open-label trial. PLoS Negl Trop Dis e2975:8
King CH, Galvani AP (2018) Underestimation of the global burden of schistosomiasis. Lancet 391:307–308
Leger E, Webster JP (2017) Hybridizations within the genus *Schistosoma*: implications for evolution, epidemiology and control. Parasitology 144:65–80
Lin DD, Liu JX, Liu YM, Hu F, Zhang YY, Xu JM, Li JY, Ji MJ, Bergquist R, Wu GL, Wu HW (2008) Routine Kato-Katz technique underestimates the prevalence of *Schistosoma japonicum*: a case study in an endemic area of the People's Republic of China. Parasitol Int 57:281–286
LoVerde PT (2019) Schistosomiasis. Adv Exp Med Biol 1154:45–70
McManus DP, Dunne DW, Sacko M, Utzinger J, Vennervald BJ, Zhou XN (2018) Schistosomiasis. Nat Rev Dis Primers 4:13
Mutapi F, Maizels R, Fenwick A, Woolhouse M (2017) Human schistosomiasis in the post mass drug administration era. Lancet Infect Dis 17:e42–e48
Nelwan ML (2019) Schistosomiasis: life cycle, diagnosis, and control. Curr Ther Res Clin Exp 91:5–9
Olveda DU, Li Y, Olveda RM, Lam AK, McManus DP, Chau TN, Harn DA, Williams GM, Gray DJ, Ross AG (2014a) Bilharzia in the Philippines: past, present, and future. Int J Infect Dis 18:52–56
Olveda DU, Olveda RM, Lam AK, Chau TN, Li Y, Gisparil AD 2nd, Ross AG (2014b) Utility of diagnostic imaging in the diagnosis and management of schistosomiasis. Clin Microbiol 3:142
Poulton K, Webster B (2018) Development of a lateral flow recombinase polymerase assay for the diagnosis of *Schistosoma mansoni* infections. Anal Biochem 546:65–71
Qian C, Zhang Y, Zhang X, Yuan C, Gao Z, Yuan H, Zhong J (2018) Effectiveness of the new integrated strategy to control the transmission of *Schistosoma japonicum* in China: a systematic review and meta-analysis. Parasite 25:54

Rollinson D, Knopp S, Levitz S, Stothard JR, Tchuem Tchuenté LA, Garba A, Mohammed KA, Schur N, Person B, Colley DG, Utzinger J (2013) Time to set the agenda for schistosomiasis elimination. Acta Trop 128:423–440

Ross AG, Chau TN, Inobaya MT, Olveda RM, Li Y, Harn DA (2017) A new global strategy for the elimination of schistosomiasis. Int J Infect Dis 54:130–137

Sanogo B, Yuan D, Zeng X, Zhang Y, Wu Z (2018) Diversity and compatibility of human schistosomes and their intermediate snail hosts. Trends Parasitol 34(6):493–510

Secor WE (2015) Early lessons from schistosomiasis mass drug administration programs. F1000Res 4:F1000 Faculty Rev-1157

Shu K, Zhang S, Han L, Lei T (2009) Surgical treatment of cerebellar schistosomiasis. Neurosurgery 64:941–944

Song LG, Wu XY, Sacko M, Wu ZD (2016) History of schistosomiasis epidemiology, current status, and challenges in China: on the road to schistosomiasis elimination. Parasitol Res 115:4071–4081

Song Z, Ting L, Kun Y, Wei L, Jian-Feng Z, Li-Chuan G, Yan-Hong L, Yang D, Qing-Jie Y, Hai-Tao Y (2018) Establishment of a recombinase-aided isothermal amplification technique to detect Schistosoma japonicum specific gene fragments. Zhongguo Xue Xi Chong Bing Fang Zhi Za Zhi 30:273–277

Tambo E, Ai L, Zhou X, Chen JH, Hu W, Bergquist R, Guo JG, Utzinger J, Tanner M, Zhou XN (2014) Surveillance-response systems: the key to elimination of tropical diseases. Infect Dis Poverty 3:17

Tchuem Tchuenté LA, Rollinson D, Stothard JR, Molyneux D (2017) Moving from control to elimination of schistosomiasis in sub-Saharan Africa: time to change and adapt strategies. Infect Dis Poverty 6:42

Tebeje BM, Harvie M, You H, Loukas A, McManus DP (2016) Schistosomiasis vaccines: where do we stand? Parasit Vectors 9:528

Toor J, Rollinson D, Turner HC, Gouvras A, King CH, Medley GF, Hollingsworth TD, Anderson RM (2020) Achieving elimination as a public health problem for *Schistosoma mansoni* and *S. haematobium*: when is community-wide treatment required? J Infect Dis 221:S525–S530

Van Dorssen CF, Gordon CA, Li Y, Williams GM, Wang Y, Luo Z, Gobert GN, You H, McManus DP, Gray DJ (2017) Rodents, goats and dogs—their potential roles in the transmission of schistosomiasis in China. Parasitology 144:1633–1642

Vicente B, López-Abán J, Muro A (2018) Field and laboratory comparative evaluation of a LAMP assay for the diagnosis of urogenital schistosomiasis in Cubal, Central Angola. Tropical Med Int Health 23:992–1001

Wang W, Liang YS, Hong QB, Dai JR (2013) African schistosomiasis in mainland China: risk of transmission and countermeasures to tackle the risk. Parasit Vectors 6:249

Wang W, Wang L, Liang YS (2012) Susceptibility or resistance of praziquantel in human schistosomiasis: a review. Parasitol Res 111:1871–1877

Wang W, Yang K (2020) Schistosomiasis and the global goals. N Engl J Med 382:1575–1576

Wang XY, He J, Juma S, Kabole F, Guo JG, Dai JR, Li W, Yang K (2019) Efficacy of China-made praziquantel for treatment of schistosomiasis haematobium in Africa: a randomized controlled trial. PLoS Negl Trop Dis e0007238:13

Wu W, Feng A, Huang Y (2015) Research and control of advanced schistosomiasis japonica in China. Parasitol Res 114:17–27

Wu W, Wang W, Huang YX (2011) New insight into praziquantel against various developmental stages of schistosomes. Parasitol Res 109:1501–1507

Wu ZL, He L (2013) Advances in traditional Chinese medicine treatment of schistosomiasis liver fibrosis. Zhongguo Xue Xi Chong Bing Fang Zhi Za Zhi 25:422–424

Xiao SH (2013) Mefloquine, a new type of compound against schistosomes and other helminthes in experimental studies. Parasitol Res 112:3723–3740

Xu J, Li SZ, Zhang LJ, Bergquist R, Dang H, Wang Q, Lv S, Wang TP, Lin DD, Liu JB, Ren GH, Yang K, Liu Y, Dong Y, Zhang SQ, Zhou XN (2020) Surveillance-based evidence: elimination of schistosomiasis as a public health problem in the Peoples' Republic of China. Infect Dis Poverty 9:63

Zhang XG, Li GX, Zhao SS, Xu FL, Wang YH, Wang W (2014) A review of dihydroartemisinin as another gift from traditional Chinese medicine not only for malaria control but also for schistosomiasis control. Parasitol Res 113:1769–1773

Zhou XN, Bergquist R, Leonardo L, Yang GJ, Yang K, Sudomo M, Olveda R (2010) Schistosomiasis japonica control and research needs. Adv Parasitol 72:145–178

Chapter 4
Schistosomiasis Control Program in Zanzibar: An Overview

Jian He and Fatma Kabole

Abstract This chapter introduced the history of schistosomiasis control in Zanzibar, then we focused on the main activities/initiatives of schistosomiasis control, introduce its background, goals, methods and findings, etc. In the end, the national plan which is associated with schistosomiasis control in Zanzibar was briefly introduced.

Keywords Schistosomiasis control · Initiatives · Plan · Zanzibar

Abbreviations

SSA	sub-Saharan Africa
MoHSW	Ministry of Health and Social Welfare
SCI	Schistosomiasis Control Initiative
GPELF	Global Program to Eliminate Lymphatic Filariasis
UNICEF	United Nations Children's Fund
NTD	Neglected Tropical Diseases
WHO	World Health Organization
MoH	Ministry of Health
SCORE	Schistosomiasis Consortium for Operational Research and Evaluation
ZEST	Zanzibar Elimination of Schistosomiasis Transmission
MDA	Mass Drug Administration
UCP-LF CAA	up-converting phosphor-lateral flow circulating anodic antigen
PZQ	praziquantel

J. He (✉)
Jiangsu Institute of Parasitic Diseases, Wuxi, Jiangsu, China
e-mail: hejian@jipd.com

F. Kabole
Ministry of Health of Zanzibar, Zanzibar, Tanzania

K. Yang, H. Mehlhorn (eds.), *Sino-African Cooperation for Schistosomiasis Control in Zanzibar*, Parasitology Research Monographs 15,
https://doi.org/10.1007/978-3-030-72165-7_4

MDG	Millennium Development Goals
NHM-SCI	Natural History Museum- Schistosomiasis Control Initiative
BCC	behaviour change communication
ZAWA	Zanzibar Water Authority
TORs	Sample Terms of Reference
IEC	Information, Education and Communication
KAP	Knowledge, Attitude and Practice
PCT	Preventive Chemotherapy Treatment
NGOs	Non-governmental organizations
HSRS	Health Sector Reform Secretariat
RGoZ	Revolutionary Government of Zanzibar

Human schistosomiasis is a public problem causing severe morbidities, which is second only to malaria in sub-Saharan Africa (SSA). Of the world's 207 million estimated cases of schistosomiasis, more than 90% are occurred in SSA, in this region, Nigeria being the first country that has the highest burden of schistosomiasis, and the United Republic of Tanzania is the second (Ross et al. 2002; Steinmann et al. 2006). As a part of Tanzania, Zanzibar is located in the west of Indian Ocean between latitudes 5 and 7 degrees south of the equator. The distance from the East Coast of Africa to it is about 30 kilometers. The total area of Zanzibar is 2654 square kilometers, which is consisted of two main islands, Unguja and Pemba, though there are several other smaller islands of which are uninhabited. Unguja is the largest island with an area of 1666 square kilometers while Pemba is the second which has an area of 988 square kilometers. 41% of Zanzibar's woody biomass is indigenous while 59% is exotic. The climate of Zanzibar is determined by the trade winds of the tropical Monsoon system. From April to October there is Southwest Monsoon, which is also called as Kusi by local people, literally meaning southerly in Swahili and Northeast Monsoon occurs from November to March, local people call it Kaskazi in Swahili. There are two rainfall pattern in nature, with a long rainy season which is from mid March to end of May (Masika), and the short one is from October to December (Vuli) (Ali et al. 2006). Pemba on average receives more rainfall than Unguja, and there is more rainfall in the western sides than the east.

Temperatures in Zanzibar are relatively high during the short dry season (January to February) with maximum mean of 32 °C. The cool season lasting from May to September which is also known as Kipupwe in Swahili, because of it is associated with periods of scattered showers. The mean annual maximum and minimum temperatures are 29.3 °C and 21.1 °C, respectively. Humidity value is around 80% due to its location, and sometimes the temperature can be as high as 40 °C when the land is braced with hot breezes.

4.1 Schistosomiasis Endemicity in Zanzibar

First documentation of urogenital schistosomiasis in Zanzibar dates to 1903, when *Schistosoma haematobium* infection symptoms were noted to be prevalent among men in Zanzibar. As early as 1972, reports emerged regarding a wide distribution of *S. haematobium* on Unguja and Pemba. The endemic region on Unguja is restricted to the northwestern and central areas, but on Pemba Island, the region spans the western, southern, central and northwest parts of the island (Fig. 4.1). The geography of disease transmission on Unguja closely follows the geographical distribution of the intermediate snail host, *Bulinus globosus*. On Pemba Island, both *Bulinus globosus* and *Bulinus nasutus* are important intermediate hosts of *S. haematobium* (Mazigo et al. 2012). There are two distinct transmission seasons of *S. haematobium*, and the time is the same as rain season (Jones 2015). Ongoing chemical control of snail hosts on both islands using molluscicides has been attempted since at least

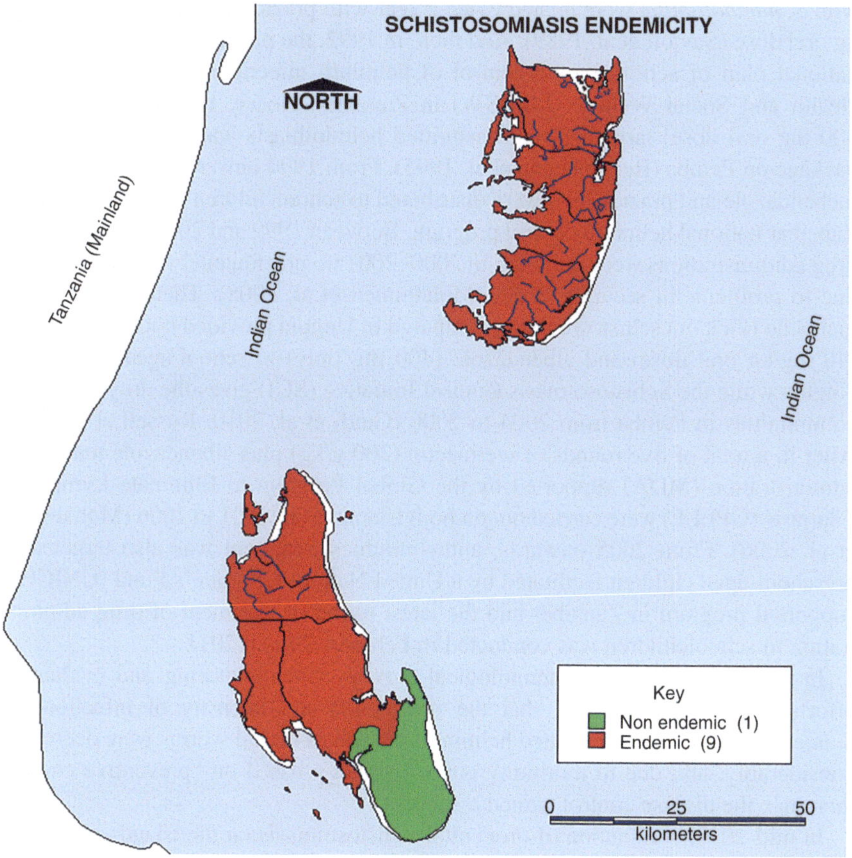

Fig. 4.1 Schistosomiasis endemicity in Zanzibar

1965, but in the last 25 years, morbidity reduction through preventative chemotherapy of school children has been the primary focus of control initiatives. In early 2012, a baseline parasitological survey was conducted in 20,000 people from 90 communities in Unguja and Pemba, Schistosoma haematobium prevalence in school children and adults was 4.3% (range: 0–19.7%) and 2.7% (range: 0–26.5%) in Unguja, and 8.9% (range: 0–31.8%) and 5.5% (range: 0–23.4%) in Pemba, respectively (Knopp et al. 2013).

4.2 Initiatives for Schistosomiasis Control in Zanzibar

4.2.1 Initiatives before 2010

The first program to control *S. haematobium* initiated in 1986 on Pemba island, with the objectives of eliminating morbidity due to *S. haematobium*. Individuals infected with *S. haematobium* were treated twice a year with praziquantel at a single 40 mg/kg oral dose (Savioli et al. 1989). And then, in 1992, the program was extended to a national plan of action for the control of helminth infections by the Ministry of Health and Social Welfare (MoHSW) in Zanzibar. Hence, mebendazole (single 500 mg oral dose) targeting soil-transmitted helminthiasis was added to the drug package on Pemba (Renganathan et al. 1995). From 1994 onwards, on both islands, mebendazole and praziquantel were distributed to schoolchildren in the frame of the Zanzibar national helminth control program. Between 1999 and 2003, anthelminthic drug administrations were irregular, in 2000–2001 no praziquantel was administered due to problems in securing drugs (Mohammed et al. 2008). Then, the Piga vita kichocho (kick out schistosomiasis) campaign in Unguja provided both praziquantel (40 mg/kg oral dose) and albendazole (400 mg only) to school-aged children in Unguja while the Schistosomiasis Control Initiative (SCI) gives the drugs to whole communities in Pemba from 2004 to 2006 (Guidi et al. 2010; Russell et al. n.d.). After it, a total of five rounds of ivermectin (200 g/kg) plus albendazole mass drug administration (MDA) supported by the Global Program to Eliminate Lymphatic Filariasis (GPELF) were carried out on both islands from 2001 to 2006 (Mohammed et al. 2006). From 2005 onwards, anthelminthics treatment was also targeted to preschool-aged children facilitated by a United Nations Children's Fund (UNICEF) supported program in Zanzibar and the latest round of treatment of drug administration to schoolchildren was conducted in February/March 2011.

In this period, many epidemiological surveys and monitoring and evaluation efforts in Zanzibar revealed that the prevalence and intensity of infections of *S. haematobium*, soil-transmitted helminths as well as filarial worms were decreased considerably, and due to morbidity control strategy based on "preventive chemotherapy", the disease control gained a success.

In mid-2010, elimination of urogenital schistosomiasis on the islands of Unguja and Pemba was committed by Zanzibar government. As main content for the commitment, Zanzibar government implemented its "National Plan" as the Zanzibar

Neglected Tropical Disease (NTD) Program in its "3-year comprehensive strategic plan to combat neglected tropical diseases in Zanzibar 2009/2011". Preventive chemotherapy by praziquantel was companied by health education and community mobilization activities to consolidate and enhance the impact of it. The National Plan is focusing on urogenital schistosomiasis areas including the districts, and zones within district (Knopp et al. 2012).

To support this National Plan, the World Health Organization (WHO) has agreed to supply sufficient quantities of praziquantel to ensure repeated rounds of preventive chemotherapy, and the Schistosomiasis Control Initiative (SCI) also helped to implement this activity at a large scale. WHO and SCI have long histories of cooperating with the Zanzibar Ministry of Health (MoH) to combat schistosomiasis and soil-transmitted helminthiasis (Savioli et al. 1989; Stothard et al. 2006).

4.2.2 Schistosomiasis Consortium for Operational Research and Evaluation (SCORE)

The Schistosomiasis Consortium for Operational Research and Evaluation (SCORE) is a program which was launched between 2008 and 2019 focusing on schistosomiasis control in SSA, it also supported an operational research project in Zanzibar (Unguja and Pemba) with the Zanzibar NTD Control Program and some other partners. It was funded by Bill & Melinda Gates Foundation at USD 23 million, which is aimed to identify some key questions and put them into a common framework geared at producing updated guidelines (Colley et al. 2020b). The project was implemented in eight sub-Saharan countries to help them combat schistosomiasis by various epidemiologic settings (King et al. 2020). Importantly, by implementing the SCORE activities, it helped to deepen the relationship between researchers and local control staff, scale up the possibility of sustainable improvements (Bergquist 2020).

Zanzibar was selected as the study site for schistosomiasis elimination in the project. The choose of it was determined by many factors, including strongly stated political support from the President of Zanzibar, clear geographic boundaries, the commitment of biannual MDA by the Ministry of Health and its partners, as well as other available resources (Colley et al. 2020b).

The goals of this project are as follows:

1. To eliminate schistosomiasis as a public health problem on Unguja in 3 years and to interrupt transmission in 5 years,
2. To control schistosomiasis throughout Pemba (prevalence $<10\%$) in 3 years and to eliminate it as a public health problem in 5 years, and
3. To learn what is effective and what are the costs, successful strategies, barriers, etc. associated with three different interventions.

This schistosomiasis elimination project was implemented in 2011 within the framework of Zanzibar national control and prevention efforts. In this period, a 5-year national schistosomiasis control strategy by the Zanzibar NTD Program with many partners, the Zanzibar Elimination of Schistosomiasis Transmission (ZEST), were committed to working together to achieve these common goals. The findings included a very low prevalence setting where most of those infected harboured very light intensities of *S. haematobium* infections, and it was also found that a low sensitivity of light infection (<5 eggs/ 10 ml urine) which was detected by the urine filtration assay to detect eggs and hematuria reagent strips to detect microhematuria (Knopp et al. 2018; Knopp et al. 2015).

Schistosome population genetics study was also carried out in Zanzibar, it examined the *schistosome* population genetic structure under multiple communitywide Mass Drug Administrations (MDAs) (four times of MDAs over a 5-year period in schistosomiasis control studies and 12 times of MDAs over a 6-year period in elimination study). The population genetic analyses found that the gene flow was at a high level, and the mixing of the parasite populations in neighbouring sites could dilute the effects of the project. Furthermore, parasite fecundity had significant inherent differences, independent of current treatment arm, but potentially of profound impact in terms of maintaining high levels of ongoing transmission in apparent PHS[1] sites.

SCORE also had some attempt to develop highly sensitive and genus-specific up-converting phosphor-lateral flow circulating anodic antigen (UCP-LF CAA) assay (Colley et al. 2020c). This assay been used many times in the SCORE studies within the ZEST program, and it showed much higher levels of *S. haematobium* than urine filtration method (Knopp et al. 2015). The project suggested this diagnose method could play an important role as a highly sensitive and specific diagnostic test and for verification of elimination of schistosomiasis.

The key findings of SCORE project were shown in Table 4.1 (Colley et al. 2020a).

4.2.3 Zanzibar Elimination of Schistosomiasis Transmission (ZEST) Project

Zanzibar Elimination of Schistosomiasis Transmission (ZEST) was an international consortium, was aimed to assist the government of Zanzibar to eliminate urogenital schistosomiasis. The consortium included the Zanzibar MoH (including the Zanzibar NTD Control Program, the Public Health Laboratory—Ivo de Carneri (PHL-IdC) Pemba), Zanzibar government agencies, WHO, SCI, the University of New Mexico and the Schistosomiasis Consortium for Operational Research and

[1]In SCORE data, MDA coverage in villages that did not respond, which were called persistent hotspots (PHSs).

Table 4.1 Summary of SCORE's six key findings and related key messages (Colley et al. 2020a)

	Key findings of SCORE	Key messages of SCORE
1	Four years of school-based treatment and/or communitywide treatment MDA with PZQ—biannually, annually, or biennially—is effective in reducing average *Schistosoma* prevalence and the intensity of infection.	Biennial MDA through school and/or communities is sufficient to reach a programmatic goal of moderate prevalence of infection. Biennial MDA is insufficient to reach low prevalence of infection.
2	No SCORE MDA regimens eliminated *Schistosoma* transmission within 5–6 years.	MDA alone will not achieve a programmatic goal of elimination of transmission in most settings.
3	All MDA regimens in the SCORE gaining control and sustaining control studies left at least 30% of the study villages in all study arms as persistent hotspots. (PHSs) these PHS villages, by definition, failed to decrease as expected in prevalence and intensity following multiple years of MDA. Monitoring outcomes to assess if villages are likely to be PHS versus responder villages is feasible as soon as a year after 2 years of annual MDAs.	A programmatic goal to gain and sustain control of schistosomiasis in all villages receiving MDA should identify persistent hotspots following 2 years of annual MDA through an epidemiologic assessment. On identification of PHSs, interventions should be adjusted, for example, more intensive MDA and/or complementary interventions to drive down the prevalence and intensity of the PHS. Continued research is needed to evaluate these options. Efforts can be maintained, adapted, or possibly decreased in responder villages.
4	If a village starts with[3] 25% prevalence, then over 4 years, annual MDA is (i) more effective at reducing prevalence and intensity than 2 years of MDA (biennial) and (ii) leaves fewer PHSs than biennial MDA.	A program will be most effective at reducing prevalence and intensity of infection and lowering the number of PHS villages with annual MDA.
5	Both control of morbidity and elimination as a public health problem goal28 based on the percent prevalence of heavy infection are achieved through MDA. Often, in areas with established moderate-to-high prevalence, these goals defined by the prevalence of heavy intensity are fulfilled even before any MDA, and, thus, need to be redefined.	Current WHO definitions of control of morbidity and elimination as a public health problem based on intensities of infection are inappropriate for determining success related to changes in infection. For programs to determine success of their efforts to control and/or eliminate schistosomiasis, new targets, which are evidence-based, urgently need to be defined.
6	Use of the point-of-care circulating cathodic antigen (POCCCA) urine assay for *S. mansoni* finds more low-intensity infections in low-to-moderate prevalence areas than the standard parasitologic methods. The POC-CCA yields some false positives, especially in areas where prevalence is extremely low (i.e., <5%).	The POC-CCA urine assay for *S. mansoni* is an appropriate tool for epidemiologic assessments (i.e., mapping and impact assessment) in areas of low-to-moderate prevalence (5–40% by Kato–Katz) to prevent undertreatment. POC-CCA cannot be used as a diagnostic tool to determine interruption of transmission (elimination).

MDA mass drug administration, *PHS* persistent hotspot, *POC-CCAs* point-of-care circulating cathodic antigen assays, *PZQ* praziquantel, *SCORE* Schistosomiasis Consortium for Operational Research and Evaluation

Evaluation (SCORE) based at the University of Georgia, the Swiss Tropical and Public Health Institute (Swiss TPH) in Basel, the Natural History Museum (NHM) in London, the London School of Hygiene and Tropical Medicine (LSHTM), with additional groups. Within the ZEST project, the work content would shift from morbidity control to comparative methods of transmission control, until local elimination of urogenital schistosomiasis. In addition to measuring the outcomes of this project, thorough documentation of the process, including lessons learned, will be crucial so that schistosomiasis control and elimination programs implemented elsewhere could benefit from the ZEST experiences.

By comparing and closely monitoring the outcome of differential intervention methods, ZEST project wanted to develop an evidence-based strategy, which can be a reference for elimination of urogenital schistosomiasis in an endemic SSA country.

Also, the goal of ZEST was to provide decisions about schistosomiasis elimination by evidence-base studies, not only for the Zanzibar NTD Control Program, but also for other settings in Africa or elsewhere that aimed for schistosomiasis elimination. Both Unguja and Pemba islands were implemented and compared snail control and behaviour change strategies as complementary measures to preventive chemotherapy in three study arms, each study group was consisting of 15 communities (shehias). A total of 45 study communities were involved on each island.

The aims of the ZEST project were as follows:

1. Eliminate schistosomiasis as public health problem in Unguja in 3 years, interrupt transmission in 5 years.
2. Control schistosomiasis in Pemba in 3 years and eliminate it as a public health problem in 5 years.
3. To identify effective behaviour change strategies with an understanding of the associated costs, motivators, triggers and barriers associated with behaviour change interventions.
4. To identify effective snail control strategies with an understanding of the associated costs, motivators, triggers and barriers associated with snail control interventions.

To reach the goal of the project, eight specific objectives, activities and milestones were proposed (Table 4.2).

The ZEST study was in the frame of SCORE and the National Plan of Zanzibar. The trial will have three study arms, each comprising 15 shehias. Hence, on each island, a total of 45 shehias was included into the study. The three study arms are as follows:

1. Treatment with praziquantel (PZQ); 2. Treatment with PZQ plus Snail control with niclosamide; 3. Treatment with PZQ plus behaviour change and community engagement (Figs. 4.2 and 4.3).

The lessons learnt and challenges identified by ZEST were showed as follows:

1. Treatment only intervention: Six rounds of treatment were implemented since 2012. It was found that drug compliance could be better when treatments were distributed to children via schools rather than via local health distributors.

Table 4.2 Eight specific objectives, activities and milestones in ZEST project

1. To assess annually the reduction in prevalence and intensity of *S. haematobium* infection according to standardized, quality-controlled methods (i.e. urine filtration and reagent strip testing) in schoolchildren aged 9–12 years within and between each study arm from year 1 to year 5.
2. To assess the reduction in the prevalence and intensity of *S. haematobium* infection as measured by urine filtration and reagent strip testing in adults in year 1 and in year 5 within and between each study arm.
3. To assess the reduction in the prevalence and intensity of *S. haematobium* infection as measured by urine filtration and reagent strip testing in first-year schoolchildren in year 1 and in year 5 within and between each study arm.
4. To assess the difference in the sero-prevalence of *S. haematobium* infection in first-year schoolchildren after 4 years of control within and between each study arm.
5. To assess annually changes in knowledge on *S. haematobium* transmission and the impact of change of behaviour on transmission.
6. To assess the impact of niclosamide on the presence of *B. globosus* in freshwater ponds and rivers.
7. To test and evaluate recently developed methods for the diagnosis of *S. haematobium* infections (such as enzyme linked immuno-sorbent assay (ELISA) for assessment of antibody levels in blood; polymerase chain reaction (PCR) for DNA detection in urine) for their sensitivity, specificity and feasibility for elimination programs.
8. To create, test and validate mathematical models for the prediction of *S. haematobium* prevalence after control interventions (to be developed with external partners).

ZEST zanzibar elimination of schistosomiasis transmission

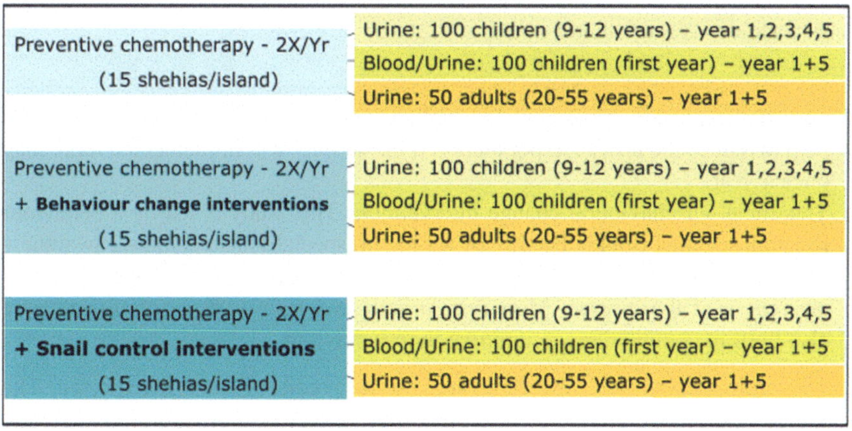

Fig. 4.2 Study arms to assess the impact of different *S. haematobium* control interventions in Zanzibar (Knopp et al. 2012)

Children receive their treatment with porridge in school which could make it more likely that they will swallow the drug.

2. Treatment + Snail Control intervention: Half of the surveyed water bodies on Unguja and Pemba had *Bulinus globosus* snails. By spraying with niclosamide in

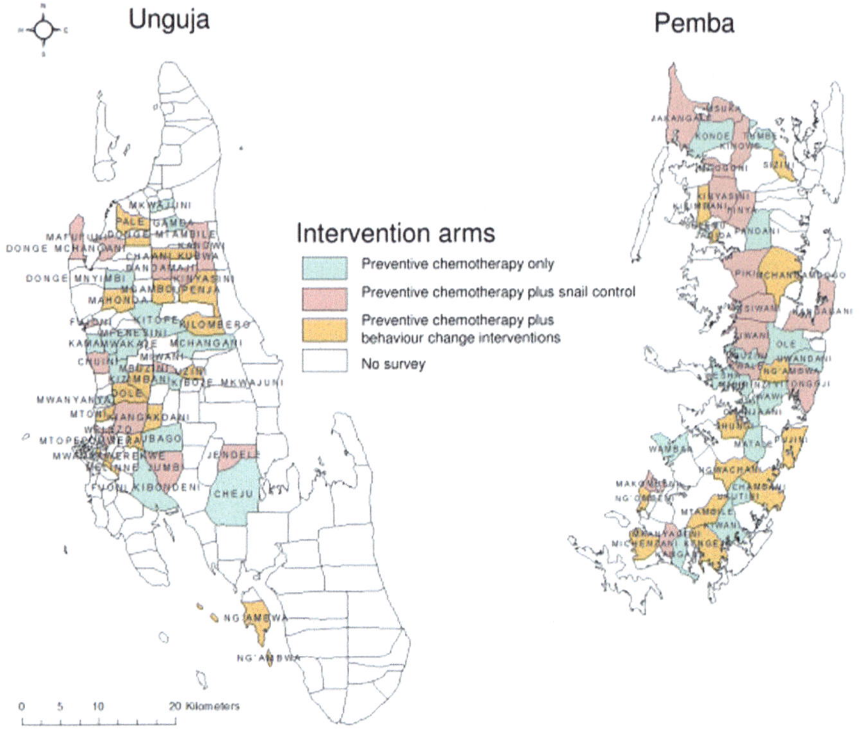

Fig. 4.3 Assignment of communities (shehias) to control interventions aiming to eliminate urogenital schistosomiasis in Zanzibar (Knopp et al. 2012)

waterbodies, a safe, biodegradable molluscicide donated by Bayer AG, it could help to reduce the host snail population if the activity was implemented more than once a year.

3. Treatment + Behavioural change intervention: This study is led by a behaviour research scientist Bobbie Person, by setting up special schistosomiasis awareness days in both public and religious (madrasssa) schools, it improved people's awareness of the disease. Moreover, urinals and safe washing platforms were also built in communities to improve their behaviour.

The project also found few challenges, some predicted some unexpected. The impact of migration could be a challenge for the effect of schistosomiasis elimination, and the Zanzibar government should be find more solutions to reduce the impact of it (Gouvras 2015).

4.3 Strategic Plans of Schistosomiasis Control in Zanzibar

4.3.1 Health Master Plan for the Control of Neglected Tropical Diseases in Zanzibar

The latest specific NTD control plan which included schistosomiasis was the Health master plan for the control of Neglected Tropical Diseases in Zanzibar 2012–2016. The development of this strategic planned to combat NTD in Zanzibar 2012–2016, with extensive efforts involving broad consultations and collaborations of various stakeholders including the NTD program of the MoH, PHL Pemba, WHO AFRO, WHO country office and WHO Zanzibar sub-office as well as NTD partners.

To control neglected tropical diseases in Zanzibar- Tanzania, the plan was an essential tool that aimed at ensuring effective plans were set forth for the implementation of sustainable NTD control programs in the Islands. This multiyear plan provided a comprehensive outlook of the 5-years implementation plan stipulating the National NTD goals, objectives and activities proposed for implementation following extensive analysis of the country's situation in various areas of NTDs control in the country. The plan was foreseen as a tool that will enhance synergies between programs and amongst Partners, but as well highlight specific vertical disease control efforts as guided by the global targets in attempt to eliminate, eradicate or control each disease at country level. The plan had clearly shown the plans and related costs required stipulating the financial gap and donors supporting various control activities.

The NTD master plan had as well included strategies for financial sustainability that linked to the health sector budgeting and planning cycles, but also encouraged strong linkages with other programs within and outside the health sector.

The thrust in this Master plan was made such that it was comprehensive enough to include both preventive chemotherapy as well as case management aspects for NTD control in Tanzania. Its approach was reflected from the national strategic priorities rather than the disease specific approach so as to ensure that integration is maximally utilized for cost effective programs as well as sustainability. Where applicable, activities have been consolidated and integration proposed with other health interventions and within the NTD program to assist solve shared problems; The plan would act as a strong base for the annual NTD work plan along the 5 years of program implementation.

A framework has been put such that it supports Partners coordination, harmonization and alignment. The plan used indicators guided by the WHO to effectively monitor and report NTD control progress at various levels of program implementation.

The Master plan had provided a description of the institutional framework used to guide its formulation. It was then followed by the main parts of the plan which included situational analysis of neglected tropical diseases and the health sector; the NTD Strategic Agenda; the Operational framework and Budget summary.

The development of this strategic plan of action was just the beginning of a process which would require key NTD stakeholders to team with the government of Zanzibar in its efforts towards attainment of Millennium Development Goals (MDG). The government of Zanzibar and the Ministry of Health in particular were committed to collaborate with these partners both within and outside Zanzibar to realize the objectives of this plan.

Data collected before the implementation of disease control interventions shows that prevalence of infection of *S. haematobium* among school-age children was about 65% in Unguja (1981) and 70% in Pemba. After a few years of irregular distribution, in 2004, prevalence of infection went down to 49.8% and 64.5% in Unguja and Pemba. After 3 years of several interventions in collaboration with Natural History Museum- Schistosomiasis Control Initiative (NHM-SCI) prevalence of Schistosomiasis decreased to 7.8% in Unguja and 8.1% in Pemba. However, recently collected data in 2007 and 2008 have shown that some pockets of high prevalence persist. According to the survey in 2011, the schistosomiasis morbidities were 7.6% and 15.1% in Unguja and Pemba, respectively.

The plan also revealed that the MDA is the major strategy of schistosomiasis control, although the geographic coverage is 100%, the percentage of No. of population is 75%, and the key partners are WHO and SCI. Case management was also implemented when doing MDA.

4.3.1.1 Overall Mission, Vision and Goal

The master plan had taken note of the importance of a clear agenda to support NTD upscale and sustainability. The vision set forth, mission and goals align with each other to support effective activity implementation and impact assessment at the end of the 5 years period. In the course of annual planning and program monitoring of this plan, milestones to be formulated will guide the trend and thus highlight any areas that will need amendment or additional support (Table 4.3).

4.3.1.2 Guiding Principles and Strategic Priorities

Priority interventions of this Master plan had been guided by the situational analysis and the country readiness for sustained control, pre-elimination and elimination. Details in each specific area have shown Zanzibar plans stipulating our readiness to

Table 4.3 Overall mission, vision and goal in health master plan for the control of neglected tropical diseases in Zanzibar

Mission	To sustain NTD control activities in all levels in the country
Vision	Zanzibar population living free from NTD infection
Goal	Control of NTD in Zanzibar that they are no longer a public health problem

NTD neglected tropical diseases

control eliminate or move towards the pre-elimination stage. Areas needing great focus include strengthening coordination mechanisms at national and sub-national level, strengthening the integration and linkages of NTD program, financial and budgetary mechanisms, strengthening integrated vector management for target NTDs, and support for operational research, documentation and evidence to guide innovative approaches to NTD program interventions.

The strategic priorities and strategic objectives are showed in Table 4.4.

Table 4.4 Strategic framework summary

Strategic priorities	Strategic objectives
Strategic priority 1: Strengthen government ownership, advocacy, coordination and partnership	Strengthen coordination mechanism for the NTD control program at national, regional, district and community level in Zanzibar Strengthen and foster partnerships for NTDs at national, regional, district and community levels in Zanzibar Enhance NTD program performance reviews and use for decision making by the revolutionary government of Zanzibar Strengthen advocacy, visibility and profile of NTD control program in Zanzibar
Strategic priority 2: Enhance planning for results, resource mobilization and financial sustainability	Support regions and districts to develop integrated annual plans for NTD control in Zanzibar Enhance resource mobilization approaches and strategies at national, regional, district and community levels in Zanzibar Strengthen the integration and linkages of NTD program and financial plans into sector-wide and national budgetary and financing mechanisms in Zanzibar Collaborate with partners to develop and update national NTD policies guidelines and tools for the Zanzibar NTD program
Strategic priority 3: Scale up access to interventions, treatment and system capacity building	Scale up preventive chemotherapy packages in Zanzibar Scale up integrated case-management-based diseases interventions and chronic care especially for lymphatic filariasis, schistosomiasis, trachoma, leprosy, human rabies prevention and other prevalent case management NTDs in Zanzibar Strengthen transmission control measures for all endemic NTDs in Zanzibar including integrated vector management, environment, and health promotion Strengthen capacity for NTD program management and implementation at national, regional and district levels
Strategic priority 4: Enhance NTD monitoring and evaluation, surveillance and operational research	Enhance monitoring of national NTD program performance and outcome at national, regional, district and community level Strengthen the surveillance of NTDs and response for epidemic prone NTDs in Zanzibar (in particular, rabies and dengue) Support operational research, documentation and evidence to guide innovative approaches to NTDs program interventions Establish integrated data management system and support impact analysis of NTD in Zanzibar as part of WHO Africa region and global NTD data management system Enhance pharmacovigilance activities linked to NTD interventions in Zanzibar Establish mechanisms for post intervention surveillance and integration within primary health care

NTD neglected tropical diseases

4.3.1.3 Measures to Be Taken in Schistosomiasis Control Plan

The plan also articulated on what the country's capacity needs are how resources will be mobilized, how potential foreseen risks will be dealt with and expected outcomes in the course of implementation (Tables 4.5 and 4.6).

Various stakeholders have been involved to have enough contributions and interests of many captured for the program's achievement and sustainability. The plan also described how the NTDs control will continue to be streamlined at the sector level to effectively establish longer term multisectoral involvement at various operating levels as already observed by the high political leaders of this country. It also showed the proposed activities to be implemented in order to achieve the strategic objectives highlighted to strengthen government ownership, advocacy, coordination and partnerships (Table 4.7).

Monitoring is the process of continuous observation and collection of data on the NTD program to ensure that the program is progressing and as planned. Evaluation is the systematic and critical analysis of the adequacy, efficiency and effectiveness of the program, its strategies as well as progress. Evaluation referred to long, mid-term and annual analysis of performance in relation to the goals, objectives and targets sets.

In the plan, monitoring and evaluation of schistosomiasis (ZEST program) will be done concurrently with Schistosomiasis monitoring.

The country has begun with pilot studies for schistosomiasis elimination with support from the ZEST, with a coalition of members who are supporting the program from 2011 onwards. ZEST two main outcomes were as follows:

1. Schistosomiasis as a public health problem eliminated in Unguja in 3 years (2012/2014).
2. Schistosomiasis disease controlled in Pemba in 3 years (2012/2014) and eliminated as a public health problem in 5 years 2012/2016).

Monitoring of schistosomiasis had therefore taken into consideration the two main outcomes of the project. Strategies would therefore be laid to effectively monitor both control and elimination plans and activities. Below are the proposed indicators.

Performance based which will check on the therapeutic and geographical coverage Geographical coverage (number of districts reached for praziquantel treatment) Therapeutic coverage (number of population eligible for preventive chemotherapy	Impact -epidemiological surveys to assess trends of infection rates humans and snail hosts Prevalence and intensity of infection with urinary schistosomiasis (microscopic, etc.) Prevalence of macrohaematuria in humans

Table 4.5 Program summary components of strategies for schistosomiasis

Program and global goal	National goals	Objectives	Interventions	Delivery channels	Target population
Schistosomiasis elimination Goal: Elimination of schistosomiasis as public health problem by 2020.	To eliminate schistosomiasis in Zanzibar by 2016	1. To establish and implement effective strategies to eliminate schistosomiasis as a public health problem in 3 years in Unguja 2. To establish and implement effective strategies to control schistosomiasis throughout Pemba in 3 years and eliminate it as a public health problem in 5 years	1. Preventive chemotherapy 2. Case management 3. Transmission control 4. Improve water supply and sanitation	1. Mass drug administration 2. Health facility management 3. School-based MDAs 4. Media programs 5. Health promotion in schools and communities 6. Advocacy	1.5 mil

MDA mass drug administration

Table 4.6 Schistosomiasis program objectives and key indicators of performance

NTD Program	Objective	Key Indictor	Baseline	Target	Milestones				
					2012	2013	2014	2015	2016
SCHISTO	1. To establish and implement effective strategies to eliminate schistosomiasis as a public health problem in 3 years in Unguja 2.To establish and implement effective strategies to control schistosomiasis throughout Pemba in 3 years and eliminate it as a public health problem in 5 years	Proportion of schools receiving PZQ in endemic districts implementing MDA in Unguja by 2016 (two rounds annual)	84 (100%) Schools	84 (100%) Schools	100%	100%	100%	100%	100%
		Proportion of districts where eligible members in communities receiving PZQ implementing MDA for schistosomiasis in Unguja by 2016	4 (100%) Districts	4 (100%) Districts	100%	100%	100%	100%	100%
		Proportion of schools receiving PZQ in endemic districts implementing MDA in Pemba by 2016	93 (100%) Schools)	93 (100%) Schools)	100%	100%	100%	100%	100%
		Proportion of districts where eligible members in communities receiving PZQ implementing MDA for schistosomiasis in Pemba by 2016	4 (100%) Districts	4 (100%) Districts	100%	100%	100%	100%	100%
		Number of schistosomiasis cases managed in health facilities in Unguja (currently *schistosoma* cases are managed at NTD centres)	0	100%	100%	100%	100%	100%	100%

		Number of schistosomiasis cases managed in health facilities in Pemba (currently *schistosoma* cases are managed at NTD centres)	0	100%	100%	100%	100%	100%	100%
		Number and type of BCC conducted for schist control in Unguja and Pemba	0	100%	100%	100%	100%	100%	100%
		Number and type of vector control activities conducted in Unguja and Pemba	0	100%	100%	100%	100%	100%	100%
		4. Number and type of collaborative activities implemented with ZAWA and the sanitation unit	0	100% of planned activities	100%	100%	100%	100%	100%

PZQ praziquantel, *MDA* mass drug administration, *NTD* neglected tropical diseases, *BCC* behaviour change communication, *ZAWA* zanzibar water authority

Table 4.7 Activities to strengthen government ownership, advocacy, coordination and partnership

Activity	Details (Sub-activities)	Timeframe	Resources needed
Strategic objective 1: Strengthen coordination mechanism for the NTD control program at national, regional, district and community level in Zanzibar			
Develop harmonized policy guidelines for implementing NTD program	Organize and support a team for the formulation of the selected policy guideline	Feb–March 2012	Conference package, stationery and printing, per diem, transport allowances
	Consultative meetings with stakeholders at all levels to review the policy drafts	April–December 2012	Refreshments,
	Printing and dissemination of the policy guideline	Jan–March 2013	Printing and dissemination funds
Formulation and periodic organization and support to the NTD steering committee	Establish/review TORs and appoint the national NTD steering committee members	Jan–Mar 2012/ Review every 2 years	In house activity
	Regularly organize national NTD steering committee meetings	Once every 6 months	Conference package, stationery and printing, per diem, transport allowances
Formulation and periodic organization and support to the NTD secretariat and smaller task force (technical advisory team) meetings	Establish the national NTD secretariat	Jan–Mar 2012 Review every 2 years	In house activity
	Regularly organize the NTD secretariat meeting	Monthly	Refreshments
	Regularly organize small task force meetings to continuously provide technical guidance to the NTD secretariat	Quarterly	Conference package, stationery and printing, per diem, transport allowances
To ensure running costs is available for the NTD secretariat activities *(Communication, Office supplies and stationeries, fuel, car maintenance)*	Participate in MOH budgeting cycles annually to have running costs expenses included in the annual plan (to be done with the strategic priority No. 2 and strategic objective No. 3)	Annually–June–July	In house
Strategic objective 2: Strengthen and foster partnerships for NTDs at national, regional, district and community levels in Zanzibar			
Forge and maintain partnership with relevant stakeholders for NTD control, elimination and eradication	Conduct mapping of national and international NTD supporting partners to also cover zoonotic diseases	Annually—June	In house
	Continuously update NTD partner information through attending coalition and NGO partner meetings	Annually	Travel allowance, per diem

(continued)

Table 4.7 (continued)

Activity	Details (Sub-activities)	Timeframe	Resources needed
Strategic objective 1: Strengthen coordination mechanism for the NTD control program at national, regional, district and community level in Zanzibar			
	Establish and facilitate review of partners/MOH MOUs through MOH legal office	As required	In house
Strategic objective 3: Enhance NTD program performance reviews and use for decision making by the revolutionary government of Zanzibar			
Conduct program review meetings with district and other key government officials at national and sub—National level (in Unguja and Pemba)	Support conduction of review meetings at district level to involve health teachers, facility staff and district officials to identify evidence-based information on the NTD program performance	Twice a year (Q2, Q4)	Conference package, stationery and printing, per diem, transport allowances
	Conduct review meetings at central level to discuss program progress (could be integrated with JAHSR etc.)	Once a year Q4	Conference package, stationery and printing, per diem, transport allowances
Conduct national NTD stakeholders meetings to review NTD performance	Conduct national NTD stakeholders meetings with national and international partners to review NTD performance in Unguja	Once a year	Conference package, stationery and printing, per diem, transport allowances
	Conduct national NTD stakeholders meetings national and international partners to review NTD performance in Pemba Island	Once a year	Conference package, stationery and printing, per diem, transport allowances
Strategic objective 4: Strengthen advocacy, visibility and profile of NTD control program in Zanzibar			
Organize meetings and or participate in other visibility activities with non-health sector and non-NTD health partners to raise NTD profile and awareness	Participate in high fora meetings to share NTD focus and achievements (house of representative, other ministries meetings, child health, malaria, immunization, military, religious)	Once a year	Travel allowance, per diem
	Organize meetings with district commissioners, religious and community leaders to advocate NTD needs and progress within the districts	Once a year	Conference package, stationery and printing, per diem, transport allowances

(continued)

Table 4.7 (continued)

Activity	Details (Sub-activities)	Timeframe	Resources needed
Strategic objective 1: Strengthen coordination mechanism for the NTD control program at national, regional, district and community level in Zanzibar			
	Organize visibility activities to advocate for NTDs through media on annual basis	Once a year	Funds
	Participate in local and international conferences related to NTD programs national and international	Once a year	Travel allowance, Perdiem
Periodically determine the knowledge, attitude, practice and acceptance of NTD control among communities in Unguja and Pemba Islands	Conduct KAP and acceptance studies on selected NTDs and their integrated control measures in endemic areas	Once in 2 years	Consultant fee, travel, perdiem
	Print and disseminate KAP study reports on integrated NTDs program to various stakeholders	Once in 2 years	Conference package, stationery and printing, per diem, transport allowances
Strengthen information, education and communication strategies for NTDs control in Zanzibar	Develop health communication strategy for integrated NTDs control program	Once a year	Funds
	Develop IEC/BCC materials and pre-test based on KAP study results	Once a year	Funds
	Print & Disseminate IEC/BCC materials on integrated NTDs in implementation (to include chronic care of LF, leprosy and zoonosis	Once a year	Funds

TORs sample terms of reference, *MDA* mass drug administration, *NTD* neglected tropical diseases, *BCC* behaviour change communication; *IEC* information, education and communication, *KAP* knowledge, attitude and practice, *JAHSR* joint annual health sector review, *NGO* non-governmental organization. Transmission control is cross-cutting activities for both vector-borne and other diseases that do not involve vectors for their transmission. In effect, these interventions compliment preventive chemotherapy and case-based intervention. Activities proposed in the plan include vector control as well as the environment measures through the package summarized below (Table 4.8). Variations in activities proposed have taken consideration of the different overlaps and differences of each transmission control approach as summarized in the intervention packages

Table 4.8 Intervention packages for transmission control

Interventions package	Methods of intervention delivery	Requirements	Other non-NTD opportunities for integration
Integrated vector control	– Snail control. – Environmental management.	– Niclosamide (molluscocide). – Clearing of grasses along water bodies banks (rivers, bridges, etc.).	
Strengthen systems to improve transmission control against NTDs	– Establishment of NTD technical team for integrated NTD transmission control that includes members from MOH, MOL MOA, ZAWA. – Conduct regular collaborative meetings addressing NTD issues. – Establish and conduct effective snail control measures for prevention of schistosomiasis.	– Regular collaborative meetings. – Appointments for NTD focal person from each ministry involved. – Encourage improvement of water supply and sanitation. – Encourage improvement of environmental management. – Encourage all necessary measures that reduce NTD transmission.	ZAWA project on country-wide improvement of water supply and latrine construction (ADB –funded project)
Community behavioural control interventions on NTDs	– Introduction of measures for behaviour changes .	– Established community committees on NTD control in communities. – Adequate provision of latrines in schools and usage. – Measure stopping children from urinating in water bodies.	
Health promotion on NTDs prevention and control	– Collaborate meetings with partners on NTD. – Development of IEC material on NTD, – Use of media TV, radio and others delivering NTD preventive and control messages.	– Advocacy for mass vaccination of dogs on control of rabies. – Several production and distribution of NTD—IEC materials (NTD posters, leaflets etc.) – Airing through media TV, radio NTD prevention and control messages .	

MDA mass drug administration, *NTD* neglected tropical diseases, *ZAWA* zanzibar water authority, *MOH* ministry of health, *MOL* ministry of lands, *MOA* ministry of agriculture; *IEC* information, education and communication, *ADB* African development bank

References

Ali AI, Schulthess F, Ogol CK, Omwega CO (2006) Effect of nitrogen fertilizer level on infestations of lepidopterous stemborers and yields of maize in Zanzibar. Annales de la Société entomologique de France 42:481–486

Bergquist NR (2020) Schistosomiasis consortium for operational research and evaluation: mission accomplished. Am J Trop Med Hyg 103(1_Suppl):1–4. https://doi.org/10.4269/ajtmh.19-0838

Colley DG et al (2020a) Contributions of the schistosomiasis consortium for operational research and evaluation (SCORE) to schistosomiasis control and elimination: key findings and messages for future goals, thresholds, and operational research. Am J Trop Med Hyg 103 (1_Suppl):125–134. https://doi.org/10.4269/ajtmh.19-0787

Colley DG, Jacobson JA, Binder S (2020b) Schistosomiasis consortium for operational research and evaluation (SCORE): its foundations, development, and evolution. Am J Trop Med Hyg 103(1_Suppl):5–13. https://doi.org/10.4269/ajtmh.19-0785

Colley DG et al (2020c) Evaluation, validation, and recognition of the point-of-care circulating cathodic antigen, urine-based assay for mapping schistosoma mansoni infections. Am J Trop Med Hyg 103(1_Suppl):42–49. https://doi.org/10.4269/ajtmh.19-0788

Gouvras A (2015) Zanzibar Elimination of Schistosomiasis Transmission (ZEST) aka Tokomeza Kichocho. http://blogs.biomedcentral.com/bugbitten/2015/07/15/zanzibar-elimination-of-schistosomiasis-transmission-zest-aka-tokomeza-kichocho/

Guidi A, Andolina C, Ame SM, Albonico M, Cioli D, Haji HJ (2010) Praziquantel efficacy and long-term appraisal of schistosomiasis control in Pemba Island. Trop Med Int Health 15 (5):614–618

Jones I (2015) An overview of schistosomiasis in Zanzibar. https://schisto.stanford.edu/pdf/Zanzibar.pdf

King CH et al (2020) Impact of different mass drug administration strategies for gaining and sustaining control of schistosoma mansoni and schistosoma haematobium infection in Africa. Am J Trop Med Hyg 103(1_Suppl):14–23. https://doi.org/10.4269/ajtmh.19-0829

Knopp S et al (2012) Study and implementation of urogenital schistosomiasis elimination in Zanzibar (Unguja and Pemba islands) using an integrated multidisciplinary approach. BMC Public Health 12(1):930

Knopp S, et al. (2013) Elimination of schistosomiasis transmission in Zanzibar: baseline findings before the onset of a randomized intervention trial

Knopp S et al (2015) Sensitivity and specificity of a urine circulating anodic antigen test for the diagnosis of schistosoma haematobium in low endemic settings. PLoS Negl Trop Dis 9(5): e0003752. https://doi.org/10.1371/journal.pntd.0003752

Knopp S et al (2018) Urogenital schistosomiasis elimination in Zanzibar: accuracy of urine filtration and haematuria reagent strips for diagnosing light intensity Schistosoma haematobium infections. Parasit Vectors 11(1):552. https://doi.org/10.1186/s13071-018-3136-6

Mazigo HD et al (2012) Epidemiology and control of human schistosomiasis in Tanzania. Parasit Vectors 5(1):1–20

Mohammed KA et al (2008) Triple co-administration of ivermectin, albendazole and praziquantel in zanzibar: a safety study. PLoS Negl Trop Dis 2(1):e171. https://doi.org/10.1371/journal.pntd.0000171

Mohammed KA, Molyneux DH, Albonico M, Rio F (2006) Progress towards eliminating lymphatic filariasis in Zanzibar: a model program. Trends Parasitol 22(7):340–344

Renganathan E et al (1995) Evolution of operational research studies and development of a national control strategy against intestinal helminths in Pemba Island, 1988–92. Bull World Health Organ 73(2):183–190

Ross AGP, Bartley PB, Sleigh AC, Olds R, Li Y, Williams GM, McManus DP (2002) Schistosomiasis. New Eng J Med 346(16):1212–1220

Russell SJ, French MD, Simba KI, María-Gloria B, David R The epidemiology and control of urinary schistosomiasis and soil-transmitted helminthiasis in schoolchildren on Unguja Island, Zanzibar. Trans R Soc Trop Med Hyg 10:1031–1044

Savioli L, Dixon H, Kisumku U, Mott K (1989) Control of morbidity due to Schistosoma haematobium on Pemba Island: program organization and management. Trop Med Parasitol 40(2):189–194

Steinmann P, Keiser J, Bos R, Tanner M, Utzinger J (2006) Schistosomiasis and water resources development: systematic review, meta-analysis, and estimates of people at risk. Lancet Infect Dis 6(7):411–425

Stothard J, Mook P, Mgeni A, Khamis I, Khamis A, Rollinson D (2006) Control of urinary schistosomiasis on Zanzibar (Unguja Island): a pilot evaluation of the educational impact of the Juma na Kichocho health booklet within primary schools. Mem Inst Oswaldo Cruz 101:119–124

Chapter 5
Sino-Africa Cooperation Project of Schistosomiasis Control: A Pathway Analysis

Jingyi Guo and Jing Xu

Abstract Schistosomiasis is still a major public health problem worldwide, which impinges on human health and depletes the development of the world economy. It is endemic in 78 countries or regions in the tropics and subtropics, while 51 of them are located in Africa. According to the WHO's estimation, more than 490 million Africans are threatened by schistosomiasis. In Africa, *Schistosoma mansoni* (*S. mansoni*) and *S. haematobium* are predominant, while *S. intercalatum* is limited to central and western Africa. The majority of cases are concentrated on Sub-Saharan Africa (SSA) where people live in poor environments with a lack of sanitation facilities and difficulties in obtaining safe water. In these countries, control and/or elimination of schistosomiasis is confronting with great challenges. Based on the current situation, there is an urgent need to map high-resolution schistosomiasis, improve diagnostic methods, expand the scope of drug treatment, and implement comprehensive approaches to improve sanitation, carry out effective hygiene education, and re-establish snail control in Africa.

Keywords Sino-Africa cooperation · Schistosomiasis control · A pathway analysis

5.1 Introduction

The People's Republic of China (P.R. China) used to have the highest disease burden of schistosomiasis caused by *S. japonicum*. The disease was widespread in twelve provinces in the south of China, mainly along the Yangtze River. National surveys conducted in the 1950s estimated that China had 11.60 million human cases and one million infected cattle. After more than 70 years of continuous national controls, the morbidity and mortality of schistosomiasis have decreased significantly. To date,

J. Guo · J. Xu (✉)
National Institute of Parasitic Diseases, Chinese Center for Disease Control and Prevention, Shanghai, China
e-mail: xujing@nipd.chinacdc.cn

K. Yang, H. Mehlhorn (eds.), *Sino-African Cooperation for Schistosomiasis Control in Zanzibar*, Parasitology Research Monographs 15,
https://doi.org/10.1007/978-3-030-72165-7_5

five provinces had eliminated schistosomiasis, and then Sichuan and Jiangsu provinces interrupted the transmission of schistosomiasis, leaving other five provinces in the stage of transmission control (both human and livestock prevalence rates are lower than 1%). During this period, China has accumulated plentiful experience and lessons in controlling schistosomiasis at different periods and conditions. These experiences may contribute to schistosomiasis control in other endemic countries in Africa.

The difference between schistosomiasis in Africa and China is primarily attributed to the intermediate hosts of schistosomes. The only snail that can transmit schistosomiasis in China, *Oncomelania hupensis*, which is amphibious, as opposed to all snail hosts in Africa. Despite this essential difference, there are many similarities in plotting, diagnostics, treatment, prevention, and management, which would prove that China's experience in schistosomiasis control could be transformed into an African scenario conducive to control activities there.

5.2 Favorable Conditions for Cooperation Between African Countries and P.R. China

5.2.1 Good Foundation for Cooperation

Since the Chinese government sent the first batch of medical rescue teams to Algeria in 1963, health cooperation between China and Africa has been strengthened by sending medical teams, establishing hospitals, training medical personnel, donating drugs and equipment, and controlling malaria. A total of 23,000 Chinese medical personnel have been allocated to 66 countries, benefiting 270 million people. In 2013, the Beijing Declaration of the China-Africa Ministerial Forum on Health Development agreed to jointly carry out pilot projects for schistosomiasis control and prevention with African countries. Through advanced courses and exchange visits by senior scientists sharing Chinese experience, cooperation between African countries and China has been carried out through two regional networks, namely, the Regional Network on Asian Schistosomiasis and other Helminth Zoonoses (RNAS+) and The Research Network on Schistosomiasis in Africa (RNSA). In 2015, to accelerate the cooperation between China and African countries, a network named Institution-based Network on China-Africa Cooperation for Schistosomiasis elimination (INCAS) was established under the support of World Health Organization and National Institute of Parasitic Disease, China CDC. It is envisaged to provide technical support in the near future for drafting national plans/strategies, training personnel, and directing control activities. In addition, in recent years, institutional cooperation between China and African countries has been previously strengthened.

5.2.2 Strong Policy Commitments and Great Achievements Gained

Over the past half century, schistosomiasis was once neglected in Africa compared to other acute infectious diseases. After recognizing the importance of schistosomiasis in public health, WHO Member States approved Resolution WHA 54.19 in 2001, which aims to achieve the goal of 75% coverage rate of praziquantel (PZQ) treatment for school-age children by 2010, which stimulated the interests in schistosomiasis control. In recent decades, many African countries have carried out national schistosomiasis control alone or together with neglected tropical diseases (NTDs), with support from key funding/donor agencies such as the World Bank, the World Vision, the US Agency for International Development (USAID), the United Kingdom Department for International Development (UKDFID), and Merck KGaA. The number of people being treated increased from 12.37 million in 2006 to 35.56 million in 2012. With Merck KGaA increasing its donation of PZQ in 2012, global donations of PZQ will exceed 250 million tablets, and an estimated 140 million people will be treated each year by 2016. In 2017, 81.82 million SAC and 16.85 million adults received treatment, and the coverage of chemotherapy in SAC reached 68% (WHO 2018). Among the population received treatment, WHO African regions accounted for 88.3%. Furthermore, many African countries are carrying out or have finished national surveys to map snail habitats and schistosomiasis endemic areas, which would help develop control strategies and effectively allocate limited resources for schistosomiasis control (19–22).

5.2.3 New Tools and Improved Capacity

To ensure the effectiveness, it is necessary to use useful tools and effective control approaches. With the recent elucidation of the schistosome genome, new immunological diagnostic methods based on recombinant and purified antigens, monoclonal antibodies against such antigens, and extraordinary sensitive and specific advanced technologies based on DNA detection have been explored and integrated into the ongoing control programs, targeting chemotherapy and monitoring snail infection. It is worth noting that some candidate vaccines have been explored and shown to be effective in the treatment of animal schistosomiasis. Geographic Information System (GIS), Global Positioning System (GPS), and Remote Sensing (RS) technologies, as well as mathematical or statistical models based on spatial, temporal, spatial-temporal analysis, could be used to draw risk maps of schistosomiasis transmission and provided guidance for policy makers. In addition, several international cooperation projects, such as the Schistosomiasis Control Initiative (SCI), the Schistosomiasis Consortium for Operational Research and Evaluation (SCORE), and CONTRAST, have initiated a multidisciplinary alliance to optimize the control and transmission surveillance of schistosomiasis in SSA.

5.2.4 Development of Social-Economics

The endemic of schistosomiasis is influenced not only by biological factors but also by other factors, such as social and economic conditions. African countries' economies have grown in the past decade, with exports growing at an annual rate of 5.2% in real terms, compared with the world average of 4.8% in 2011.Their per capita development value reached US $1000 for the first time in 2013. In the meantime, the living conditions in Africa have improved. In 2012, the safe water coverage rate of SSA reached 64%, while overall sanitation coverage rate of SSA grew slowly, from 25% in 1990 to 30% in 2012. Moreover, health system reforms have been strengthened or innovated in some countries, including the enactment of health-related laws, the establishment or amelioration of clinics that provide primary health care, the development of health information systems for health services monitoring, the classification of medical services, and the nationalization of health laboratories, thereby making health service more accessible.

5.3 Limited Factors for Cooperation Between African Countries and P.R. China

5.3.1 Different Epidemiological Characteristics

Although *Schistosoma* species that attack humans have similar life cycles and modes of transmission, there are always some differences in their epidemic characteristics in intermediate and definitive hosts, clinical manifestations, and seasons of transmission. *S. japonicum* is the only schistosome species in China; it is a zoonotic infection that affects more than 40 mammals. There is only one species of snail as intermediate host, i.e. the amphibious and dioecious *Oncomelania hupensis* snail. However, in Africa, *S. mansoni, S. haematobium*, and *S. intercalatum* are transmitted by aquatic and androgynous snails, *Biomphalaria spp., Bulinus spp., Bulinus globosus*, respectively. Among them, schistosomiasis japonica, schistosomiasis mansoni, and schistosomiasis intercalatum are characterized by symptoms of intestinal schistosomiasis, which may cause abdominal pain and bloody diarrhea. While Schistosomiasis haematobium is manifested as an urinary system disease, which could lead to severe damage to the bladder, ureter, and kidney, with common symptoms of hematuria.

5.3.2 Low Chemotherapy Coverage

In most endemic countries, prophylactic chemotherapy is the primary intervention, and coverage is a key issue in controlling efficacy. As of 2012, 44 countries in Africa needed chemotherapy, but prophylactic chemotherapy only covered 13.6% of the population. Many people are at high-risk due to the gap between drug donation and demand, as well as inadequate delivery capability of drugs. School-aged children are often treated with chemotherapy because of high-intensity infections, while other high risk groups, including fishermen, boatmen, irrigation workers, and housewives, account for a large proportion of cases and have difficulty accessing treatment. In addition, treatment gap existed in the infants and pre-school aged children due to lack of pediatric praziquantel. Therefore, in addition to school-based mass drug administration (MDA) as the main drug supply route in Africa, it is necessary to supplement additional drug delivery approaches such as community-based approach to increase the coverage of chemotherapy.

5.3.3 Inadequate Water Supply and Poor Sanitation

Access to safe water and improved sanitation could interrupt transmission routes by reducing typical exposure to infected water and environmental pollution from eggs in order to prevent the persistence and re-emergence of schistosomiasis. WHA 65.21 proposed the importance of using safe water, improving environmental sanitation and public health education as the measures to control schistosomiasis, but these measures are basically dependent on economic development. By 2012, 89% of the world's population had access to improved drinking water, but more than 700 million people still had no access to improved drinking water sources, 50% of whom resided in SSA. In fact, as of 2012, improved sanitation coverage in SSA was only 30%, and 644 million people had no access to improved sanitation. The Millennium Development Goals of safe water, sanitation and hygiene are far from being achieved. Inequality in economic development, donor funding, and national or regional access to drinking water and sanitation services also affect the control of schistosomiasis.

5.3.4 Lack of Perfect Health System

Despite a nearly threefold increase in donations from western developed countries, from $1.5 billion in 1990 to 4.24 billion in 2011, Africa's fragile health systems have not greatly improved, as evidenced by the outbreak of Ebola virus contagion in 2014. In most African countries, lack of health insurance systems, funding, medical personnel, and well-functioning medical information systems is key issues of health care, and many of these countries do not possess public health basic facilities or the necessary resources to implement schistosomiasis control. The migration of experienced health workers from Africa to other countries was not conducive to the effective functioning of the health system in Africa.

5.3.5 Insufficient Government Coordination Capacity

Due to the natural, social, and biological factors will impact on the transmission of schistosomiasis, a comprehensive approach based on multi-sectoral cooperation would be conducive to control activities. At present, however, most African countries lack such a mechanism. Chemotherapy can reduce the prevalence and incidence rates of schistosomiasis and inhibit the transmission from reservoir hosts to intermediate hosts snails. Mollusciciding and environmental modification can kill snails. Health education, access to clean water, and enhanced sanitation may prevent human and/or snail infections. Some studies observed that inadequate knowledge of behavioral problems such as seeking medical treatment and hygienic habits to influence the prevalence of schistosomiasis, while the construction of water conservancy and irrigation projects did not consider the impacts on schistosomiasis, resulting in the spread of schistosomiasis. Therefore, in the context of ensuring political commitment, a leadership group composed of policy makers in the health, education, agriculture, water conservancy, and other sectors of governments could be established. The formulation of national strategic plans and the establishment of long-term cooperation mechanisms could help mobilize resources from all aspects of society and strengthen the comprehensive intervention measures of schistosomiasis control.

5.4 Cooperation Priorities Between China and Africa in Schistosomiasis Control

With the development of society and science, China's strategies for schistosomiasis control have changed, with the intervention measures adjusted and reorganized in turn. A variety of products, technologies, and well-experienced specialists have ensured the success of schistosomiasis control. These experiences cover various aspects that may be suitable for implementation in African countries to speed up the control process there.

5.4.1 To Improve Coordination at Different Administrative Levels

In China, achievements in schistosomiasis control were based on close cooperation among National Health Commission, State Forestry Administration, Ministry of Water Resources, Ministry of Agriculture, Ministry of Finance, Ministry of Land and Resources, and National Development and Reform Commission. In addition to this, the State Council on Schistosomiasis Control had responsibility for coordination among various sectors. The health bureaus, Centers for Diseases Control

(CDCs), and hospitals at all levels assisted in management and implementation of disease control activities. On the one hand, China could propagate its experience in cooperation mechanisms and capacity-building of the management systems through organizing training course. On the other hand, international advocacy workshops and stakeholder meetings at the national or regional levels are also necessary.

5.4.2 To Adopt New and Effective Snail Control Methods

Since schistosomes have to complete their life cycle through their intermediate host, eliminating snails would interrupt the transmission of schistosomiasis in theory. China has developed and implemented a variety of environmental modifications, through wetland reclamation, new ditches digging and old ditches filling, the transition from paddy fields to dry grain, etc., to integrate projects in land resources, agriculture, water conservancy, forestry, and other sectors. In addition, various molluscicides have been developed adapting to local conditions in different snail-infected areas, such as 50% niclosamide ethanolamine salt wettable powder (or other formulations, such as suspensions, granules, retarders, etc.) and plant molluscicides, such as tea seed distilled saponins (TDS) which are low toxic to fish. Annual production capacity of niclosamide could reach 1000 tons in Chinese companies, and if the prequalification of this molluscicide is approved by WHO, China could export the quality-assured molluscicides to other endemic countries at lower prices. Training courses and exchanges with senior scientists to share experience and technologies about snail control could offer assistance to effectively control snails in African countries.

5.4.3 To Integrate High-Quality Diagnostic Tools

Accurate diagnosis in areas where schistosomiasis is endemic can help to obtain the data on the distribution, prevalence, and infection intensity of schistosomiasis, and the diagnostic methods should vary from different stages of schistosomiasis control. Chinese scientists developed a variety of diagnostic tools, especially immunodiagnostic assays. Over the past decade, State Food and Drug administration (SFDA) has approved more than ten quality-assured immunoassays, playing different roles in the Chinese control programs or research activities. In China, under different schistosomiasis control thresholds, appropriate diagnostic tools have been selected to guide treatment strategies and identify elimination. A diagnostic network was established through the construction of provincial reference laboratories and county-level sentinel monitoring laboratories to monitor the quality of diagnostics and to provide scientific evidence for schistosomiasis surveillance and response in areas where schistosomiasis transmission had been interrupted. Some kits produced in China have high sensitivity and specificity for the diagnosis of other schistosomiasis;

therefore, these kits could be provided for African countries with competitive prices or free of charge. Training programs or academic activities could share the experience in developing diagnostic assays, selecting appropriate diagnostic tools, and managing diagnostic quality control.

5.4.4 To Strengthen Disease Mapping

Disease mapping plays an important role in schistosomiasis control and can provide base information when drawing up control programs or carrying out activities. Since the 1950s, China has conducted nationwide surveys to understand the distribution of schistosomiasis. According to the data gathered from annual survey and surveillance between 2002 and 2010, the latest version of the China schistosomiasis transmission atlas (2nd edition) was issued in 2012, which present the thematic maps and charts of schistosomiasis transmission in China. Annual schistosomiasis risk mapping and snail geographical distribution model were conducted at the national, regional, or county level to guide policy decision-making. Geospatial tools (such as GIS, RS, geostatistics, etc.) can be used to better describe, understand, and predict the distribution of schistosomiasis. With skilled technical personnel and extensive experience in disease mapping, China could host communication and advanced training courses of disease mapping. In countries where there are currently no schistosomiasis distribution maps, technical support could be provided to finalize the mapping.

5.4.5 To Improve Surveillance Systems

Rapid, effective response monitoring and early warning mechanism are the premise of efficient management of infectious diseases. China has always paid attention to monitoring the endemic tendency of schistosomiasis in the fixed surveillance sites and evaluating the transmission risk along the border between the non-endemic and endemic areas to ensure effective management of schistosomiasis. Schistosomiasis information management system which was built in 2010 and used nationwide provided valuable source data at the village level to support national management. In the meantime, new technologies such as predictive modeling, xeno-monitoring tools based on LAMP, GIS, and RS have also been used for schistosomiasis monitoring and surveillance. Chinese scientists could disseminate their skills and experience in schistosomiasis monitoring and surveillance to African countries by visiting or providing training courses in the form of distance learning, basic and advanced courses. With the help of Chinese scientists, an early warning system could be established by tracking information about snail species density.

5.4.6 To Expand the Coverage of Chemotherapy

PZQ is a cheap drug against all schistosome species with high-efficiency and low toxicity. Currently, the amount of existing donated PZQ is far from the actual demand in African countries. Numerous cases could not be treated due to existing insufficient drug delivery systems. Up to now, Nanjing Pharmaceutical Factory produced 50 tons of raw materials and 100 million PZQ tablets (200 mg/tablet) per year, it increased the dosage of PZQ while make it easier to swallow. A randomized control trial conducted in Zanzibar proved that the praziqantel produced by China didn't differ in cure rate for schistosomiasis cases with that produced by Merck KGaA. Besides, community-based drug delivery system that distributes drugs by local hospitals or village doctors in collaboration with CDC or schistosomiasis stations ensures the high coverage of chemotherapy in P.R. China. In order to expand the coverage of chemotherapy, China could export improved PZQ tablets to other endemic countries at lower prices. Government and non-governmental organizations (NGOs) could also donate PZQ to close the existed gap. At the same time, China's experience could be disseminated through regional training programs for pharmaceutical administrators and healthcare workers in African countries to improve their drug delivery skills. Chinese scientists could also support the operation of reporting systems for drug delivery facilities.

5.5 The Way Forwards

The cooperation between China and African countries is on its way step by step. At the fifth China-Africa Cooperation Forum in July 2012, P.R. China made a promise to increase its investment continuously, extend cooperation to fields such as health care and disease control as well as carry out joint pilot projects in the field of public health. In 2013, the Beijing Declaration was produced during the Ministerial Forum of China-Africa Health Development in 2013, extending its cooperation to health care and disease control. Schistosomiasis, malaria, and HIV were given priority for pilot studies in Africa and Asia through multilateral cooperation. On 21 May 2014, a memorandum of understanding was signed by the governments of China, Zanzibar and the WHO in Geneva. Under the framework of cooperation, a China-Zanzibar collaborating project aiming at accelerating elimination of schistosomiasis in Zanzibar was financed by Chinese government in August 2016. Tremendous achievements had been obtained and the experience and lessons are summarizing and distilling.

In September 2018, President Xi Jinping announced at the Beijing summit of the China-Africa Cooperation Forum that China will implement the "health campaign" in the next three years. Collaboration on schistosomiasis control and prevention is one of the topics. A series of international communication and field investigations will be conducted to draft and formulate feasible and scientific cooperation proposals which should be met the needs of African countries. These cooperation programs aim to decrease the disease burden caused by schistosomiasis in selected countries, increase the self-protection awareness of African people in endemic areas, and strengthen the capacity of public health service. Appropriate and suitable strategies and interventions will be formulated adaptable to African situations by transferring the Chinese experience and technologies distilled or developed during the process of schistosomiasis control. The institutional-based Network on China-Africa cooperation of schistosomiasis (INCAS), which was coordinated by the National Institute of Parasitic Diseases (NIPD) and the WHO, would help to strengthen cooperation between basic and operational research institutions, and then accelerate the control process of schistosomiasis in Africa. During 2013 and 2014, cooperation of Chinese and African institutions had been further strengthened by memorandums of understanding concerning Sino-Africa collaboration. Through this network, memorandums aiming to strengthening the cooperation on schistosomiasis control and research were signed between the Public Health Laboratory in Pemba Island, which is part of the semi-autonomous of Zanzibar in Tanzania, and Jiangsu Institute of Parasitic Diseases, the Blue Nile Institute in Sudan, and the National Institute of Parasitic Diseases, China CDC, respectively.

It cannot be ignored that challenges still exist when facing different environments in new collaborations, for example, lack of experience and specialists related to global health in China. Also, the endemicity of schistosomiasis in African countries presents huge heterogeneity. Field investigations to explore the real situation of the local situation and close communication should be conducted, to understand the real endemic status of schistosomiasis, real requirements of local people, existing gaps, health system conducting schistosomiasis control activities, etc. At present, international agencies and major national donor organizations have cooperated with African countries in schistosomiasis control. Cooperating with them is expected to be a shortcut to facilitate the Sino-Africa cooperation on schistosomiasis control and elimination programs.

Further Reading

Asaolu SO, Ofoezie IE (2003) The role of health education and sanitation in the control of helminth infections. Acta Trop 86:283–294

Assare RK, Knopp S, N'guessan NA, Yapi A, Tian-Bi YN, Yao PK et al (2014) Sustaining control of schistosomiasis mansoni in moderate endemicity areas in western Côte d'Ivoire: a SCORE study protocol. BMC Public Health 14:1290

Bergquist R, Johansen MV, Utzinger J (2009) Diagnostic dilemmas in helminthology: what tools to use and when? Trends Parasitol 25:151–156

Berriman M, Haas BJ, LoVerde PT et al (2009) The genome of the blood fluke *Schistosoma mansoni*. Nature 460:352–358

Brooker S, Kabatereine NB, Gyapong JO, Stothard JR, Utzinger J (2009) Rapid mapping of schistosomiasis and other neglected tropical diseases in the context of integrated control programs in Africa. Parasitology 136:1707–1718

Chen MG, Feng Z (1999) Schistosomiasis control in China. Parasitol Int 48:11–19

Chen M (1999) Progress in schistosomiasis control in China. Chin Med J 112:930–933
Chen YY, Liu JB, Huang XB, Cai SX, Su ZM, Zhong R et al (2014) New integrated strategy emphasizing infection source control to curb schistosomiasis japonica in a marshland area of Hubei Province, China: findings from an eight-year longitudinal survey. PLoS One 9:e89779
Chitsulo L, Engels D, Montresor A, Savioli L (2000) The global status of schistosomiasis and its control. Acta Trop 77:41–51
Cioli D, Pica-Mattoccia L, Basso A, Guidi A (2014) Schistosomiasis control: praziquantel forever? Mol Biochem Parasitol 195:23–29
Clements AC, Firth S, Dembelé R et al (2009) Use of Bayesian geostatistical prediction to estimate local variations in *Schistosoma haematobium* infection in western Africa. Bull World Health Organ 87:921–929
Connell J, Zurn P, Stilwell B, Awases M, Braichet JM (2007) Sub-Saharan Africa: beyond the health worker migration crisis? Soc Sci Med 64:1876–1891
Curti E, Kwityn C, Zhan B, Gillespie P, Brelsford J, Deumic V et al (2013) Expression at a 20L scale and purification of the extracellular domain of the *Schistosoma mansoni* TSP-2 recombinant protein: a vaccine candidate for human intestinal schistosomiasis. Hum Vaccin Immunother 9:2342–2350
Da'dara AA, Li YS, Xiong T, Zhou J, Williams GM, McManus DP et al (2008) DNA-based vaccines protect against zoonotic schistosomiasis in water buffalo. Vaccine 26:3617–3625
Dang H, Xu J, Li SZ et al (2014) Monitoring the transmission of *Schistosoma japonicum* in potential risk regions of China, 2008–2012. Int J Environ Res Public Health 11:2278–2287
Dorkenoo AM, Bronzan RN, Ayena KD et al (2012) Nationwide integrated mapping of three neglected tropical diseases in Togo: countrywide implementation of a novel approach. Tropical Med Int Health 17:896–903
Fenwick A, Rollinson D, Southgate V (2006) Implementation of human schistosomiasis control: challenges and prospects. Adv Parasitol 61:567–622
Fenwick A, Rollinson D, Southgate V (2006) Implementation of human schistosomiasis control: challenges and prospects. Adv Parasitol 61:567–622
Fenwick A, Webster JP, Bosque-Oliva E, Blair L, Fleming F, Zhang Y et al (2009) The Schistosomiasis Control Initiative (SCI): rationale, development and implementation from 2002–2008. Parasitology 136:1719–1730
Frenk J, Horton R (2006) Evidence for health-system reform: a call to action. Lancet 368:3–4
Green A (2013) Kenya's new president promises health system reform. Lancet 381:1348–1349
Hodges M, Dada N, Wamsley A et al (2011) Improved mapping strategy to better inform policy on the control of schistosomiasis and soil-transmitted helminthiasis in Sierra Leone. Parasit Vectors 4:97
Huang YX, Manderson L (2005) The social and economic context and determinants of schistosomiasis japonica. Acta Trop 96:223–231
King CH (2009) Toward the elimination of schistosomiasis. N Engl J Med 360:106–109
Lei ZL, Zheng H, Zhang LJ et al (2014) Endemic status of schistosomiasis in People's Republic of China in 2013. Chin J Schisto Contr 26:591–597 (in Chinese)
Liu P, Guo Y, Qian X, Tang S, Li Z, Chen L (2014) China's distinctive engagement in global health. Lancet 384:793–804
Moendeg KJ, Angeles JM, Goto Y et al (2015) Development and optimization of cocktail-ELISA for a unified surveillance of zoonotic schistosomiasis in multiple host species. Parasitol Res 114:1225–1258
Musuva RM, Awiti A, Omedo M, Ogutu M, Secor WE, Montgomery SP et al (2014) Community knowledge, attitudes and practices on schistosomiasis in western Kenya-The SCORE Project. Am J Trop Med Hyg 90:646–652
National Health and Family Planning Commission (2013) Beijing declaration of the ministerial forum of China-Africa Health Development. http://www.nhfpc.gov.cn/gjhzs/s3590/201308/da8ad62e487a481f987e631e1318c6fc.shtml
NIPD (2012) Atlas of schistosomiasis transmission in China. China Map Publishing House, Beijing

Obe AF (2015) Praziquantel: do we need another antischistosoma treatment? Future Med Chem 7:677–680

Odhiambo GO, Musuva RM, Atuncha VO, Mutete ET, Odiere MR, Onyango RO et al (2014) Low levels of awareness despite high prevalence of schistosomiasis among communities in Nyalenda informal settlement, Kisumu city, western Kenya. PLoS Negl Trop Dis 8:e2784

OECD (2013) Aid to the health, 2013. http://www.oecd.org/dac/stats/health. Accessed 6 Jan 2015

Pontes LA, Dias-Neto E, Rabello A (2002) Detection by polymerase chain reaction of *Schistosoma mansoni* DNA in human serum and feces. Am J Trop Med Hyg 66:157–162

Protasio AV, Tsai IJ, Babbage A et al (2012) A systematically improved high quality genome and transcriptome of the human blood fluke *Schistosoma mansoni*. PLoS Negl Trop Dis 6:e1455

Porter J, Webbe G, Sturrock RF (1993) Human schistosomiasis. CAB International, Wallingford UK

Qin ZQ, Xu J, Feng T et al (2013) Strategic thinking of the construction of national schistosomiasis laboratory network in China. Chin J Schisto Contr 25:329–32 (in Chinese)

Raso G, Matthys B, N'Goran EK, Tanner M, Vounatsou P, Utzinger J (2005) Spatial risk prediction and mapping of *Schistosoma mansoni* infections among schoolchildren living in western Cote d'Ivoire. Parasitology 131:97–108

Richards FO Jr, Eigege A, Miri ES, Jinadu MY, Hopkins DR (2006) Integration of mass drug administration programmes in Nigeria: the challenge of schistosomiasis. Bull World Health Organ 84:673–676

Riveau G, Deplanque D, Remoue F, Schacht AM, Vodougnon H, Capron M et al (2012) Safety and immunogenicity of rSh28GST antigen in humans: phase 1 randomized clinical study of a vaccine candidate against urinary schistosomiasis. PLoS Negl Trop Dis 6:e1704

Rollinson D, Knopp S, Levitz S et al (2013) Time to set the agenda for schistosomiasis elimination. Acta Trop 128:423–440

Rollinson D, Knopp S, Levitz S, Stothard JR, Tchuem Tchuente LA, Garba A et al (2013) Time to set the agenda for schistosomiasis elimination. Acta Trop 128:423–440

Rozendaal JA (1997) Vector control. Methods for use by individuals and communities. WHO, Geneva

Sekwat A (2003) Health financing reform in sub-Saharan Africa: major constraints, goals, and strategies. J Health Care Finance 29:67–78

Shaw V (2005) Health information system reform in South Africa: developing an essential data set. Bull World Health Organ 83:632–636

Simoonga C, Utzinger J, Brooker S et al (2009) Remote sensing, geographical information system and spatial analysis for schistosomiasis epidemiology and ecology in Africa. Parasitology 136:1683–1693

Steinmann P, Keiser J, Bos R, Tanner M, Utzinger J (2006) Schistosomiasis and water resources development: systematic review, meta-analysis, and estimates of people at risk. Lancet Infect Dis 6:411–425

Stothard JR, Chitsulo L, Kristensen TK, Utzinger J (2009) Control of schistosomiasis in sub-Saharan Africa: progress made, new opportunities and remaining challenges. Parasitology 136:1665–1675

Stothard JR, Kabatereine NB, Tukahebwa EM et al (2006) Use of circulating cathodic antigen (CCA) dipsticks for detection of intestinal and urinary schistosomiasis. Acta Trop 97:219–228

Stothard JR, Sousa-Figueiredo JC, Betson M, Bustinduy A, Reinhard-Rupp J (2013) Schistosomiasis in African infants and preschool children: let them now be treated! Trends Parasitol 29:197–205

Stothard JR, Sousa-Figueiredo JC, Betson M, Green HK, Seto EY, Garba A et al (2011) Closing the praziquantel treatment gap: new steps in epidemiological monitoring and control of schistosomiasis in African infants and preschool-aged children. Parasitology 138:1593–1606

Sturrock HJ, Pullan RL, Kihara JH, Mwandawiro C, Brooker SJ (2013) The use of bivariate spatial modeling of questionnaire and parasitology data to predict the distribution of *Schistosoma haematobium* in Coastal Kenya. PLoS Negl Trop Dis 7:e2016

Tchuem Tchuente LA, Kamwa Ngassam RI, Sumo L et al (2012) Mapping of schistosomiasis and soil-transmitted helminthiasis in the regions of centre, east and west Cameroon. PLoS Negl Trop Dis 6:e1553

Tong Q, Chen R, Zhang Y et al (2015) A new surveillance and response tool: risk map of infected *Oncomelania hupensis* detected by Loop-mediated isothermal amplification (LAMP) from pooled samples. Acta Trop 141:170–177

Tong Q, Chen R, Zhang Y, Yang G, Kumagaid T, Shimogawarad R et al (2015) A new surveillance and response tool: risk map of infected *Oncomelania hupensis* detected by Loop-mediated isothermal amplification (LAMP) from pooled samples. Acta Trop 141:170–177

UNCTAD (2013) Economic development in Africa Report, 2013. http://unctad.org/en/pages/PublicationWebflyer.aspx?publicationid=604

Utzinger J, Raso G, Brooker S et al (2009) Schistosomiasis and neglected tropical diseases: towards integrated and sustainable control and a word of caution. Parasitology 136:1859–1874

Utzinger J, Bergquist R, Shu-Hua X, Singer BH, Tanner M (2003) Sustainable schistosomiasis control – the way forward. Lancet 362:1932–1934

Utzinger J, Raso G, Brooker S, De Savigny D, Tanner M, Ornbjerg N et al (2009) Schistosomiasis and neglected tropical diseases: towards integrated and sustainable control and a word of caution. Parasitology 136:1859–1874

Wang LD, Utzinger J, Zhou XN (2008) Schistosomiasis control: experiences and lessons from China. Lancet 372:1793–1795

Wang XY, He J, Juma S, Kabole GJG, Dai JR, Li W, Yang K (2019) Efficacy of China-made praziquantel for treatment of *Schistosomiasis haematobium* in Africa: a randomized controlled trial. PLoS Negl Trop Dis:13e0007238

WHO (2012) Key issues and challenges arising out or the Sixty-fifth World Health Assembly and the 130th and 131st sessions of the WHO Executive Board. Elimination of schistosomiasis (WHO 65.21). Sixty-fifth Session SEA/RC65/17. Yogyakarta, Indonesia

WHO (2013) Schistosomiasis: progress report 2001–2011 and strategic plan 2012–2020 [Online]. World Health Organization, Geneva

WHO (2014) Schistosomiasis: number of people receiving preventive chemotherapy in 2012. Wkly Epidemiol Rec 89:21–28

WHO (2018) Schistosomiasis and soiltransmitted helminthiases: numbers of people treated in 2017. Wkly Epidemiol Rec 93:681–692

WHO/UNICEF (2014) Progress on drinking water and sanitation, 2014. http://www.who.int/water_sanitation_health/publications/2014/jmp-report/en/

Worku L, Damte D, Endris M, Tesfa H, Aemero M (2014) *Schistosoma mansoni* infection and associated determinant factors among school children in Sanja Town, northwest Ethiopia. J Parasitol Res 2014:792536

Xu J, Xu JF, Li SZ et al (2015) Integrated control programmes for schistosomiasis and other helminth infections in P.R. China. Acta Trop 141:332–341

Xu J, Guo JG, Wu XH, Zeng XJ, Yang WP, Yang GB et al (2009) Efficacy and adverse effects of film coated praziquantel for treatment of schistosomiasis japonica. Chin J Prev Med 43:718–722

Yang GJ, Li W, Sun LP et al (2010) Molluscicidal efficacies of different formulations of niclosamide: result of meta-analysis of Chinese literature. Parasit Vectors 3:e84

Yang GJ, Vounatsou P, Zhou XN, Utzinger J, Tanner M (2005) A review of geographic information system and remote sensing with applications to the epidemiology and control of schistosomiasis in China. Acta Trop 96:117–129

Yang K, Sun LP, Huang YX et al (2012) A real-time platform for monitoring schistosomiasis transmission supported by Google Earth and a web-based geographical information system. Geospat Health 6:195–203

Yang GJ, Vounatsou P, Tanner M, Zhou XN, Utzinger J (2006) Remote sensing for predicting potential habitats of *Oncomelania hupensis* in Hongze, Baima and Gaoyou lakes in Jiangsu province, China. Geospat Health 1:85–92

Yang K, Sun LP, Huang YX, Yang GJ, Wu F, Hang DR et al (2012) A real-time platform for monitoring schistosomiasis transmission supported by Google Earth and a web-based geographical information system. Geospat Health 6:195–203

Yuan Y, Xu XJ, Dong HF, Jiang MS, Zhu HG (2005) Transmission control of schistosomiasis japonica: implementation and evaluation of different snail control interventions. Acta Trop 96:191–197

Zhang Z, Fang R, Yu B et al (2013) Field evaluation of a novel plant molluscicide "Luo-wei" against *Oncomelania hupensis* snails II Molluscicidal effect in the field of lake areas in Hanchuan City, Hubei Province. Chin J Schisto Contr 25:481–84 (in Chinese)

Zhang ZY, Xu DZ, Zhou XN, Zhou Y, Liu SJ (2005) Remote sensing and spatial statistical analysis to predict the distribution of *Oncomelania hupensis* in the marshlands of China. Acta Trop 96:205–212

Zhao GM, Zhao Q, Jiang QW, Chen XY, Wang LY, Yuan HC (2005) Surveillance for schistosomiasis japonica in China from 2000 to 2003. Acta Trop 96:288–295

Zhou XN, Guo JG, Wu XH et al (2007) Epidemiology of schistosomiasis in the People's Republic of China, 2004. Emerg Infect Dis 13:1470–1476

Zhou XN, Olveda R, Sripa B, Yang GJ, Leonardo L, Bergquist R (2015) From gap analysis to solution and action: the RNAS(+) model. Acta Trop 141:146–149

Zhou Y, Wang Z, Zhang B et al (2013) Field evaluation of a novel plant molluscicide "Luo-wei" against the snail *Oncomelania hupensis* III molluscicidal effect by spraying method in hilly regions. Chin J Schisto Contr 25:495–497

Zhou XN, Bergquist R, Tanner M (2013) Elimination of tropical disease through surveillance and response. Infect Dis Poverty 2:1

Zhou XN, Wang LY, Chen MG, Wu XH, Jiang QW, Chen XY et al (2005) The public health significance and control of schistosomiasis in China – then and now. Acta Trop 96:97–105

Zhou XN, Xu J, Chen HG, Wang TP, Huang XB, Lin DD et al (2011) Tools to support policy decisions related to treatment strategies and surveillance of Schistosomiasis japonica towards elimination. PLoS Negl Trop Dis 5:e1408

Zhou XN, Xu J, Chen HG, Wang TP, Huang XB, Lin DD et al (2011) Tools to support policy decisions related to treatment strategies and surveillance of Schistosomiasis japonica towards elimination. PLoS Negl Trop Dis 5:e1408

Zhu R, Qin ZQ, Feng T, Dang H, Zhang LJ, Xu J (2013) Assessment of effect and quality control for parasitological tests in national schistosomiasis surveillance sites. Chin J Schisto Contr 25:11–15 (in Chinese)

Zhu YC (2005) Immunodiagnosis and its role in schistosomiasis control in China: a review. Acta Trop 96:130–136

Zhu N (2013) China to strengthen cooperation with Africa on health [Online]. Xinhua Newspaper, Beijing. http://news.xinhuanet.com/english/china/2013-08/16/c_132637291

Zhu YC, Socheat D, Bounlu K, Liang YS, Sinuon M, Insisiengmay S et al (2005) Application of dipstick dye immunoassay (DDIA) kit for the diagnosis of schistosomiasis mekongi. Acta Trop 96:137–141

Zhu Y, Hassen S, He W, Cao G (2006) Preliminary study on detection of schistosomiasis mansoni with dipstick dye immunoassay (DDIA) kit. Chin J Schisto Contr 18:419–421

Chapter 6
China-Zanzibar Cooperation Project of Schistosomiasis Control: Study Design

Kun Yang, Saleh Juma, Fatma Kabole, Jiagang Guo, Amadou Garba, Jian He, Jian-Rong Dai, Wei Li, and Jian-Feng Zhang

Abstract Schistosomiasis is a public health problem mostly in Africa. In 2014, World Health Organization (WHO) signed a tripartite memorandum of understanding (MOU) with China and Tanzania-Zanzibar paving the way for piloting interventions for schistosomiasis elimination in Zanzibar. The study has been designed including the transmission pattern of schistosomiasis; the effectiveness studies of praziquantel from China and mass drug administration (MDA) for suitable method of human treatment; biological study of snails and effectiveness of novel molluscicides from China in laboratory and field for the suitable method of snail control; KAP study and health education design for suitable method of behavior interventions by health education. The integrated measures will be implemented after baseline surveillance. At last, the effectiveness of the integrated measure will be evaluated by the index of snail density, prevalence, and KAP of human.

Keywords Schistosomiasis · Haematobium · Elimination · Pilot project · Zanzibar · China

---

K. Yang (✉) · J.-R. Dai · W. Li · J.-F. Zhang
Jiangsu Institute of Parasitic Diseases, Key Laboratory on Technology for Parasitic Disease Prevention and Control, Ministry of Health, Jiangsu Provincial Key Laboratory on the Molecular Biology of Parasites, Wuxi, China
e-mail: yangkun@jipd.com

S. Juma · F. Kabole
Ministry of Health of Zanzibar, Zanzibar, United Republic of Tanzania

J. Guo · A. Garba
Department of Control of Neglected Tropical Diseases, World Health Organization, Geneva, Switzerland

J. He
Global Health Center, Jiangsu Institute of Parasitic Diseases, Wuxi, Jiangsu, China

K. Yang, H. Mehlhorn (eds.), *Sino-African Cooperation for Schistosomiasis Control in Zanzibar*, Parasitology Research Monographs 15,
https://doi.org/10.1007/978-3-030-72165-7_6

Abbreviations

KAP	Knowledge attitudes and practices
MDA	Mass drug administration
MOU	Memorandum of understanding
NTD	Neglected tropical diseases
SCORE	Schistosomiasis Consortium for Operational Research and Evaluation
WHO	World Health Organization
WHOPES	WHO Pesticide Evaluation Scheme
ZEST	Zanzibar Elimination of Schistosomiasis Transmission

6.1 Background

Schistosomiasis is a global public health problem, with 78 endemic countries and regions in Asia, South America, Middle East, and Africa. The report of World Health Organization (WHO) shows that at least 206.5 million people required preventive treatment for schistosomiasis in 2016, out of which more than 88 million people were reported to have been treated (Ross et al. 2017; Lo et al. 2017). Schistosomiasis was one of the most serious parasitic diseases in the People's Republic of China with a documented history of over 2100 years. During the mid-1950s, at the beginning of the national control program, schistosomiasis was endemic in 12 provinces, with estimated 11.6 million people and 1.2 million cattle infected, and an area of 14,300 km^2 infested by the intermediate host snail. Up to now, China has been successful with control and elimination of schistosomiasis. According to the actual situation and the medium- and long-term planning goals, China put forward a new comprehensive control strategy (Wang et al. 2017; Li et al. 2005). Many scientific and technological achievements obtained the popularization and application, provided the strong technical support for the goal of comprehensive control of schistosomiasis. By the end 2015, all endemic counties reached the standard of under control of schistosomiasis, namely, the prevalence of human was less than 1% (Liu et al. 2016; Shi et al. 2016; Xu et al. 2016c; Basch 1986; Zhang and Wong 2003). China developed many prevention experts and accumulated rich experience of the comprehensive control of schistosomiasis [10] (Song et al. 2017).

In 2012, the World Health Assembly adopted a resolution on elimination of schistosomiasis (WHA65.21) (Savioli et al. 2017). WHO is in the process of finalizing a revised global strategy for schistosomiasis including control of morbidity, elimination as a public health problem, and interruption of schistosomiasis transmission as the last step (Zoni et al. 2016; Savioli et al. 2015; Fenwick and Jourdan 2016). In 2014, WHO signed a tripartite memorandum of understanding (MOU) with the China and Zanzibar paving the way for the start of a pilot study to support the schistosomiasis elimination program in Zanzibar. Our pilot study is designed to master the transmission pattern of schistosomiasis, formulate the control strategies and measures based on the local conditions, research studies, and the China's experience to eliminate schistosomiasis.

6.2 The Aim of the Pilot Study

The aim is to promote the process of schistosomiasis elimination in Pemba island of Zanzibar.

1. To master the transmission pattern of schistosomiasis based on studies in the laboratory and field.
2. To explore the suitable method of human treatment by effectiveness studies of praziquantel from China and coverage rate of mass drug administration (MDA).
3. To explore the suitable method of snail control by the biological study of snails and effectiveness study of molluscicides from China.
4. To explore the suitable method of health education by KAP study and design for health education material from China.
5. To evaluate the effectiveness of the integrated measure by the index of snail density, prevalence, and KAP of a human.
6. To improve and build the capacity of the schistosomiasis control and explore the cooperation mechanism to eliminate schistosomiasis in Africa.

6.3 Study Method and Design

The study is designed as a controlled trial combined with the planned preventive chemotherapy conducted by the Zanzibar NTD Program from 2017 to 2019, and the timeline is shown in Fig. 6.1.

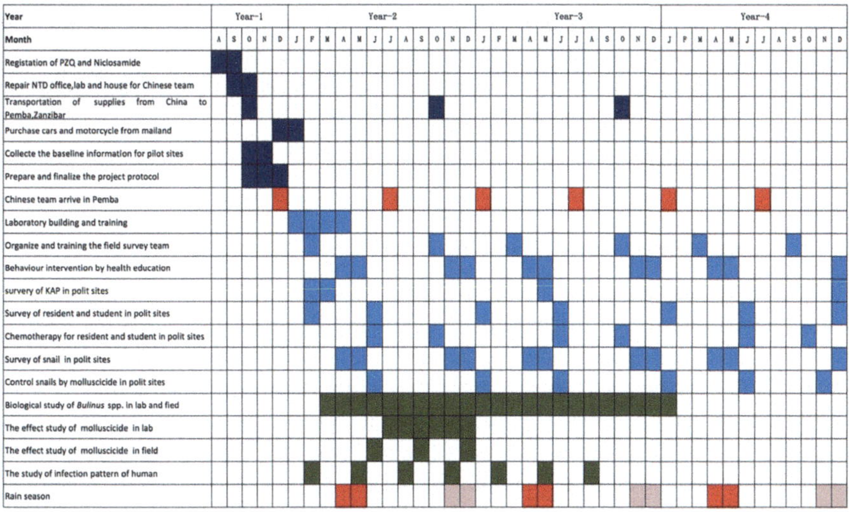

Fig. 6.1 The timeline of the pilot study to control and eliminate schistosomiasis in Zanzibar based on China-Africa collaboration

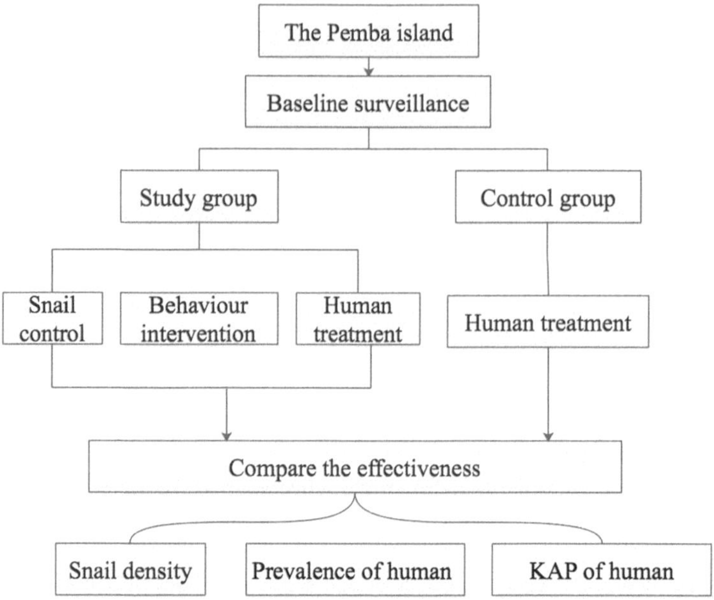

Fig. 6.2 The structure of the pilot study to control and eliminate schistosomiasis in Zanzibar based on China-Africa collaboration

During the course of the implementation there will be annual evaluation of the study to understand the progress of the study and if any adjustment of the interventions will be needed. This study will have two study arms: the control group is only human treatment by the national plan of twice yearly preventive chemotherapy, the study group is human treatment plus snail control and behavior intervention, the effectiveness index includes the prevalence of human, snail density, and KAP of human (Fig. 6.2).

6.3.1 Study Area and Population

Zanzibar is composed of two islands, Unguja and Pemba, situated off the eastern coast of Tanzania mainland and it lies between latitude 4 and 5-degree south of equator (http://zanzibar.go.tz/). There are two administrative regions and more than 100 shehias in Pemba island. The estimated total resident population for Pemba was 511,576 in 2010. The average number of school-age children per study household was 1.9, and there was wide variation in the proportion of non-enrolled children (19–84%) between shehias (Montresor et al. 2001). There is annual dry and rainy season, and the average annual temperature ranges between 23 °C and 28 °C. The main economic activities include seawater fishing and cash crop production.

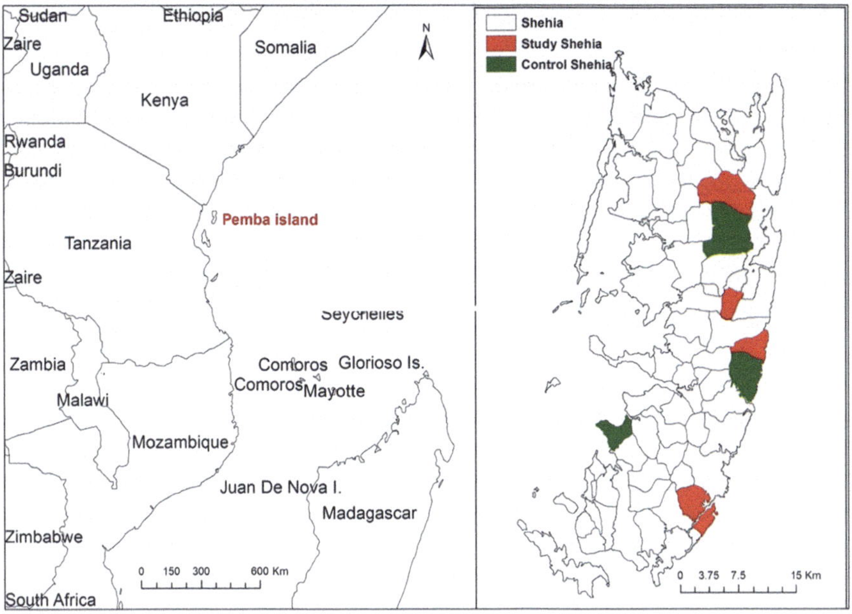

Fig. 6.3 The location of the pilot study to control and eliminate schistosomiasis in Zanzibar based on China-Africa collaboration

Zanzibar is endemic for *S. haematobium*. The distribution of schistosomiasis on Unguja and Pemba was first described in 1925. The intermediate host is *Bulinus spp* snail (Goatly and Jordan 1965; Knopp et al. 2013). In 1975, they estimated that 60% of the Pemba population had urinary schistosomiasis and about 10% of those infected had severe morbidity. By 2004, the infection rate of schistosomiasis was about 64.5% (Stothard et al. 2009; French et al. 2007; Rollinson et al. 2005). In 2011, the infection rate was 9% in Pemba island, before the onset of large-scale schistosomiasis control interventions implemented in the frame of the Zanzibar Elimination of Schistosomiasis Transmission (ZEST) project (Knopp et al. 2013).

In this study, 7 shehias of Pemba island are selected as study and control group. The study group includes 4 shehias, namely Tangani, Wingwi, Kiyuyu, and Uwandani. The control group includes 3 shehias, namely Wambaa, Vitongoji, and Shengejuu (Fig. 6.3). There are permanent and temporary water bodies (pond and stream) in the study and control region. The estimated resident population for study shehia is 4000 and the proportion of the female and less than 15 years old is about 61.9% and 45.8%, respectively.

6.3.2 Human Treatment by Preventive Chemotherapy

According to the National Plan of the Zanzibar MoH, preventive chemotherapy with praziquantel (40 mg/kg) against schistosomiasis will be conducted to all children aged >3 years and all adults throughout the islands. Severely sick people and pregnant women will be excluded from preventive chemotherapy interventions. Their serial studies including the transmission pattern and effectiveness of praziquantel from China, etc., will be implemented to explore an suitable method of human treatment. The coverage rate of chemotherapy should reach more than 90% in the study group (Fig. 6.4).

6.3.2.1 Baseline Investigation of *S. haematobium* Infection

The baseline investigation of *S. haematobium* infection will try best to cover all residents and students, and at least the coverage rate will be more than 80%. The urine samples will be collected every day from 10 am to 2 pm detected with the naked eye, and then tested by membrane filter method, the infection rate and intensity of *S. haemotobium* will be calculated. According to the egg count per 10 ml urine sample, infection density will be classified into four grades: negative (0 eggs/10 ml), light (1–50 eggs/10 ml) and heavy (>50/10 ml) (WHO 2002). Each day, 10% of the samples will be re-checked by the controller, results should include negative or positive, egg counts, and infection density. The original results

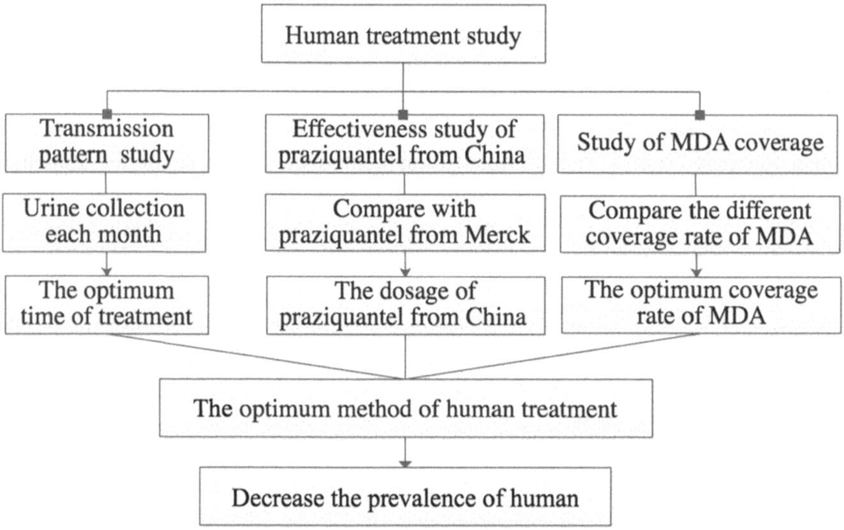

Fig. 6.4 The human treatment of the pilot study to control and eliminate schistosomiasis in Zanzibar based on China-Africa collaboration

will be compared with the control results. If the discrepancy rate (false negative, false positive, the inconsistency of infection density) is more than 20% of the re-check samples, the original technician should be trained, qualified to carry out the detection.

6.3.2.2 Transmission Pattern of *S. haematobium* Infection

One hundred students (50 boys and 50 girls) grade from one to three, and one hundred adult residents (50 males and 50 female) aged from 20 to 55 will be selected. All participants will take mass drug administration (MDA) by praziquantel before the pilot study. Then, the *S. haematobium* will be detected with a standard method for these enrolled students and residents every two months from January to December 2017. The infection rate and density of *S. haematobium* will be calculated every two months to find the transmission pattern of schistosomiasis haematobium in Pemba island, Zanzibar. During the study, all enrolled students and residents will not follow national programs' biannual MDA.

6.3.2.3 The Assessment of China's Praziquantel Against *S. haematobium*

A randomized controlled trial will be used to compare the efficacy of China-made versus WHO-PQ Praziquantel for treatment of *Schistosoma haematobium* in 2017. The study will propose to conduct an open-label, randomized trial to evaluate the comparative efficacy of China-made Praziquantel versus WHO Praziquantel in the treatment of 200 people infected with *S. haematobium*. To do this the investigators will screen about 4000 people by examination of urine for schistosome eggs. Eligible participants will be randomized (1:1) to receive a single dose of Chinese-made and WHO-Prequalified (WHO-PQ) Praziquantel. Four weeks after treatment, the participants will be assessed for cure and egg reduction. At the end of the study, all participants who will still be excreting *S. haematobium* eggs will be treated with WHO-PQ praziquantel again. This trial is registered with ClinicalTrials.gov, number NCT03133832.

6.3.3 *Snail Control by a Novel Molluscicides*

Figure 6.5 shows the study structure of snail control by a novel molluscicides including distribution pattern and biological characteristic of *Bulinus* spp. snail, and the molluscicidal effect of China's novel niclosamide in field and laboratory. At last, the suitable method of snail control will be identified to control and eliminate *Bulinus* spp. snail in Zanzibar.

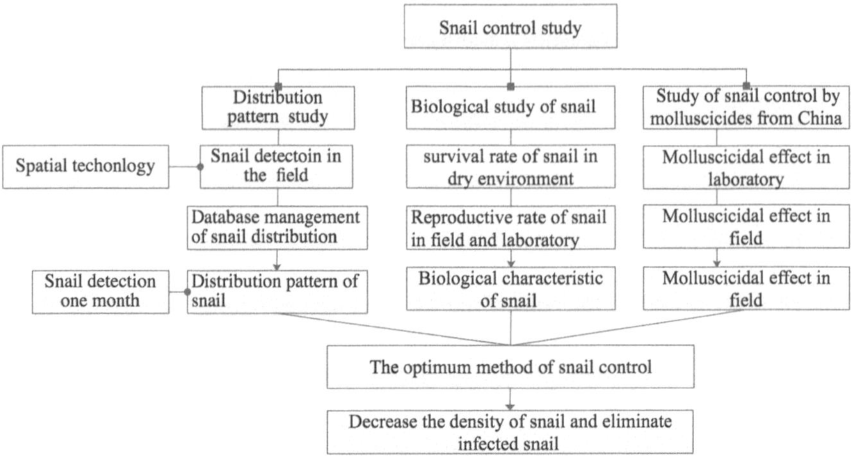

Fig. 6.5 The snail control of the pilot study to control and eliminate schistosomiasis in Zanzibar based on China-Africa collaboration

6.3.3.1 Field Investigation of the Distribution of *Bulinus* spp.

All aquatic snails will be captured in suspected habitats of *Bulinus* spp. each month of 2017, and the number of *Bulinus* spp. will be counted after morphological identification. The distribution and density of *Bulinus* spp. will be investigated, and the infection will be detected by microscopy. The suspected habitat will be positioned with GPS (Garmin, USA). The hydrographic and environmental characteristics of the investigated areas will be collected during the snail survey, and the length, width, and water flow velocity of the river and pond will be measured. In addition, the fluctuation of the water areas will be also investigated. The database of hydrological and snail status will be established, and the distribution of *Bulinus* spp. will be mapped using Google Earth. The snail distribution area and the range will be investigated in both dry and rainy seasons, and the distribution, reproduction, and spread of *Bulinus* spp. will be analyzed.

6.3.3.2 Biological Study of *Bulinus* spp. Snail in Laboratory

The *Bulinus* spp. snails captured in the field will be bred and passaged in the laboratory, and egg production, egg hatching time, time from snail development to oviposition, duration of oviposition, and lifespan will be tested at various temperatures. The reproduction rate, the cumulative growth and development temperatures will be estimated to investigate the reproduction and growth pattern of snails.

6.3.3.3 The Molluscicidal Efficacy in Killing *Bulinus* spp. Snails in Laboratory and Field

The efficacy of novel molluscicides developed in China (26% metaldehyde and niclosamide and 5% powder of niclosamide ethanolamine salt granules) killing snails will be assessed, while WHO-PQ 70% wettable powder of niclosamide ethanolamine salt (WPN) serves as control. The compounds are formulated into solutions with effective concentrations of 0.5, 0.25, 0.125, 0.063, 0.032, and 0.016 mg/L, respectively. Each solution is transferred to a 100 ml beaker. Then, 10 2-month-aged snails are placed in each beaker, which is covered with stainless steel net to prevent snails escaping from the solution. Following immersion for 24 and 48 hrs, the solution is removed, and the snails are washed with dechlorinated water, and bred in water for 48 h. The mortality and median lethal concentration (LC_{50}) are estimated.

Based on hydrographic survey and epidemiological investigation of snails, a small ditch or canal is divided into several segments, and the volume of the water body is measured. The amount of the China's molluscicide is estimated according to the drug dose, and immersion test is performed to evaluate the molluscicidal effect. After immersion for 1, 3, and 7 d, snails are captured and stored in various containers. All snails are tested for death in the laboratory, the snail mortality or reduction of snail density is estimated after two weeks. Following immersion for 1 or 3 months, the long-term molluscicidal effect is assessed. In each experiment, the air temperature and climate are recorded, the study area is protected to prevent human and animal contacting by warning mark.

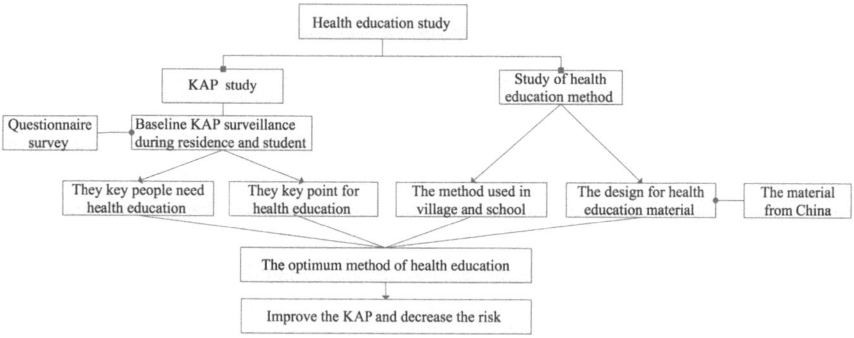

Fig. 6.6 The health education of the pilot study to control and eliminate schistosomiasis in Zanzibar based on China-Africa collaboration

6.3.4 *Behavior Interventions by Health Education*

Figure 6.6 shows the study structure of behavior interventions by health education. Baseline KAP (knowledge attitude and practice) of residents and students will be carried out to find the key people who especially need behavior intervention. Then, health education will be designed to raise the awareness of children and adults in schools and communities, and creates community participation against schistosomiasis transmission. Health education materials will be designed based on the local culture and the epidemic situation of the schistosomiasis and translated into the local language of Swahili.

6.3.4.1 Historical Data Collection and Questionnaire Survey

The historical data of last 5 years including demographic, economic and hydrographic characteristics, distribution of *Bulinus* spp. snail populations, *S. haematobium* infection of human, and the implementation of control interventions will be collected from local station of schistosomiasis control. The questionnaire survey will be performed in 50 households that are randomly selected from each pilot, and household- and individual-level information will be collected. The household-level information involves the composition of the family, economic level, the style of water use, and children's school enrollment, and the individual-level information involves age, gender, occupation, production and lifestyle, piss test place, *S. haematobium* infection, and treatment. In addition, the household will be located with GPS.

6.3.4.2 Knowledge, Attitude, and Practice (KAP) About Schistosomiasis

One class from each grade in a primary school and 100 adults will be randomly selected in the pilot area, and the investigation on the KAP about schistosomiasis will be carried out. The medical record of KAP includes the medical history and treatment. In the knowledge part, the host and transmission of schistosomiasis, as well as knowledge of infection source will be included. In attitude and practice part, attitudes toward disease prevention, access to praziquantel and doctor, peeing in ponds or rivers, and other water contact activities like swimming, washing, and passing will be questioned.

6.4 Statistical Analysis

Assessment indexes mainly include the infection rate and intensity of the schistosomiasis in different groups of people, the distribution and density of *Bulinus* snails, and the score of KAP in different groups of people.

Descriptive, correlation, regression analysis, cost-effectiveness/efficiency analysis, and spatial analysis method will be used (1) to analyze the epidemic and transmission pattern of schistosomiasis, access the feasibility of schistosomiasis control and elimination, (2) to screen comprehensive control measures in different levels of infection, (3) to analyze the cost-effectiveness/efficiency of the control measures, and (4) to access the control efficiency of comprehensive control measures through differential analysis.

6.5 Beneficiaries and Stakeholders

The management of the pilot study will be meeting with stakeholders responsible for distribution of water supply as well as community leaders in the respective 7 shehias where the study will be implemented.

The interventions set forth for this study will be applied shehia wide and not targeted to a specific group of individuals, except for mass treatment of schistosomiasis which will be provided as per national NTD program plans (this excludes children <3 year and individuals with certain disease conditions). So overall, this study will benefit people living in pilot shehia of Pemba island. The people from the identified shehia will be benefited by being prevented with schistosomiasis. The implementation of the study will not cause any individual or group of individuals to lose anything. However, the molluscicide may transient kills the fishes in the fresh water.

6.6 Discussion

Schistosomiasis is a serious public health problem mostly in Africa. In Zanzibar, the efforts to control schistosomiasis especially in Pemba have begun since 1986 and during that time the prevalence of infection based on hematuria was 54.1%. This was progressively reduced to 12.8% following implementation of school-based treatment with praziquantel (PZQ). However, the program was somewhat interrupted with irregular availability of drug, PZQ, and the infection prevalence bounced back to 63.0% in 2004 (Knopp et al. 2013, 2016; Pennance et al. 2016; Bergquist et al. 2017a). Following this, in the same year (2004) the program was revived by reintroduction of intense treatment through mass drug administration (MDA) which led to marked reduction in prevalence which in 2007 fell to 18% based on both haemastix testing and egg detection (Stothard et al. 2009). This encouraging progress attracted more partners to assist Zanzibar in tackling the disease, for example, the ZEST program that was implemented from 2011 to 2017 and focussed on MDA, snail control, and behavior change interventions across Unguja and Pemba in 90 shehias (Knopp et al. 2013). The government of the People's Republic of China in collaboration with WHO has agreed to work together with Government of Zanzibar through Ministry of Health with the objective of elimination of schistosomiasis as a public health problem through implementation of integrated approach using mollusciciding, MDA, and behavioral change intervention (Xu et al. 2016a, b). This is the first cooperation between three sides to control schistosomiasis in Africa, and the responsibility and coordination of each side is very explicit. China government will support the pilot study including identifying resources for implementation and purchase of commodities and equipment, deploying technical experts to Zanzibar, training of staff in Zanzibar, and supervising study implementation. Zanzibar will implement the study by putting in place a policy framework for elimination, providing the required infrastructure and personnel, coordinating research activities and priorities, and identifying and coordinating other potential partners who could contribute to the collaboration. WHO provide technical guidance and oversight for the study through identification and support of independent experts, coordination of partners by convening timely meetings and with a view to preventing no-overlapping activities, assistance in the pre-qualification of medicines for the treatment of school-children and communities, assistance in the assessment and evaluation of molluscicides (pesticides) for the control of vector snails to meet WHOPES (WHO Pesticide Evaluation Scheme) standards.

In general, different control strategies for this infection such as improved sanitation and hygiene, increased access with clean water, health education and information campaigns, and mollusciciding have shown to be effective in interrupting infection transmission (Abou-El-Naga 2018; Phillips et al. 2017, 2018). In most cases, they are rarely undertaken concurrently in developing countries where the burden of diseases is more apparent. In the past decade chemotherapy based control was the most advocated measure because of its simplicity in the implementation and its effectiveness in reverting morbidity related to schistosomiasis. However, in the recent past, following endorsement of WHA resolution 65.21, endemic countries were urged considering changing of approach from morbidity control to elimination. This was found feasible due to an increased commitment from donor countries and pharmaceutical companies as well as the experience learned from other countries such as China where schistosomiasis has been successfully eliminated (Lo et al. 2017; Xu et al. 2016a).

Schistosomiasis is a water-related disease, i.e. its transmission can only happen where surface water that could allow survival and development of aquatic snails is available. Lack of safe water supply exacerbated by socio-economic factors enhances individuals to use alternative sources of water such as ponds, streams, lake, or river which in turn put them at risk of acquisition of schistosomiasis. Thus involvement and commitment of water authority to ensure availability of safe water supply could minimize schistosomiasis transmission (Grimes et al. 2014, 2015;

Krauth et al. 2015; Evan Secor 2014; Bergquist et al. 2017b). In this proposed study, the program will collaborate with water authority to ensure safe water supply is available in project areas.

Recently, there an international consortium, ZEST including Schistosomiasis Consortium for Operational Research and Evaluation (SCORE) was committed to assisting the government of Zanzibar in its efforts to eliminate urogenital schistosomiasis (Pennance et al. 2016; Celone et al. 2016). However, there are some differences with the other project and study. Firstly, the Chinese expert team will continuously work in the field of Pemba island accompany with African colleagues and improve and build the capacity of the schistosomiasis control. Second, China and most countries in Africa are developing countries, and condition including economy and health system are similar, the successful experience and produce from China will be easily imported and used in Africa. In so doing, it increases awareness of what is happening in their respective community (Utzinger et al. 2013; Korte et al. 1986; Collins et al. 2012). In this pilot study, we will make sure we engage the communities where the project will be implemented through regular sensitization meetings.

This pilot study will utilize new formulation of molluscicide—which has been extensively used in China for the control of snails during the schistosomiasis elimination campaigns (Xia et al. 2014; Dai et al. 2008, 2014). However, as most pesticides do, the new molluscicide might have a negative impact on aquatic life, i.e. may kill small freshwater fish. But this effect is only transient and that fishes or any aquatic animals that might be killed as a result of the application of the new molluscicide repopulate immediately after application stops and hence no long-term impact (fish or whatsoever will not permanently disappear) (Liu et al. 2017; Li and Wang 2017; Coelho and Caldeira 2016). Indeed, there is no scientific evidence on the bioaccumulation and magnification of the new molluscicide and hence there is no risk of development of any medical condition or disease because of its application in freshwater (King and Bertsch 2015; Lo et al. 2018).

6.7 Conclusions

In conclusion, given Chinese experience and expertise with schistosomiasis, particularly with controlling its transmission, we hope that these studies will provide clues and evidence that the collaboration with African countries can prompt implementation of comprehensive strategy and improve possibility to eliminate schistosomiasis in African countries.

References

Abou-El-Naga IF (2018) Towards elimination of schistosomiasis after 5000 years of endemicity in Egypt. Acta Trop 181:112–121

Basch PF (1986) Schistosomiasis in China: an update. Am J Chin Med 14:17–25

Bergquist R, Utzinger J, Keiser J (2017a) Controlling schistosomiasis with praziquantel: how much longer without a viable alternative? Infect Dis Poverty 6:74

Bergquist R, Zhou XN, Rollinson D, Reinhard-Rupp J, Klohe K (2017b) Elimination of schistosomiasis: the tools required. Infect Dis Poverty 6:158

Celone M, Person B, Ali SM, Lyimo JH, Mohammed UA, Khamis AN, Mohammed YS, Mohammed KA, Rollinson D, Knopp S (2016) Increasing the reach: involving local Muslim religious teachers in a behavioral intervention to eliminate urogenital schistosomiasis in Zanzibar. Acta Trop 163:142–148

Coelho P, Caldeira RL (2016) Critical analysis of molluscicide application in schistosomiasis control programs in Brazil. Infect Dis Poverty 5:57

Collins C, Xu J, Tang S (2012) Schistosomiasis control and the health system in P.R. China. Infect Dis Poverty 1:8

Dai JR, Wang W, Liang YS, Li HJ, Guan XH, Zhu YC (2008) A novel molluscicidal formulation of niclosamide. Parasitol Res 103:405–412

Dai JR, Li Y, Wang W, Xing Y, Qu G, Liang Y (2014) Sensitivity of *Oncomelania hupensis* to niclosamide: a nation-wide survey in China. Int J Environ Res Public Health 11:3086–3095

Evan Secor W (2014) Water-based interventions for schistosomiasis control. Pathog Glob Health 108:246–254

Fenwick A, Jourdan P (2016) Schistosomiasis elimination by 2020 or 2030? Int J Parasitol 46:385–388

French MD, Rollinson D, Basanez MG, Mgeni AF, Khamis IS, Stothard JR (2007) School-based control of urinary schistosomiasis on Zanzibar, Tanzania: monitoring micro-haematuria with reagent strips as a rapid urological assessment. J Pediatr Urol 3:364–368

Goatly KD, Jordan P (1965) Schistosomiasis in Zanzibar and Pemba. East Afr Med J 42:1–9

Grimes JE, Croll D, Harrison WE, Utzinger J, Freeman MC, Templeton MR (2014) The relationship between water, sanitation and schistosomiasis: a systematic review and meta-analysis. PLoS Negl Trop Dis 8:e3296

Grimes JE, Croll D, Harrison WE, Utzinger J, Freeman MC, Templeton MR (2015) The roles of water, sanitation and hygiene in reducing schistosomiasis: a review. Parasit Vectors 8:156

King CH, Bertsch D (2015) Historical perspective: snail control to prevent schistosomiasis. PLoS Negl Trop Dis 9:e0003657

Knopp S, Person B, Ame SM, Mohammed KA, Ali SM, Khamis IS, Rabone M, Allan F, Gouvras A, Blair L, Fenwick A, Utzinger J, Rollinson D (2013) Elimination of schistosomiasis transmission in Zanzibar: baseline findings before the onset of a randomized intervention trial. PLoS Negl Trop Dis 7:e2474

Knopp S, Person B, Ame SM, Ali SM, Muhsin J, Juma S, Khamis IS, Rabone M, Blair L, Fenwick A, Mohammed KA, Rollinson D (2016) Praziquantel coverage in schools and communities targeted for the elimination of urogenital schistosomiasis in Zanzibar: a cross-sectional survey. Parasit Vectors 9:5

Korte R, Schmidt-Ehry B, Kielmann AA, Brinkmann UK (1986) Cost and effectiveness of different approaches to schistosomiasis control in Africa. Trop Med Parasitol 37:149–152

Krauth SJ, Musard C, Traore SI, Zinsstag J, Achi LY, N'Goran EK, Utzinger J (2015) Access to, and use of, water by populations living in a schistosomiasis and fascioliasis co-endemic area of northern Cote d'Ivoire. Acta Trop 149:179–185

Li H, Wang W (2017) Apropos: critical analysis of molluscicide application in schistosomiasis control programs in Brazil. Infect Dis Poverty 6:54

Li YS, Zhao ZY, Ellis M, McManus DP (2005) Applications and outcomes of periodic epidemiological surveys for schistosomiasis and related economic evaluation in the People's Republic of China. Acta Trop 96:266–275

Liu Y, Zhou YB, Li RZ, Wan JJ, Yang Y, Qiu DC, Zhong B (2016) Epidemiological features and effectiveness of schistosomiasis control programme in mountainous and hilly region of the People's Republic of China. Adv Parasitol 92:73–95

Liu Y, Zhong B, Wu ZS, Liang S, Qiu DC, Ma X (2017) Interruption of schistosomiasis transmission in mountainous and hilly regions with an integrated strategy: a longitudinal case study in Sichuan, China. Infect Dis Poverty 6:79

Lo NC, Addiss DG, Hotez PJ, King CH, Stothard JR, Evans DS, Colley DG, Lin W, Coulibaly JT, Bustinduy AL, Raso G, Bendavid E, Bogoch II, Fenwick A, Savioli L, Molyneux D, Utzinger J, Andrews JR (2017) A call to strengthen the global strategy against schistosomiasis and soil-transmitted helminthiasis: the time is now. Lancet Infect Dis 17:e64–e69

Lo NC, Gurarie D, Yoon N, Coulibaly JT, Bendavid E, Andrews JR, King CH (2018) Impact and cost-effectiveness of snail control to achieve disease control targets for schistosomiasis. Proc Natl Acad Sci U S A 115:E584–E591

Montresor A, Ramsan M, Chwaya HM, Ameir H, Foum A, Albonico M, Gyorkos TW, Savioli L (2001) School enrollment in Zanzibar linked to children's age and helminth infections. Tropical Med Int Health 6:227–231

Pennance T, Person B, Muhsin MA, Khamis AN, Muhsin J, Khamis IS, Mohammed KA, Kabole F, Rollinson D, Knopp S (2016) Urogenital schistosomiasis transmission on Unguja Island, Zanzibar: characterisation of persistent hot-spots. Parasit Vectors 9:646

Phillips AE, Gazzinelli-Guimaraes PH, Aurelio HO, Ferro J, Nala R, Clements M, King CH, Fenwick A, Fleming FM, Dhanani N (2017) Assessing the benefits of five years of different approaches to treatment of urogenital schistosomiasis: a SCORE project in Northern Mozambique. PLoS Negl Trop Dis 11:e0006061

Phillips AE, Gazzinelli-Guimarães PH, Aurelio HO, Dhanani N, Ferro J, Nala R, Deol A, Fenwick A (2018) Urogenital schistosomiasis in Cabo Delgado, northern Mozambique: baseline findings from the SCORE study. Parasit Vectors 11:30

Rollinson D, Klinger EV, Mgeni AF, Khamis IS, Stothard JR (2005) Urinary schistosomiasis on Zanzibar: application of two novel assays for the detection of excreted albumin and haemoglobin in urine. J Helminthol 79:199–206

Ross AG, Chau TN, Inobaya MT, Olveda RM, Li Y, Harn DA (2017) A new global strategy for the elimination of schistosomiasis. Int J Infect Dis 54:130–137

Savioli L, Fenwick A, Rollinson D, Albonico M, Ame SM (2015) An achievable goal: control and elimination of schistosomiasis. Lancet 386:739

Savioli L, Albonico M, Colley DG, Correa-Oliveira R, Fenwick A, Green W, Kabatereine N, Kabore A, Katz N, Klohe K, LoVerde PT, Rollinson D, Stothard JR, Tchuem Tchuenté LA, Waltz J, Zhou XN (2017) Building a global schistosomiasis alliance: an opportunity to join forces to fight inequality and rural poverty. Infect Dis Poverty 6:65

Shi L, Li W, Wu F, Zhang JF, Yang K, Zhou XN (2016) Epidemiological features and control progress of Schistosomiasis in waterway-network region in the People's Republic of China. Adv Parasitol 92:97–116

Song L, Wu X, Ning A, Wu Z (2017) Lessons from a 15-year-old boy with advanced schistosomiasis japonica in China: a case report. Parasitol Res 116:1787–1791

Stothard JR, French MD, Khamis IS, Basáñez MG, Rollinson D (2009) The epidemiology and control of urinary schistosomiasis and soil-transmitted helminthiasis in schoolchildren on Unguja Island, Zanzibar. Trans R Soc Trop Med Hyg 103:1031–1044

Utzinger J, Brattig NW, Kristensen TK (2013) Schistosomiasis research in Africa: how the CONTRAST alliance made it happen. Acta Trop 128:182–195

Wang X, Wang W, Wang P (2017) Long-term effectiveness of the integrated schistosomiasis control strategy with emphasis on infectious source control in China: a 10-year evaluation from 2005 to 2014. Parasitol Res 116:521–528

WHO (2002) Prevention and control of schistosomiasis and soil-transmitted helminthiasis. World Health Organ Tech Rep Ser 912:i–vi, 1–57, back cover

Xia J, Yuan Y, Xu X, Wei F, Li G, Liu M, Li J, Chen R, Zhou Z, Nie S (2014) Evaluating the effect of a novel molluscicide in the endemic schistosomiasis japonica area of China. Int J Environ Res Public Health 11:10406–10418

Xu J, Bergquist R, Qian YJ, Wang Q, Yu Q, Peeling R, Croft S, Guo JG, Zhou XN (2016a) China-Africa and China-Asia collaboration on Schistosomiasis control: a SWOT analysis. Adv Parasitol 92:435–466

Xu J, Yu Q, Tchuenté LA, Bergquist R, Sacko M, Utzinger J, Lin DD, Yang K, Zhang LJ, Wang Q, Li SZ, Guo JG, Zhou XN (2016b) Enhancing collaboration between China and African countries for schistosomiasis control. Lancet Infect Dis 16:376–383

Xu J, Steinman P, Maybe D, Zhou XN, Lv S, Li SZ, Peeling R (2016c) Evolution of the national schistosomiasis control programmes in the People's Republic of China. Adv Parasitol 92:1–38

Zanzibar. http://zanzibar.go.tz/

Zhang W, Wong CM (2003) Evaluation of the 1992–1999 World Bank Schistosomiasis control project in China. Acta Trop 85:303–313

Zoni AC, Catalá L, Ault SK (2016) Schistosomiasis prevalence and intensity of infection in Latin America and the Caribbean countries, 1942–2014: a systematic review in the context of a regional elimination goal. PLoS Negl Trop Dis 10:e0004493

Chapter 7
Epidemiological Survey of Schistosomiasis in Zanzibar

Liang Shi, Saleh Juma, Fatma Kabole, Jian He, Yu-Zheng Huang, Wei Li, Jian-Feng Zhang, and Kun Yang

Abstract A cross-sectional survey of schistosomiasis was carried out in Pemba Island, Zanzibar from 2017 to 2019. The overall infection rate was 1.84% (353/19,194, 95% CI: 1.65%–2.03%). A total of 73.53% of shehia (50/68) were found to be infected with schistosomiasis, and the highest infection rate reached 18.77%. The infection rate of children aged 6–18 was 2.76%. The student infection rate was 2.2%, higher than other occupations. The infection rate of primary school students is 2.41%. There were statistically significant differences in the infection rate among regions, age groups, occupations, and education levels. Snail habitats were mainly found in stream (79.22%). Positive snail habitats were mainly found in stream (51.43%). Control measures must be strengthened in high infection areas and groups. Snail monitoring and control should be intensified in the future.

Keywords Schistosomiasis · Snails · Zanzibar · Pemba Island · Cross-sectional survey

7.1 Introduction

Schistosomiasis, caused by a trematode worm belonging to genus *Schistosoma*, is a potentially debilitating disease and can result in both acute and chronic infections. It occurs in tropical and subtropical countries and remains a major burden for public

L. Shi (✉) · Y.-Z. Huang · W. Li · J.-F. Zhang · K. Yang
Jiangsu Institute of Parasitic Diseases, Key Laboratory on Technology for Parasitic Disease Prevention and Control, Ministry of Health, Jiangsu Provincial Key Laboratory on the Molecular Biology of Parasites, Wuxi, China
e-mail: shiliang@jipd.com

S. Juma · F. Kabole
Ministry of Health of Zanzibar, Zanzibar, United Republic of Tanzania

J. He
Global Health Center, Jiangsu Institute of Parasitic Diseases, Wuxi, Jiangsu, China

K. Yang, H. Mehlhorn (eds.), *Sino-African Cooperation for Schistosomiasis Control in Zanzibar*, Parasitology Research Monographs 15,
https://doi.org/10.1007/978-3-030-72165-7_7

health in these regions. The World Health Organization (WHO) estimates that currently at least 218 million individuals require preventive treatment for schistosomiasis. The global burden of schistosomiasis has been estimated at 3.31 (95% confidence interval (CI): 1.70–6.26) million disability-adjusted life years (DALYs). In Africa alone, it is estimated that some 200 million people are infected with the blood fluke of the genus schistosomiasis. It negatively impacts on the health and wellbeing of mainly rural dwellers in Zanzibar. The World Health Organization (WHO) recently put forward an ambitious goal for the year 2020: to control schistosomiasis globally. Interruption of transmission and elimination of schistosomiasis are encouraged whenever resources allow. Examples of where schistosomiasis has been successfully controlled or even eliminated using integrated measures in People's Republic of China (*S. japonicum*) were used as an example. The Zanzibar archipelago, part of the United Republic of Tanzania, has been identified as a candidate area, where schistosomiasis elimination might be achieved. But the prevalence of schistosomiasis is complicated, there are still many challenges in Zanzibar.

After careful consideration, with the help of WHO and Zanzibar ministry of health, the China aid Schistosomiasis Control project selected Pemba Island in Zanzibar to eliminate schistosomiasis. In this chapter, we report the disease status and blister snail status baseline survey in Pemba Island.

7.2 Method

7.2.1 Study Area and Population

The Zanzibar archipelago includes the two large islands of Unguja and Pemba. Pemba is divided into four districts, which are further subdivided into smaller administrative units, known as shehias. The local administration in the shehias is governed by the community leader (sheha). According to the 2002 census, Pemba consists of 72 shehias. The majority of the population is Muslim. We surveyed 68 shehias in Pemba mainland.

7.2.2 Field Procedures

Details of the study surveys are provided elsewhere between April 2017 and Feb 2020. We conducted a baseline parasitological survey assessing *S. haematobium* infection in adults and children. In each study shehia, the sheha was invited to answer a set of standard questions about the demographics, sanitary infrastructure, and water availability and use in the shehia. Moreover, 50 randomly selected households in each shehia were visited by a member of a 3–10 headed trained interviewer team. Participation included answering standard questions about demographics such as age, occupation, and risk factors potentially associated with *S. haematobium* transmission. Moreover, all participants were invited to submit a urine sample right after the questionnaire interview between 09:00 and 18:00 hours. Urine samples were transferred to the laboratory in Pemba Island Tropical Disease Office Laboratory immediately after collection.

7.2.3 Laboratory Procedures

Urine samples were processed either the same day or stored in the fridge until the next morning. *S. haematobium* eggs by filtering 10 ml of urine through a polycarbonate filter that was quantitatively examined under a microscope by experienced laboratory technicians.

7.2.4 Snail Survey

Investigate all the fresh water environment suitable for blister snails on Pemba Island in Zanzibar. During each Shehia survey, under the guide of local residents who are familiar with the environment, search for freshwater bodies in the jurisdiction. According to the distribution range and characteristics of freshwater bodies, the freshwater body environment is divided into: temporary ponds and permanent ponds. Permanent streams and temporary streams: In permanent and temporary ponds, according to the survey results of the basic conditions of the water body, the ponds are divided into several survey units along a circle with a distance of 15 m. Before the survey of each unit, trained staff use Garmin etrex 30x device (Kansas City, KS, USA) to mark the starting point of each unit, and record the coordinate point number and latitude and longitude. The staff used tweezers to collect *B. globosus* in the unit. The collection range was concentrated on the borders of water and vegetation and other places where snails were relatively concentrated. The captured *B. globosus* was placed in a 50–100 ml plastic cup containing fresh water in the unit (one for each unit). The plastic cup is brought back to the laboratory for testing, and the coordinate point number of the starting point of the unit is marked with a Mark pen on the lid of the plastic cup. After the investigation of each unit, fill in the snail capture information. The number of survey units for each pond is not less than 3, the survey time for each unit is 15 minutes, and the survey area does not exceed 15 m^2. In permanent and temporary streams, according to the survey results of the basic conditions of the water body, starting from the source of the stream, the stream is divided into several survey units with an interval of 100 m. The specific investigation requirements are the same as pond. All captured *B. globosus* were counted and checked for infection in the vector laboratory of NTD on Pemba Island.

7.2.5 *Infectious Snail Detection*

Place the blister snail on a glass slide, use another thick glass slide to gently crush the blister snail, then add a drop of dechlorinated water on the blister snail and place the blister snail under a dissecting mirror (×10 times) or a microscope (4×objective lens, 10×eyepiece), use a dissecting needle to pull apart the shell, shred the vesicular snail digestive gland and other soft tissues in turn, and observe one by one to find *Schistosoma elegans* cercariae and/or *Ascomycetes* are infective blister snails.

Every time the dissecting needle fiddles with the soft tissue of the snail, it should be cleaned in time to prevent cercaria contamination. The snail infection rate and the density of positive snails were calculated in units of water.

7.3 Results

7.3.1 *Disease Survey*

7.3.1.1 General Situation

A total of 68 shehia and 19,194 residents were surveyed in Pemba Island, which distributed in Chakechake, Micheweni, Mkoani, and Wete 4 districts. The composition ratios of shehia number in the 4 survey districts were 27.94% (19/68), 19.12% (13/68), 29.41% (20/68), and 23.53% (16/68), and the population composition ratios in the 4 survey districts were 28.92% (5550/19,194), 18.18% (3489/19,194), 28.96% (5559/19,194), and 23.94% (4596/19,194). With an average age of 23.23 years old among 19,194 residents, males accounted for 48.60% (9321/19,194), females accounted for 51.40% (9873/19,194), male to female ratio was 0.94:1; those younger than 5 years old accounted for 11.79% (2263/19,194), 6–18 years old accounted for 41.67% (7998/19,194); those over 19 years old accounted for 46.54% (8933/19,194); farmers accounted for 18.21% (3495/19,194), fishermen accounted for 1.80% (345/19,194), students accounted for 33.89% (6505/19,194), workman accounted for 1.11% (213/19,194), businessmen accounted for 2.42% (465/19,194), other accounted for 42.57% (8171/19,194); the number of people who did not receive education accounted for 33.29% (6389/19,194), primary school education accounted for 39.42% (7566/19,194), secondary school education accounted for 26.11% (5012/19,194), high school education accounted for 0.29% (56/19,194), college education accounted for 0.89% (171/19,194). Results shown in Figs. 7.1, 7.2, and 7.3 and Table 7.2.

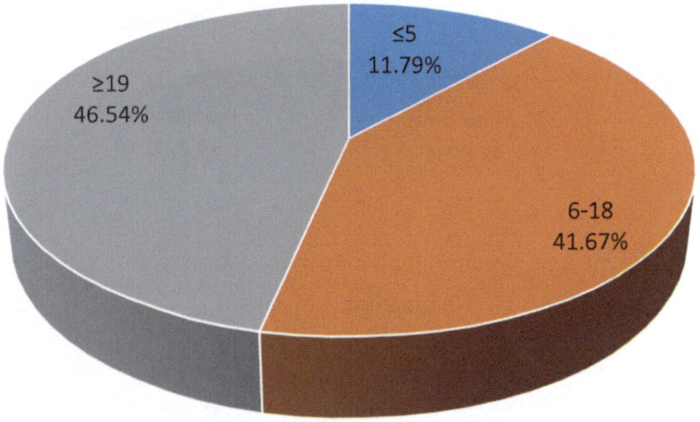

Fig. 7.1 Composition ratio of different age groups

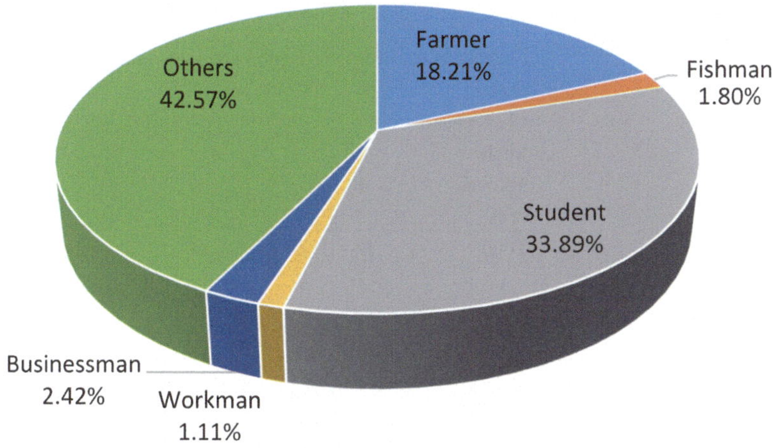

Fig. 7.2 Composition ratio of different occupations

7.3.1.2 District Schistosomiasis Infection

A total of 353 cases of schistosomiasis were detected in 4 districts across the island, with an infection rate of 1.84% (353/19,194, 95% CI: 1.65%–2.03%). The infection rate from Chakechake district was the highest, 2.59% (95% CI: 2.18%–3.01%), and the overall difference in the infection rate among the 4 regions was statistically significant ($P \leq 0.05$). A total of 73.53% of shehia (50/68) were found to be infected with schistosomiasis, of which 48% (24/50) had an infection rate of 0–1%, and 36% of shehia (18/50) had an infection rate of 1–5%, 16% shehia (8/50) infection rate >5%. Results shown in Table 7.1, Fig. 7.4.

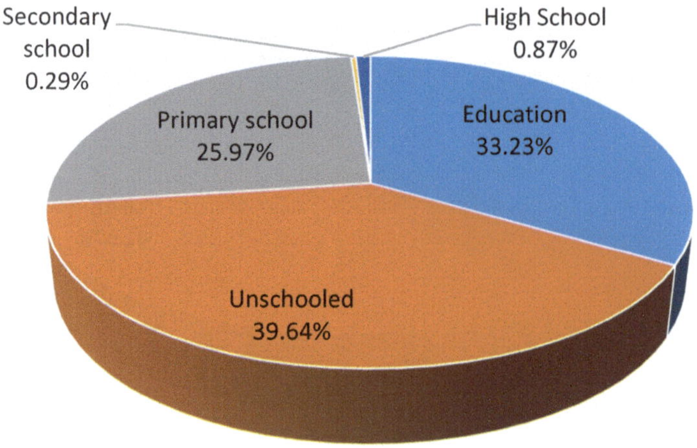

Fig. 7.3 Composition ratio of different education levels

Table 7.1 District prevalence of infection with schistosomiasis

Number of schistosomiasis infection rates in the shehia					
District	0	0–1%	1–5%	>5%	Total
Chakechake	4	8	4	3	19
Micheweni	5	5	2	1	13
Mkoani	6	4	6	4	20
Wete	3	7	6	0	16
Total	18	24	18	8	68

7.3.1.3 Population Schistosomiasis Infection

The infection rates of men and women were 1.92% and 1.76%, respectively, and the difference was not statistically significant ($P = 0.223$); the infection rates of age ≤ 5, 6–18, and ≥ 19 were 0.13%, 2.76%, and 1.44%, respectively. Statistical significance ($P \leq 0.05$); The infection rates of farmers, fishermen, students, workers, businessmen, and others were 1.60%, 1.74%, 2.23%, 0.47%, 0.43%, 1.75%, 1.41%, and 0.62%, respectively. There was statistical significance ($P \leq 0.05$); the infection rates of unschooled, Primary school, and Secondary school were 1.67%, 2.41%, and 1.28%, respectively. Results shown in Table 7.2 and Fig. 7.5.

7.3.2 Snails Survey

7.3.2.1 General Survey Situation

A total of 628 water environments were investigated in 68 shehia of the Pemba main island, of which streams accounted for 71.82% and ponds accounted for 28.18%.

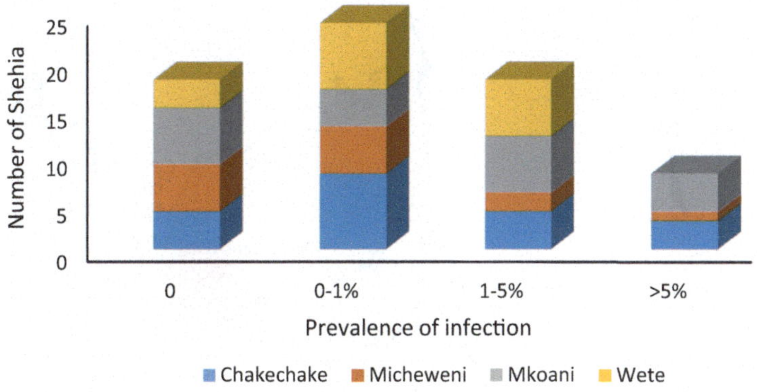

Fig. 7.4 District prevalence of infection with schistosomiasis

Table 7.2 Prevalence of infection with *Schistosoma haematobium*

Variable	*N* (%)	*n*	%	*P*-value
District				
Chakechake	5550	144	2.59%	$P \leq 0.05$
Micheweni	3489	38	1.09%	
Mkoani	5559	111	2.00%	
Wete	4596	60	1.31%	
Sex				
Male	9321	179	1.92%	$P = 0.223$
Female	9873	174	1.76%	
Age group				
≤5	2263	3	0.13%	$P \leq 0.05$
6–18	7998	221	2.76%	
≥19	8933	129	1.44%	
Occupation				
Farmer	3495	56	1.60%	$P \leq 0.05$
Fisherman	345	6	1.74%	
Student	6505	145	2.23%	
Worker	213	1	0.47%	
Businessman	465	2	0.43%	
Others	8171	143	1.75%	
Education				
Unschooled	6389	107	1.67%	$P \leq 0.05$
Primary school	7566	182	2.41%	
Secondary school	5012	64	1.28%	
High school	56	0	0.00%	
College	171	0	0.00%	

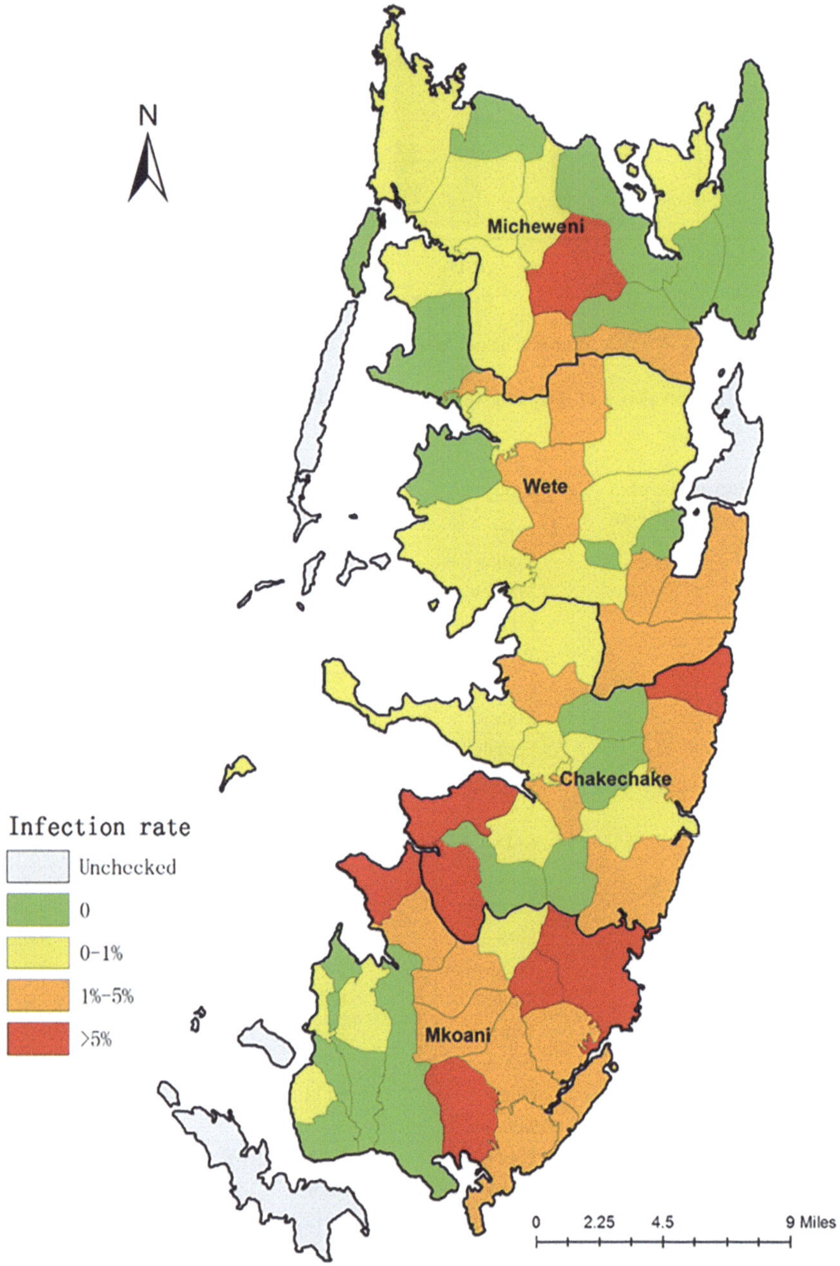

Fig. 7.5 Shehia prevalence of infection with schistosomiasis

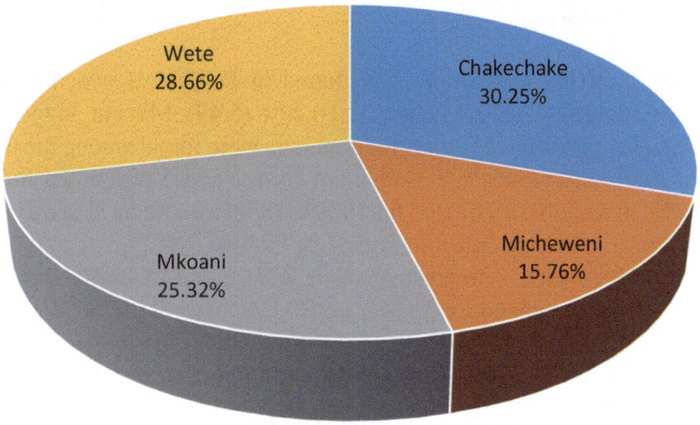

Fig. 7.6 Composition ratio of environment in different districts

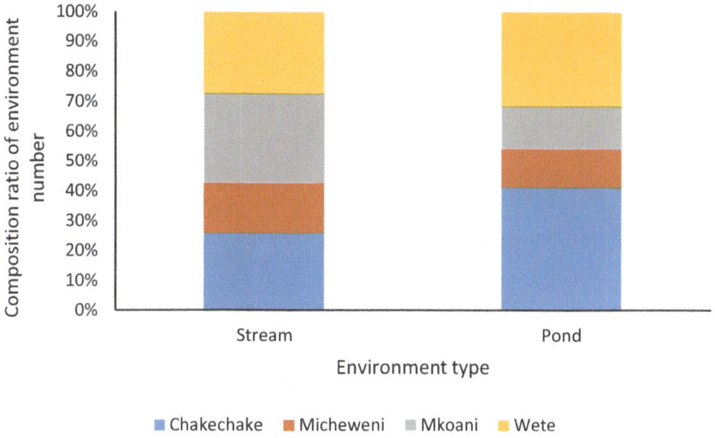

Fig. 7.7 Composition ratio of different types of environments

Among 4 districts, Chakechake accounted for 30.25% (190/628), Micheweni accounted for 15.76% (99/628), Mkoani accounted for 25.32% (159/628), and Wete accounted for 28.66% (180/628). In the stream environment type, the composition ratios of Chakechake, Micheweni, Mkoani, and Wete are 25.94%, 16.85%, 29.71%, and 27.49%. In the type of pond environment, the composition ratios of Chakechake, Micheweni, Mkoani, and Wete were 41.24%, 12.99%, 14.12%, and 31.64%. Results shown in Figs. 7.6 and 7.7.

7.3.2.2 Snail Results

72.06% (49/68) shehia with snails were found in Pemba, of which Chakechake accounted for 34.69% (17/49), Micheweni 10.20% (5/49), Mkoani 34.69% (17/49), and Wete 20.41% (10/49). 255 snail environments were found among the 68 shehia. Number of snail environments in shehia between 1 and 5 accounted for 67.35% (33/49), between 6 and 10 accounted for 16.33% (8/49), between 11 and 20 accounts for 16.33% (8/49). Results shown in Figs. 7.8, 7.9, and 7.10.

32.65% (16/49) shehia with positive snails were found in shehia with snails, of which Chakechake accounted for 31.25% (5/16), Micheweni 6.25% (1/16), Mkoani 37.50% (6/16), Wete 25.00% (4/16). 35 positive snail environments were found among the 68 shehia. Number of positive snail environments in shehia between

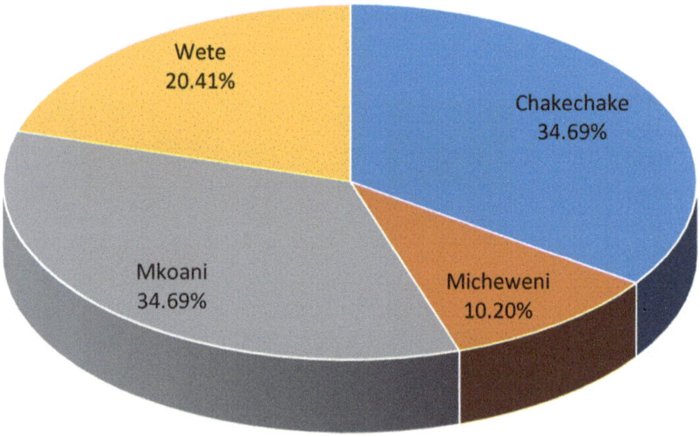

Fig. 7.8 Composition ratio of snail districts

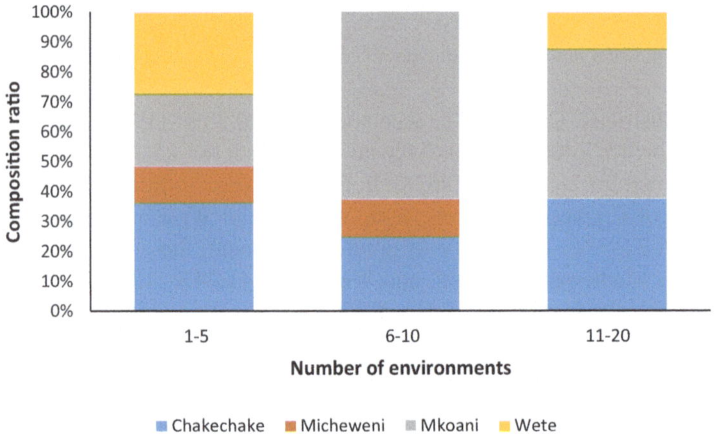

Fig. 7.9 Composition ratio of the number of snails in different districts

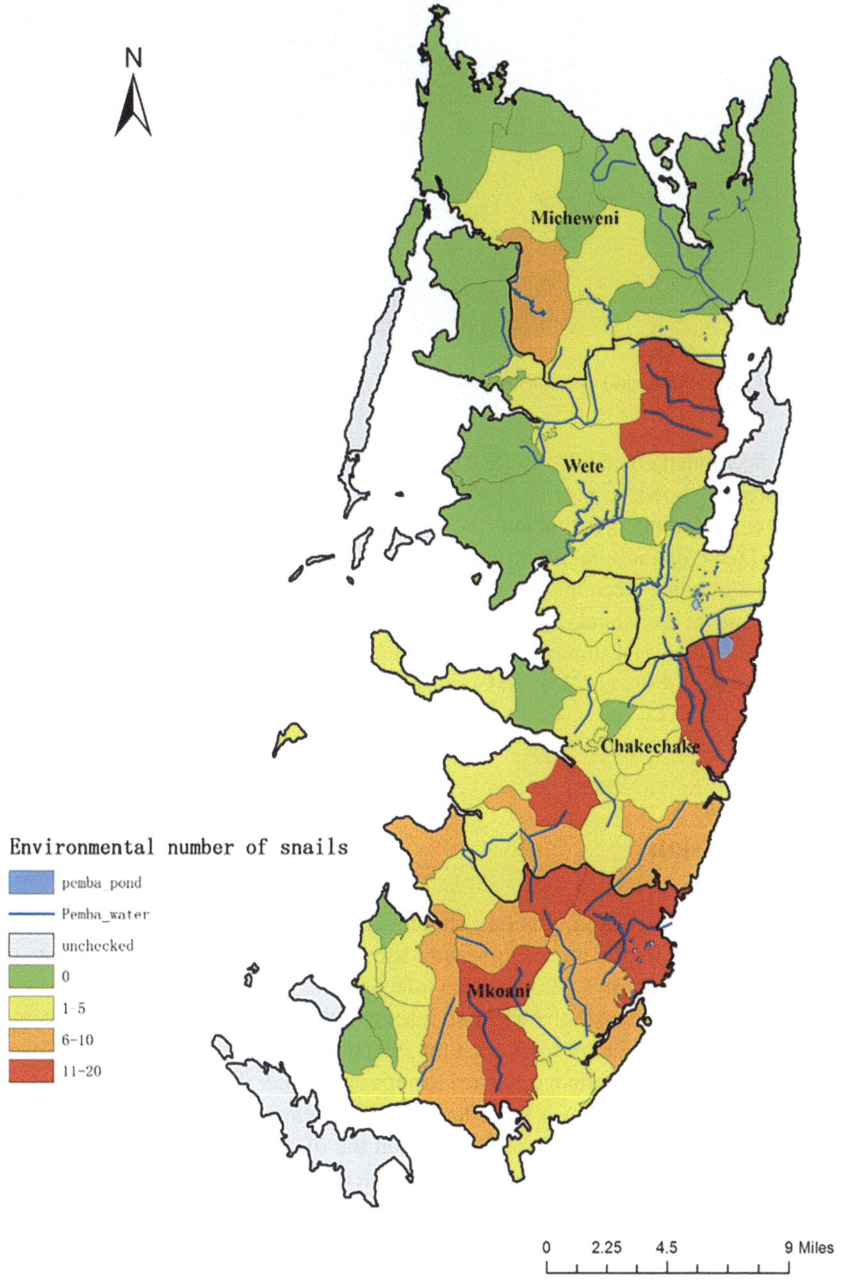

Fig. 7.10 Shehia snail distribution

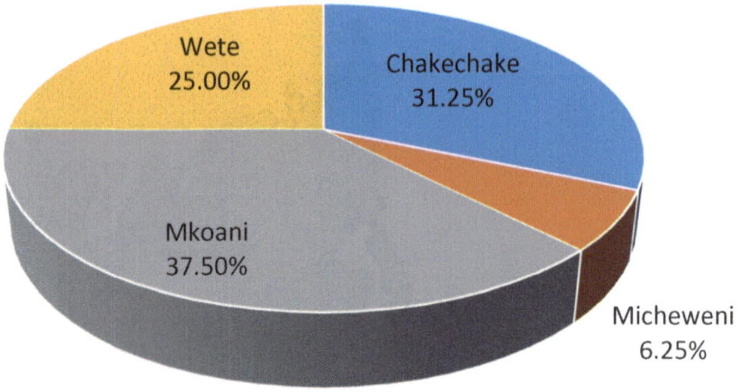

Fig. 7.11 Composition ratio of positive snail districts

1 and 5 accounted for 67.35% (15/16), between 6 and 10 accounted for 16.33% (1/16). Results shown in Figs. 7.11 and 7.12.

255 snail breeding environments were found in 68 shehia of the island, accounting for 40.61% of the total environment, of which streams accounted for 32.17% (202/628) and ponds accounted for 8.44% (53/628). Among the 255 snail breeding environments on the island, 35 positive snail environments were found, accounting for 13.73% of the total number of snail breeding environments, of which streams accounted for 51.43% (18/35), and ponds accounted for 48.57% (17/35). Results shown in Table 7.3.

7.4 Discussion

The epidemic of schistosomiasis on Pemba Island has caused a serious disease and economic burden. The results of this study showed that the overall infection rate of the island was 1.84%, 38% of the shehia infection rate exceeded the overall level of the island, and the highest infection rate reached 18.77%.

The results of this survey found that the infection rate of children aged 6–18 was 2.76%, which was higher than other age groups and higher than the overall infection rate of schistosomiasis across the island. The student infection rate was 2.23%, higher than other occupations, and higher than the overall infection rate of schistosomiasis on the island. The infection rate of elementary school education is 2.41, which is higher than other education, and higher than the overall infection rate of schistosomiasis in the whole island. However, no infected persons have been detected with high school and university education.

It is also widely acknowledged that students or adolescents, people using natural freshwater, and those pursuing specific occupations that expose them to open freshwater bodies (e.g., farmers) are at an elevated risk of Schistosomiasis infection. Therefore, future surveillance and control of the disease will need to pay special attention to these factors.

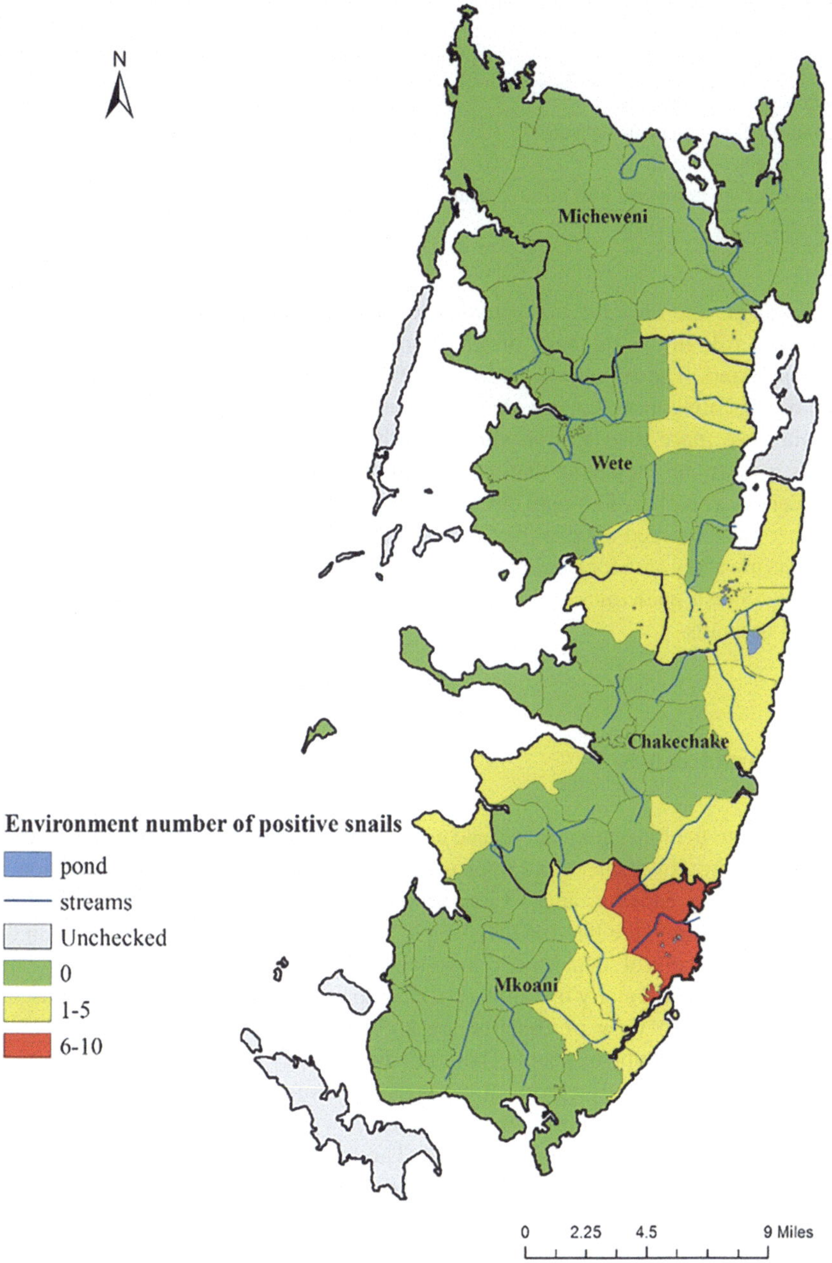

Fig. 7.12 Shehia positive snail distribution

Table 7.3 Survey of snails

Survey item	Environment type		
	Stream	Pond	Total
Number of environments	451	177	628
Number of environments with snails	202	53	255
Total number of snails	29,552	4341	33,893
Number of environments with positive snail	18	17	35
Number of positive snails	35	68	103

Snails are the only intermediate host of schistosomiasis and play an extremely important role in the epidemic of schistosomiasis. Snails are widely distributed on Pemba Island. The investigation found 72.06% shehia with snails were found in Pemba indicating that snail is widely distributed on the island. More than 30% of the shehia have more than 5 environments with snails. 32.65% shehia with positive snails were found indicating that the snail environment and positive snails distribution density is higher. The snail environment number and positive snail environment number has exceeded 30% in the Chakechake and Mkoani district reveal that these districts are a key area for snail control. Therefore, to effectively control the spread of schistosomiasis epidemics, it is necessary to control and monitor the growth area and spread of snails.

7.5 Conclusions

According to the formation factors and epidemic characteristics of epidemic areas on the island, the government and relevant departments should pay great attention to schistosomiasis endemic areas. First, strengthen the monitoring of snails and the source of infection, the intermediate host of schistosomiasis, and cooperate with the local disease control department to establish an epidemic monitoring network information, and regularly organize personnel to monitor the snail status and disease status: Second, in view of endemic areas children (age 6–18) incidence is high; therefore, these groups of people should be used as key disease monitoring targets. The third is to promptly carry out pathogenic examinations for people who have been engaged in production, living, and contact with infected water, and promptly treat patients with schistosomiasis found. The fourth is to kill snails with drugs in areas with positive blisters. Fifth, it is necessary to strengthen the research and application of new technologies for schistosomiasis, strengthen the publicity and education of schistosomiasis knowledge, establish a disease inspection and monitoring system in health centers, and grasp the epidemic information in a timely manner to achieve early detection, early diagnosis, early treatment, and early control of the epidemic.

References

Abe EM, Tambo E, Xue J et al (2020) Approaches in scaling up schistosomiasis intervention towards transmission elimination in Africa: leveraging from the Chinese experience and lessons. Acta Trop 208:105379

Chen J, Ding W, Li Z et al (2020) From parasitic disease control to global health: new orientation of the National Institute of Parasitic Diseases, China CDC. Acta Trop 201:105219

Hailegebriel T, Nibret E, Munshea A (2020) Prevalence of *Schistosoma mansoni* and *S. haematobium* in snail intermediate hosts in Africa: a systematic review and meta-analysis. J Trop Med 2020:8850840

Hotez PJ, Pecoul B, Rijal S et al (2016) Eliminating the neglected tropical diseases: translational science and new technologies. PLoS Negl Trop Dis 10(3):e0003895

Knopp S, Stothard JR, Rollinson D et al (2013) From morbidity control to transmission control: time to change tactics against helminths on Unguja Island, Zanzibar. Acta Trop 128(2):412–422

Lackey EK, Horrall S (2020) Schistosomiasis (*Schistosoma haematobium*). In: StatPearls. StatPearls Publishing LLC, Treasure Island (FL)

Lustigman S, Prichard RK, Gazzinelli A et al (2012) A research agenda for helminth diseases of humans: the problem of helminthiases. PLoS Negl Trop Dis 6(4):e1582

Mazigo HD, Nuwaha F, Kinung'hi SM et al (2012) Epidemiology and control of human schistosomiasis in Tanzania. Parasit Vectors 5:274

McManus DP, Dunne DW, Sacko M et al (2018) Schistosomiasis. Nature reviews. Disease Primers 4(1):13

Murray CJ, Vos T, Lozano R et al (2012) Disability-adjusted life years (DALYs) for 291 diseases and injuries in 21 regions, 1990–2010: a systematic analysis for the Global Burden of Disease Study 2010. Lancet (London, UK) 380(9859):2197–2223

Qian MB, Chen J, Bergquist R et al (2019) Neglected tropical diseases in the People's Republic of China: progress towards elimination. Infect Dis Poverty 8(1):86

Ren Guanghui LY (2015) African schistosomiasis. People's Medical Publishing House, Beijing

Rollinson D, Knopp S, Levitz S et al (2013) Time to set the agenda for schistosomiasis elimination. Acta Trop 128(2):423–440

Rujeni N, Morona D, Ruberanziza E et al (2017) Schistosomiasis and soil-transmitted helminthiasis in Rwanda: an update on their epidemiology and control. Infect Dis Poverty 6(1):8

Steinmann P, Keiser J, Bos R et al (2006) Schistosomiasis and water resources development: systematic review, meta-analysis, and estimates of people at risk. Lancet Infect Dis 6 (7):411–425

WHO (2012) Accelerating uncork to tropical diseases, Geneva, p 42

Yang Kun LS (2020) Appropriate technology for schistosomiasis control and elimination. People's Medical Publishing House, Beijing

Yi-Ting L, Guang-Hui R, You-Sheng L et al (2018) Global burden and challenges of parasitic diseases in Africa. Chinese J Schistosomiasis Control 30(2):226–231

Chapter 8
Mapping of Schistosomiasis Haematobia in Zanzibar

Mingzhen He

Abstract Schistosomiasis remains a serious health problem in Africa. Although a strong, coordinated agenda for research on this disease has been in place for the last 50 years in Zanzibar, data storage, retrieval of survey data, and management remain problem areas. We investigated the use of Google Earth (GE) in conjunction with a handheld, global positioning system as a pilot project for managing schistosomiasis control. In this way, risk areas can be surveyed and followed up by visualizing both the distribution of human infections and that of the intermediate snail host together with environmental information. A platform with three spatial databases was created: (1) Distribution of infected humans; (2) Distribution of the intermediate snail host in ponds (infected and not infected specimens); (3) Distribution of the intermediate snail host in streams (infected and non-infected specimens). The GE spatial database increased the efficiency of follow-up case treatment as well as snail control and contributed also to the discovery of previously unknown areas in need of snail control. We conclude that this platform is advantageous not only by being useful for management and visualization of spatial data, but also because it is easy to operate and available free of charge. Also spatial autocorrelation and spatial scan statistics were used to detect the spatial characteristics of the distribution of *B. globosus*. The results indicated that there was spatial correlation between the snail distribution globally, there were both positive correlation and negative correlation in the local snail sites, and there existed spatial clustering areas of snail sites.

Keywords Google Earth · Geographic Information System · Schistosomiasis · Snail · Database · Spatial autocorrelation · Spatial analysis · Spatial autocorrelation · Spatial scan statistics

M. He (✉)
Changzhou Center for Disease Control and Prevention, Changzhou, P. R. China

K. Yang, H. Mehlhorn (eds.), *Sino-African Cooperation for Schistosomiasis Control in Zanzibar*, Parasitology Research Monographs 15,
https://doi.org/10.1007/978-3-030-72165-7_8

8.1 Background

Schistosomiasis, one of the poverty-related, neglected tropical diseases (NTDs), is caused by trematode parasites of the genus *Schistosoma*, which reside in the abdominal capillaries of the infected human definitive host. Different kinds of snails play the role of intermediate host depending on the species of the parasite. So far, many areas in sub-Saharan Africa still have a high schistosomiasis prevalence with many people presenting with high intensities of disease due to constant reinfection (Olsen et al. 2015). The World Health Organization (WHO) estimates that more than 90% of all humans requiring treatment for this disease currently live in Africa (WHO 2018). In Zanzibar (including the islands of Unguja and Pemba), urogenital schistosomiasis is still a considerable public health problem with high prevalence (Knopp et al. 2013; Pennance et al. 2016). Schistosomiasis on these islands is urogenital which is caused by *Schistosoma haematobium* relying on the intermediate host snail *Bulinus globosus* (Allan et al. 2013).

In 2012, WHO set the goal to interrupt transmission of schistosomiasis in selected countries of the African region by 2025 (WHO 2013). Much work and research have been carried out to reach this goal in Zanzibar. Prevention and control on the island of Pemba was once the focus of strong control efforts (Savioli et al. 1989; Savioli and Mott 1989; Lwambo et al. 1997; Guidi et al. 2010); however, although disease burdens have been alleviated through preventative chemotherapy together with additional control interventions, transmission has not been interrupted on a large scale (Knopp et al. 2012, 2013, 2016). Regrettably, collected data are dispersed and scattered and cannot be effectively utilized since the local schistosomiasis control institution, the NTD office, has neither access to easily available information about the distribution of *B. globosus* nor to environmental data associated with this infection.

At the World Health Assembly on May 21st 2014, the Government of Zanzibar, WHO, and China jointly signed a memorandum of understanding to cooperate in the control of schistosomiasis in Zanzibar. This multilateral cooperation resulted in the decision to conduct a pilot project with the aim of eliminating schistosomiasis in this part of Tanzania. In August 2016, Jiangsu Institute of Parasitic Disease in Wuxi, China officially took on this China-aid project for schistosomiasis elimination in Zanzibar. The project was stationed on Pemba Island and a systematic disease investigation and snail survey were initiated together with the local NTD office. It was felt that the work should focus on the spatial component to establish effective data collection and management for schistosomiasis control using geographical information systems (GIS) based on the Google Earth (GE) platform (Google Inc., Mountain View, CA, USA), which is a free software displaying geographic data.

Elliott and Wartenberg (2004) pointed out the importance of spatial epidemiology and a review by Bergquist and Rinaldi (2010) noted the usefulness of remote sensing (RS) for health-related research on endemic diseases. GE integrates the functions of RS, GIS, and global positioning system (GPS) technologies to present users with satellite-generated aerial imagery that varies from the global scale to the local (Lisle

2006). Combined with GIS software, *e.g.*, ArcGIS (ESRI, Redlands, CA, USA), and network technology, GE is an excellent approach to spatial epidemiology as various files used for the collection of data, *e.g.*, Excel and shapefiles. Such files can easily be reformatted and incorporated as keyhole markup language (KML) files, a commonly used format pioneered by Keyhole Inc., which was acquired by Google in 2004. The shapefile is a popular geospatial vector data format developed by ESRI for interoperability between various software products. Human residences as well as streams and ponds are spatially represented as points, lines, and polygons, respectively. A large number of different annotations describing these vectors, such as name, environmental variables, infection status, *etc*., can be added as annotations.

The GE virtual globe approach incorporates aerial images, digital terrain, and a large amount of basic geographic information, the utilization of which is simple and convenient. Collected data can be accurately recorded and zooming makes it possible to present geographically limited areas as flat maps where the landscape scenes can be intuitively understood with distances and areas derived in a straightforward way. The spatial data management methods applicable to basic schistosomiasis control work on Pemba were explored by adopting the GE platform to establish a spatial database for the demonstration of the distribution of human schistosomiasis infections and the presence of the intermediate snail host in the various water bodies available on the island. The specific aim of this work was to improve management of schistosomiasis control in Zanzibar by investigating the spatial relation between infected humans and *B. globosus* habitats.

Due to the influence of the micro environment factors in local areas, the attraction of the same species, the limitation of species diffusion, and other internal factors, the spatial distribution of biological population has a correlation. The intermediate snail host of *Schistosoma haematobium* is also spatially related (Cheng et al. 2016). Data attributes and spatial location were both considered in spatial autocorrelation, of which the theoretical basis is that the adjacent data in space usually have higher similarity than the remote data Geographic Information System (GIS) and spatial analysis technology provide support for the study of spatial distribution characteristics of snails. Moran's I and Getis-Ordgeneral G indicator can be used to test the relationship between a spatial variable and its adjacent space. Spatial clustering analysis is an important method for epidemiological study of disease causes or risk factors. It can detect whether the observation variables have regional clustering, and determine the specific location and scope of the clustering area, so as to reflect the distribution rule and epidemic trend of the disease. In this study, the snail sites of 50 shehias on Pemba were taken as the study objects, spatial autocorrelation and spatial clustering analysis were used to detect the spatial characteristics of the distribution of *B. globosus*.

8.2 Research Materials and Methods

8.2.1 *Time and Area Under Study*

S. haematobium prevalence and snail status on Pemba Island was surveyed from February 2017 to November 2018. The island is divided into four districts that are further divided into about 120 smaller administrative areas, referred to as shehias, which can contain several villages varying in population size and household number. The total area of Pemba Island is 984 km^2, and the average area of a shehia is 8.2 km^2.

8.2.2 *Data Records and Generation of Spatial Coordinates*

The data consisted of geographic information, such as the whereabouts of human residences, ponds, and streams. All findings regarding these entities, such as diagnostic results and snail data, *etc*., were named attribute data and given unique identification numbers (ID), for example, for the human cases the record included name, house location, laboratory test results, and treatment (Shehia no./Village no./Household no./Person no.).

8.2.2.1 Human Data

Schistosomiasis-related data included the infection rate at each shehia and individual case investigation of positive cases. There are around 3000 people in each shehia on Pemba Island. The whole population were surveyed in four pilot shehias, *i.e.* Mtangani, Kiyuyu, Wingwi, and Uwandani. In other shehias, about 200 residents (including at least 30 children aged 9–12 years) from 50 households were randomly selected and surveyed. Urine samples were collected and 10 ml of each sample was filtered and subjected to microscopy by experienced laboratory technicians who recorded all *S. haematobium* eggs present for each individual. All this information were recorded in an Excel database.

8.2.2.2 Snail Data

Snail survey was conducted after the rainy season. All water bodies in the project area were investigated. The snail data included the distribution of *B. globosus* in the environment inhabited by humans and included the number of surveyed sites (ponds and streams), the total number of snails, and the number of infected ones. The ponds were encoded as Shehia no./Village no./Serial no./Pond type, while the streams were only labeled with its serial no. according to order of investigation. The sites were surveyed for snails at distance gaps of 30 and 100 m, respectively. At each identified site, presence of *B. globosus* was searched for 15 min over a maximum 15 m^2 area. The borders of the water bodies and their vegetation, where snails are most likely to be found, were more intensively searched than other areas. *B. globosus* was preliminary identified upon shell morphology, placed in screw top plastic containers with freshwater, and taken to the laboratory for microscopic examination of cercariae after crushing. A handheld Garmin etrex30x device (Kansas City, KS, USA) was used to specify the coordinates of all findings. All data were saved as KML files through BaseCamp software, version 4.5.2 (Garmin).

8.2.3 Annotating Attribute Data

Positive human cases were recorded as a point on the map and the total number counted, while the length of the streams, given as lines, and the surface area of the ponds, given as polygons, were measured using the GE properties window and recorded in the Excel database.

Water bodies without snails were given blue colour, those with presence of non-infected snails yellow, and those with presence of infected snails red. These spatial, qualitative pieces of information, each identified by its ID, were saved as shapefiles, reformatted as KML files and subsequently imported into GE Pro, version 7.1.8.3036 (Google).

8.2.4 The Spatial Database

The spatial, qualitative pieces of information, each identified by its ID, were collected in a GIS database, reformatted by ArcGIS, version 10.2.2 (ESRI) as KML files and subsequently imported into GE Pro, version 7.1.8.3036 (Google) to display the distribution of positive human cases in their environment and the snails (infected and uninfected) in theirs (ponds and streams). In this way, a GIS database containing the coordinates of all human positive cases found, including the status of the surrounding ponds and streams was established. Digital maps and shapefiles of Pemba Island were obtained from the NTD office and associated with the collected schistosomiasis data with reference to the shehias so that the GIS database could tell the schistosomiasis infection rate in all the different shehias on the island. Managed this way, all geographic and attribute data would be available and possible to visualize by using the software and clicking on the object in question.

8.2.5 The Nearest Neighbor Analysis

According to the nearest neighbor analysis method, first the distance between each snail site and its nearest site was measured, and the average value was calculated, then the similarity between the actual average distance and the expected average distance assumed to be random distribution was estimated, and the Z value was returned. If the value of Z is negative and smaller, the distribution tends to be clustering distribution mode, on the contrary, it tends to be discrete distribution mode. ArcGIS software was used to conduct the nearest neighbor analysis.

8.2.6 Spatial Autocorrelation Analysis

The correlation of the same variable in different spatial positions is a measure of the correlation degree of the attribute values of spatial units. In practical analysis, indicators such as global and local Moran's I, Geary's C and General G are often used to reflect spatial autocorrelation. Using ArcGIS software, the global and local spatial autocorrelation of snail sites were analyzed by global Moran's I, General G and local Moran's I (LISA).

8.2.6.1 Global Spatial Autocorrelation

Refers to the similarity of different spatial units in a general spatial pattern in terms of certain characteristics. The global Moran's I is a commonly used indicator to measure the global spatial autocorrelation, and calculated by the following equation:

$$I = \frac{n\sum_{i=1}\sum_{j=1} w_{ij}(x_i - \overline{x})\left(x_j - \overline{x}\right)}{\sum_{i=1}\sum_{j=1} w_{ij}\sum_{i=1}(x_i - \overline{x})^2}$$

In the equation, n is the number of spatial units, x_i and x_j represent the attribute values in i and j points, and w_{ij} is the spatial weight matrix. The range of Moran's I is $[-1, 1]$, the theoretical value $E(I) = -1/(n-1)$. The significance level of spatial autocorrelation can be tested by the standardized statistic $Z(I)$. The formula of $Z(I)$ is that $Z(I) = (I - E(I))/\sqrt{Var(I)}$, $Var(I)$ is the theoretical variance of Moran's I. When the observation value is larger than the theoretical value and has statistical significance, it suggests existing spatial positive autocorrelation (high value aggregation or low value aggregation), and there is spatial negative autocorrelation when the observation value is smaller than the theoretical value. The larger the absolute value is, the greater the spatial autocorrelation is.

Because Moran's *I* cannot judge whether the spatial data is a high value aggregation or a low value aggregation, Getis and Ord proposed the general *G* indicator. The general *G* indicator uses the distance weight and requires the attribute value of the spatial unit to be positive. The calculation formula is as follows:

$$G(d) = \frac{\sum_{i=1}\sum_{j=1} w_{ij}(d)x_i x_j}{\sum_{i=1}\sum_{j=1} x_i x_j}$$

where *d* is the distance, $w_{ij}(d)$ is the distance weight between *i* and *j* points. Similar to the global Moran's *I*, the general *G* can also be standardized: $Z(G) = (G(d) - E(G(d)))/Var(G(d))$, positive $Z(G)$ indicates high value aggregation, negative $Z(G)$ indicates low value aggregation.

The above two global autocorrelation statistics can only reflect the spatial autocorrelation on the whole, but not which spatial units have similar attributes. To understand the correlation between an area and its surrounding area, the local spatial autocorrelation statistics was needed.

8.2.6.2 Local Spatial Autocorrelation

Describe the similarity between a spatial unit and its surrounding unit. LISA (local indicators of spatial autocorrelation) is a commonly used method to test the local spatial autocorrelation. The calculation formula is as follows:

$$I_i = \frac{n^2}{\sum_i\sum_j w_{ij}} \frac{(x_i - \bar{x})\sum_j w_{ij}(x_j - \bar{x})}{\sum_j (x_j - \bar{x})^2}$$

where I_i is the LISA value of point *i*, and the definition of other parameters is the same as those of global Moran's *I*. According to LISA values and related observations, local spatial correlation patterns can be divided into four types: H-H (high-high, high value is surrounded by high value), L-L (low-low, low value is surrounded by low value), L-H (low-high, low value is surrounded by high value), H-L (high-low, high value is surrounded by low value), in which H-H, L-L are spatial positive correlation, L-H, H-L are spatial negative correlation.

8.2.7 *Spatial Scan Statistics*

Spatial scan statistics is a commonly used method to analyze spatial clustering. In this method, a circular (cylindrical) scanning window is established to calculate the theoretical incidence of each scanning window, and then a generalized likelihood function is constructed according to the actual incidence and the theoretical incidence. The Monte Carlo iterative simulation method is used to test the likelihood ratio. The greater the likelihood ratio, the more likely it is to be an aggregation region.

Based on the snail data of each surveyed point, space-time permutation distribution model was used for spatiotemporal clustering analysis. In this method, different time and area are scanned by dynamic cylindrical window. The bottom of the cylinder corresponds to a certain geographical area, and the height of the cylinder corresponds to a certain time length. In the process of modeling, only the number of live snails, survey time, longitude and latitude coordinates of each surveyed point are needed.

SaTScanversion 7.0.3 (Martin Kulldorff) is used for spatial scan statistics, and ArcGIS is used to realize the visualization of analysis results.

8.2.8 Staff Training

The Chinese team trained the local NTD staff on the operation of GE and how to use a handheld GPS device to specify the coordinates of all findings, save tracks, and mark points, lines, and polygons, how to import GPS data into GE as well as how to measure distances and how to calculate areas. The staff was also trained to use ArcGIS software manipulating ArcGIS shapefile vector features of interest such as points, lines, and polygons, here representing human infections, streams, and ponds, respectively, and converting this format into KML files to be fed into the GE software. They were also trained in associating attribute data related to the spatial data collected so as to create spatial databases.

8.3 Results

8.3.1 Geographic Information Database of Schistosomiasis Control Based on Google Earth and ArcGIS

8.3.1.1 Surveys

From Feb 2017 to Nov 2018, a total of 50 shehias were investigated for *S. haematobium* infection. A total of 20,263 urine samples were collected and 617 were found to be positive corresponding to an average positive rate of 3.04%. Figure 8.1 shows the prevalence rates in the surveyed 50 shehias on Pemba Island.

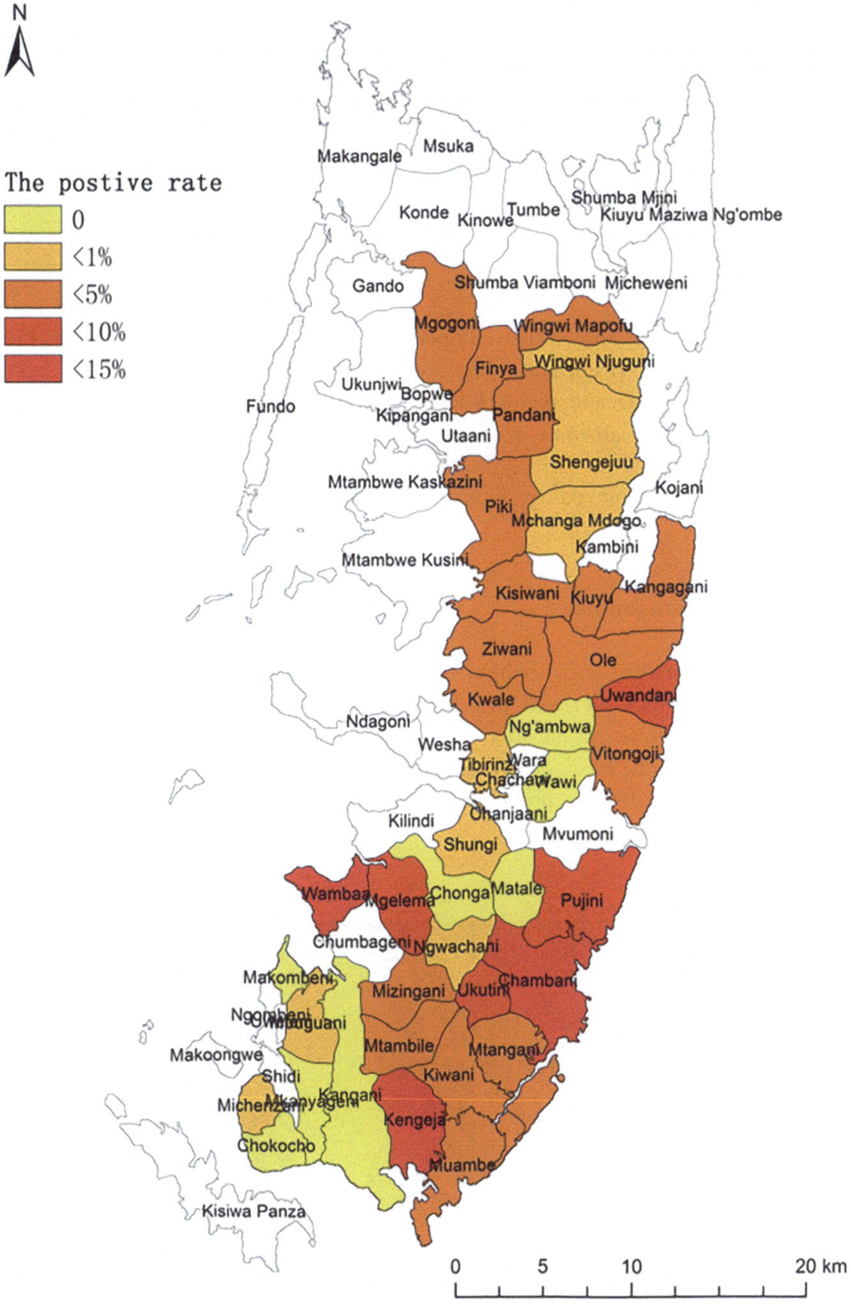

Fig. 8.1 Prevalence of *S. haematobium* in 50 surveyed shehias on Pemba Island

8.3.1.2 Spatial Databases

Three spatial databases were created by means of GE and they included distribution of infected humans (191 persons in 8 shehias); distribution of the intermediate snail host in ponds (all the ponds surveyed in 21 shehias); and distribution of the intermediate snail host in streams (all the streams surveyed in 50 shehias) (Fig. 8.2).

8.3.1.3 Infected Humans

After diagnosis, it was shown that 191 persons in the 8 shehias were infected. The coordinates of these residences were available in the database that had been established and the ID and name, householder name, egg counts, and treatment were recorded in the attribute data for each positive case. They could then be visualized and marked on GE as shown in Fig. 8.3. The attribute information of each case could be displayed by clicking on the point in question.

8.3.1.4 Snail Distribution

A total of 144 ponds in the 21 shehias were surveyed for *B. globosus* (Fig. 8.4). The total area of the ponds surveyed was 2,587,347 m^2. Although we found 46 ponds with *B. globosus* covering an area of 1,437,200 m^2 only 15 (836,492 m^2) were harbored infected with *B. globosus*. Figure 8.5 shows ponds without snails, ponds with non-infected snails, and ponds with infected snails. The basic information and

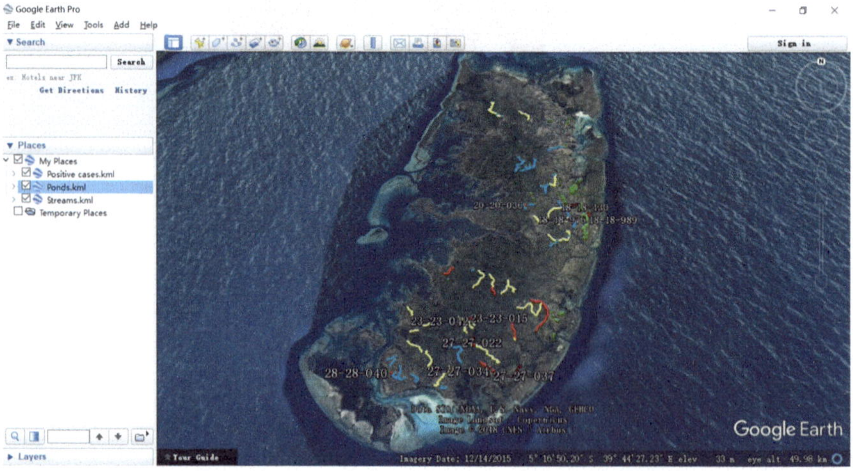

Fig. 8.2 Display of how three different spatial databases can be simultaneously visualized via Google Earth

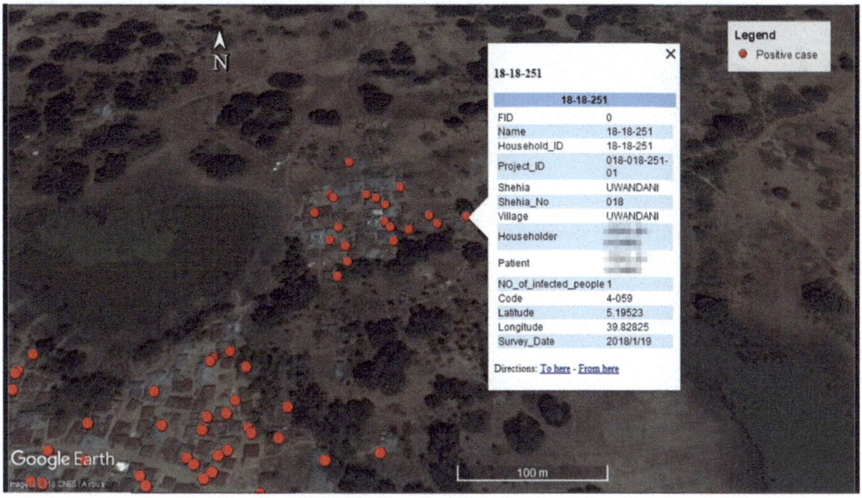

Fig. 8.3 Distribution of positive human cases and display of annotated information of cases

snail information for each pond could be displayed by clicking on that pond shown in GE.

In the shehia of Uwandani, four infected ponds were found in August 2017, and nearly 200 positive cases were found in the whole population survey. The location of human cases and infected waterbodies was displayed on GE, and we found there were no infected ponds nearby some human cases. Then the waterbodies and snail situation were investigated again, and another two infected ponds surrounding the human cases were found (Fig. 8.6).

A total of 138 streams in the 50 shehias were surveyed as shown in Fig. 8.4. The total length of the streams surveyed was 312.8 km. We found 59 streams with *B. globosus*, out of which 9 harbored infected snails. Figure 8.7 shows streams without snails, streams with snails, and streams with infected snails. The basic information and snail information for each stream could be displayed by clicking on that stream in the software.

8.3.2 *Spatial Characteristics of Snail Distribution*

8.3.2.1 Distribution of Snail Sites

The spatial analysis was based on the single surveyed site. From Feb 2017 to Nov 2018, the distribution of *B. globosus* was investigated in 50 shehias, and 795 snail sites were found (Fig. 8.8).

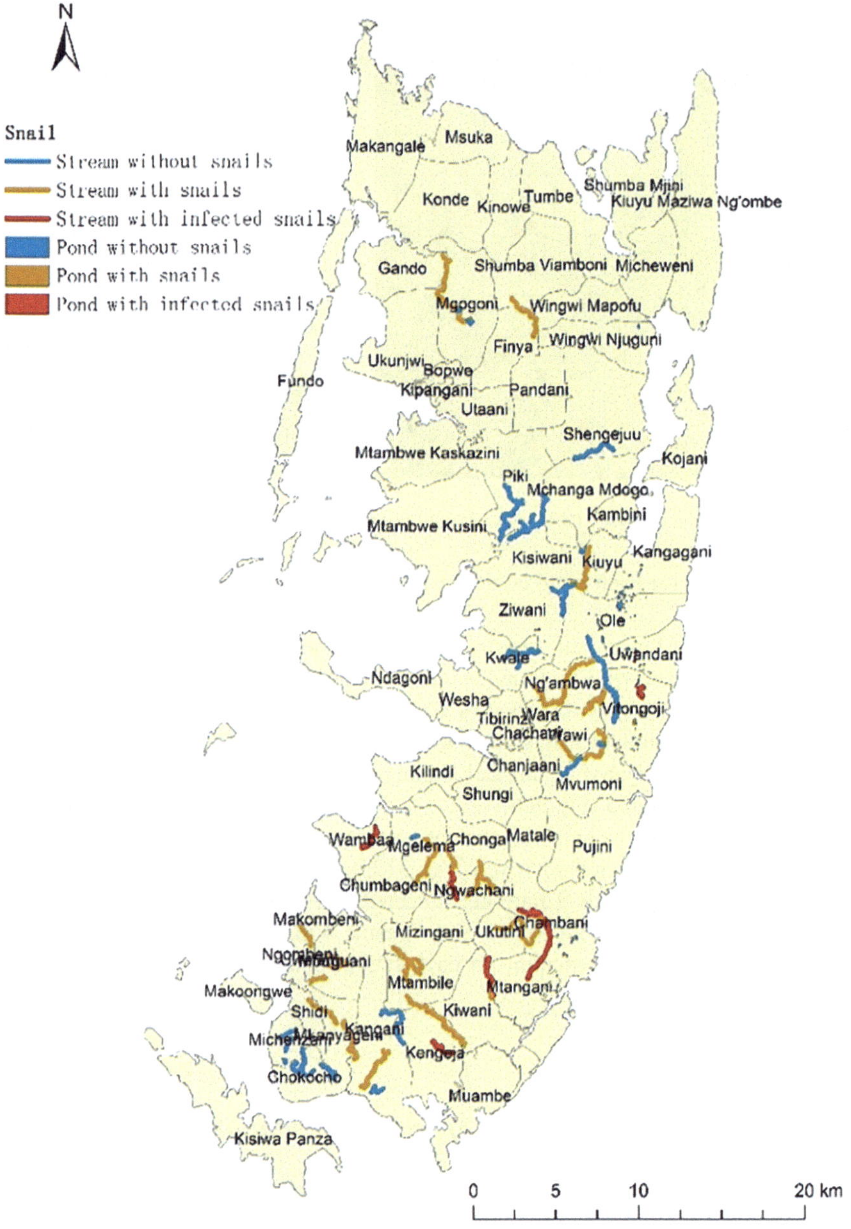

Fig. 8.4 Distribution of *B. globosus* in ponds and streams surveyed in 50 shehias on Pemba Island

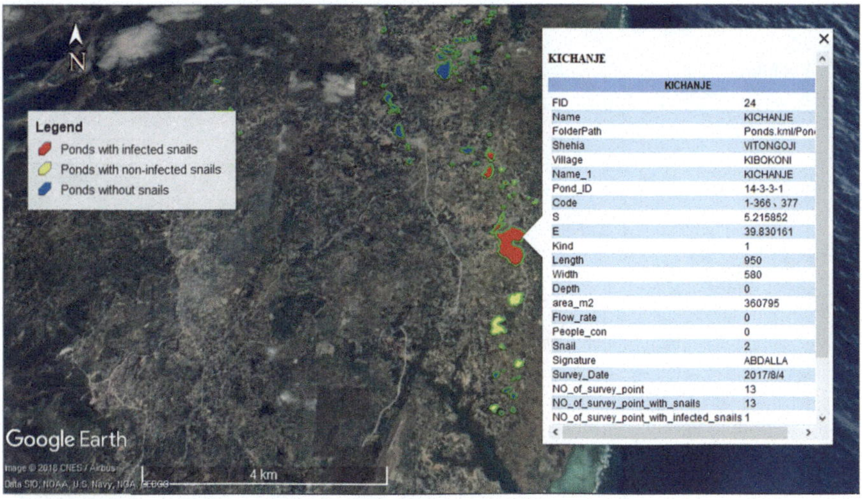

Fig. 8.5 Distribution of *B. globosus* in ponds and display of annotated information of ponds

8.3.2.2 The Nearest Neighbor Analysis

According to the nearest neighbor analysis, the ratio of the actual average distance to the expected average distance is 0.25, $z = -40.47$ ($P < 0.05$), which indicated that the spatial distribution of snail sites is a clustering distribution pattern.

8.3.2.3 Global Spatial Autocorrelation

The global Moran's I indicator of the number of live snails in 50 shehias was calculated. The observed value was 0.013, and the theoretical value was −0.001 ($P < 0.05$), which indicated that the spatial distribution of *B. globosus* had a positive autocorrelation on the whole.

The general G indicator was also calculated, the observed value was 0.462, and the theoretical value was 0.342, $z = 6.37$ ($P < 0.05$), indicating that there was a high value clustering area of *B. globosus* on the whole.

8.3.2.4 Local Spatial Autocorrelation

Local spatial autocorrelation analysis showed that the LISA value of 161 snail sites was statistically significant ($P < 0.05$), among which 84, 10, 67, and 0 were in the four modes of H-H, H-L, L-H, and L-L, respectively (Fig. 8.9). The snail habitats of H-H pattern is located in shehias including Kilindi, Shungi, Chonga, Ngwachani, Chumbageni, Mizingani, Mtambile, Ukutini, Chambani, Kengeja, Kangani, which

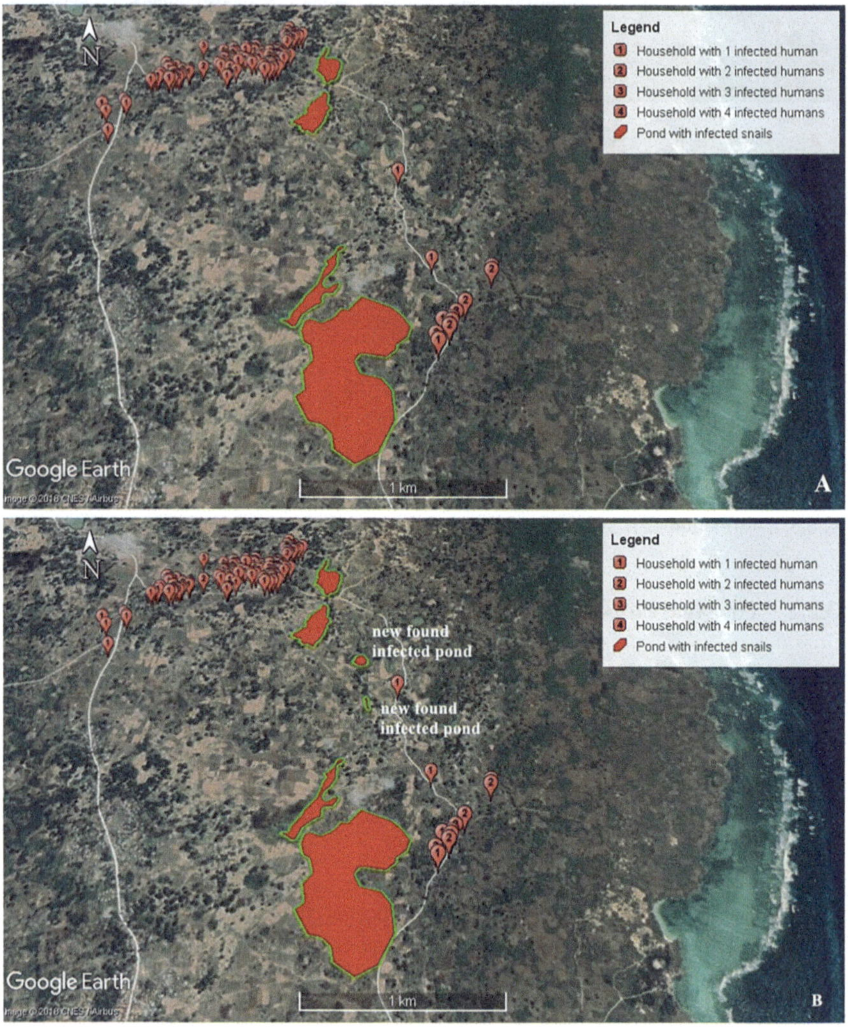

Fig. 8.6 Two new infected ponds found in Uwandani by analyzing the location of human cases and infectious ponds

suggested that these shehias were densely distributed areas of snails and should be taken as key areas of snail control.

Fig. 8.7 Distribution of *B. globosus* in streams and display of annotated information of streams

8.3.2.5 Spatial Clustering Analysis

A total of 8 clustering areas of snail sites were detected by using spatial scan statistics method ($P < 0.05$), as shown in Table 8.1. The buffer of clustering areas was shown in Fig. 8.10.

8.3.3 Training

After training and work practice, all NTD staff easily grasped the use of GPS and GE. They could utilize GPS devices to specify coordinates and save tracks and use GE to import GPS data. They could also measure distances and areas and use ArcGIS to convert shapefile vector files to KML files, create spatial databases, and make maps.

8.4 Significance and Prospects

Many diseases show spatial clustering (Elliott and Wartenberg 2004; Kirby et al. 2017), including parasitic diseases (Barbosa et al. 2014; Hundessa et al. 2016), other infectious diseases (Smith et al. 2015; Lai et al. 2018), and also non-communicable diseases like cancer (Goodman et al. 2014; Sun et al. 2015). Schistosomiasis transmission is focal in nature (Manyangadze et al. 2016) and schistosomiasis cases are consequently also spatially clustered as shown in many studies (Brooker

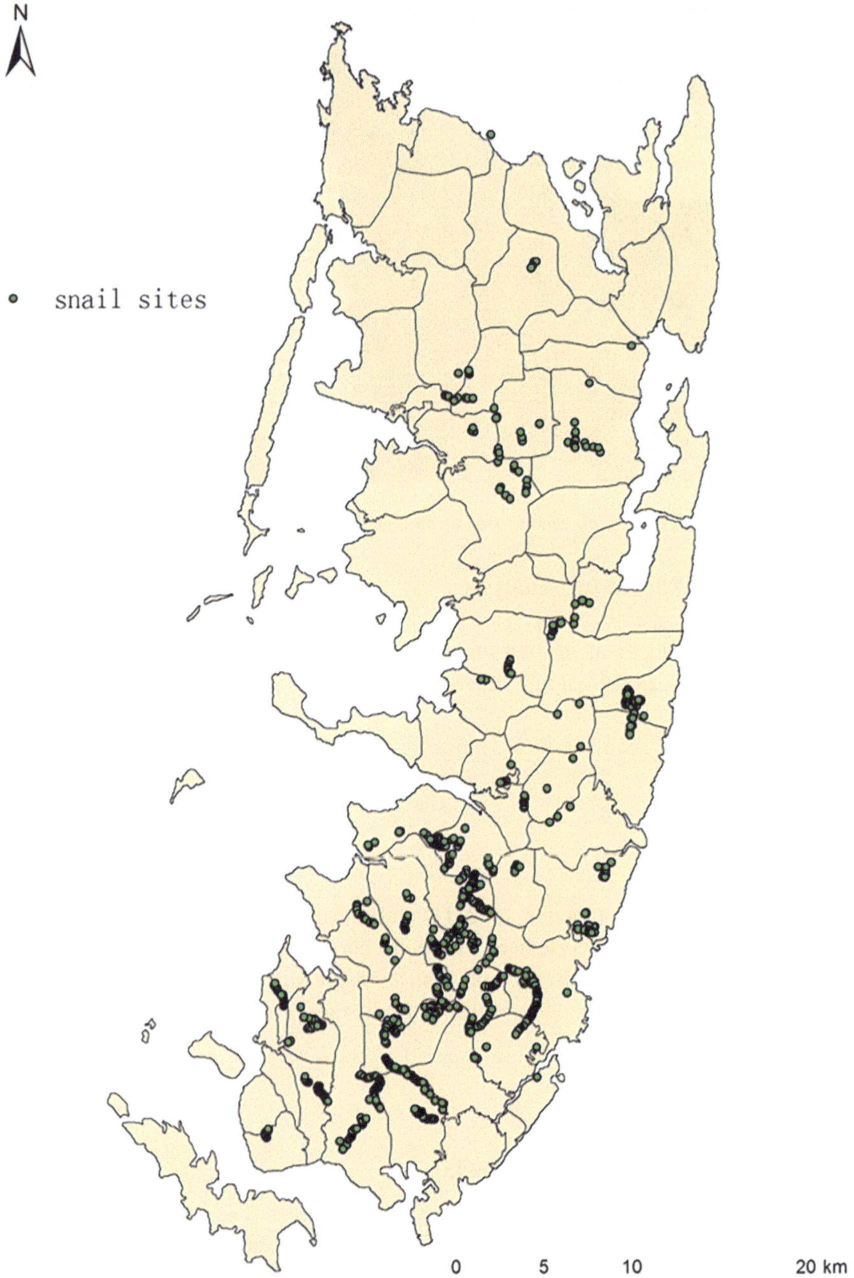

Fig. 8.8 Distribution of snail sites in 50 shehias of Pemba Island

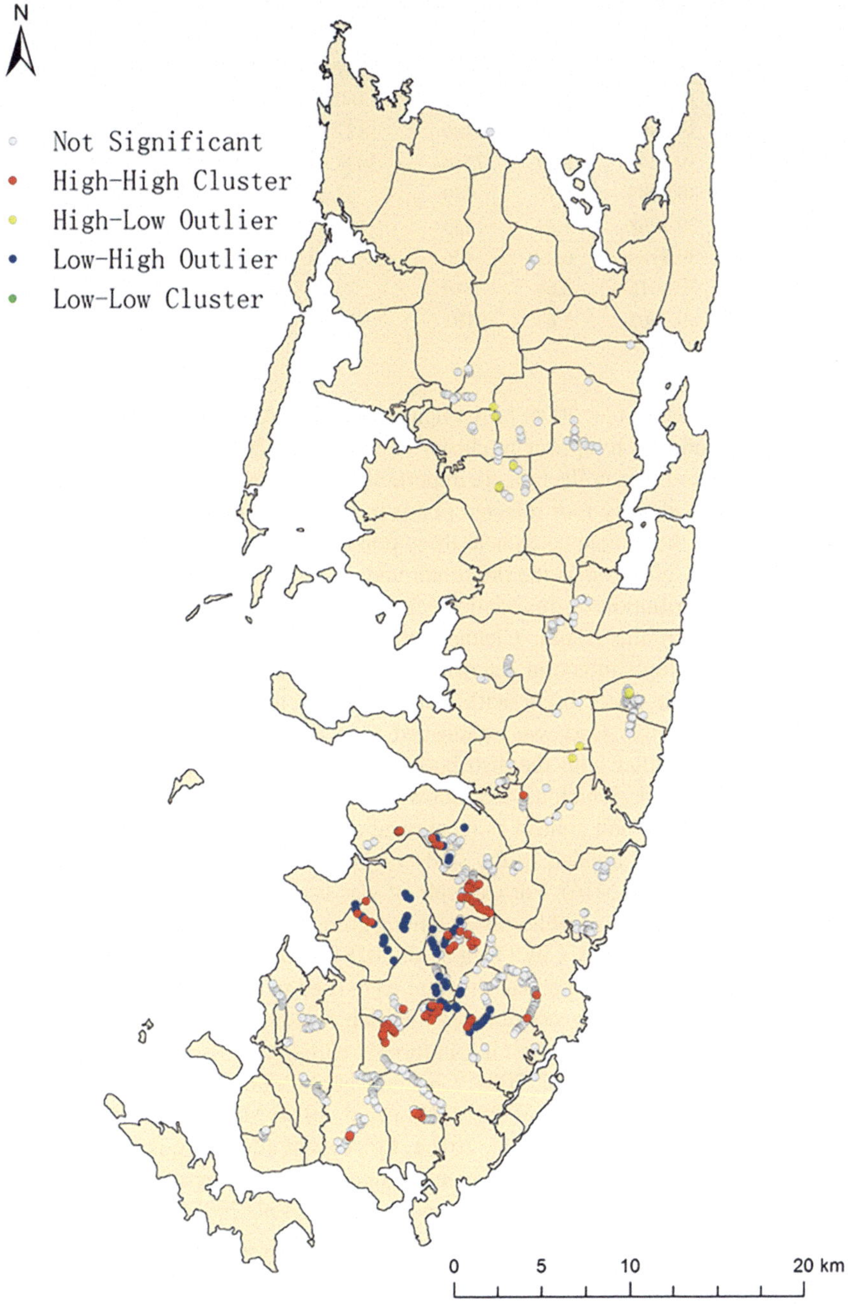

Fig. 8.9 Distribution of LISA values of snail sites in 50 shehias of Pemba Island

Table 8.1 Results of spatial clustering analysis on the snail sites in Pemba Island

No	Latitude	Longitude	Radius (km)	Observed	Expected	ODE (obs/exp)	Test statistics	*P* value
1	−5.3680	39.7727	3.68	3452	460.34	7.50	4128.43	0.001
2	−5.2848	39.7361	3.79	6406	1887.68	3.39	3715.38	0.001
3	−5.4198	39.6440	10.28	7346	3038.87	2.42	2562.72	0.001
4	−5.1963	39.8293	0.70	979	33.73	29.03	2368.10	0.001
5	−5.3594	39.7268	1.51	1533	143.67	10.67	2274.54	0.001
6	−5.2870	39.8184	6.27	706	26.15	27.00	1655.25	0.001
7	−5.3073	39.6916	2.69	849	77.10	11.01	1275.36	0.001
8	−5.3223	39.7605	1.64	2191	776.07	2.82	895.69	0.001

et al. 2006; Clements et al. 2006; Simoonga et al. 2008). Also, the intermediate snail host has the tendency to appear in cluster (Cheng et al. 2016; Gao et al. 2014). It follows that spatial data collection are important for schistosomiasis control. Indeed, this helped us to identify four infected ponds in Uwandani that had not been found until we located the human cases near these ponds.

On Pemba, *B. globosus* is the only intermediate host of *S. haematobium*, so the geographical distribution of this snail species determines the spatial distribution of schistosomiasis on the island. Clennon et al. (2004) have already shown that households with high infection intensities of *S. haematobium* were significantly clustered around a water body with infected snails near the southern coast of Kenya. Following this lead, we established a spatial database of distribution of positive cases, together with the distribution of *B. globosus* in ponds and streams based on GPS and GE, which helped following up cases as well as guide us to other key areas in need of snail control as mentioned above.

GE is increasingly applied to the study of infectious diseases (Escamilla et al. 2014; Simonsen et al. 2016). For example, it was used to map HCV infections by home, social, and sexual neighborhoods among community-recruited men who have sex with men in New York City (Tieu et al. 2018). In another study, Google Earth imagery and geographical analysis software were used to develop a sampling frame by digitizing and assigning coordinates for each household within the catchment area (Escamilla et al. 2014). The GE platform is particularly useful for research on schistosomiasis (Clennon et al. 2004; Clements et al. 2006; Simoonga et al. 2008, 2009; Yang et al. 2012; Sun et al. 2014; Gao et al. 2014; Manyangadze et al. 2015; Wang et al. 2014, 2016; Xiao et al. 2016a, b). A great advantage is that GE users need not be computer specialists since its operation is straightforward and the software easy to apply, *e.g.*, for map creation (Lefer et al. 2008) which is the first step. As long as the computer is connected to Internet, any map can be rapidly displayed. The application of GE in grass-root schistosomiasis control institutions is therefore a useful platform for the storage and display of spatial data (Wang et al. 2014; Xiao et al. 2016a). Combined with GIS software, such as ArcGIS, GE is also reliable for surveillance and forecasts (Xiao et al. 2016b) and can be implemented for

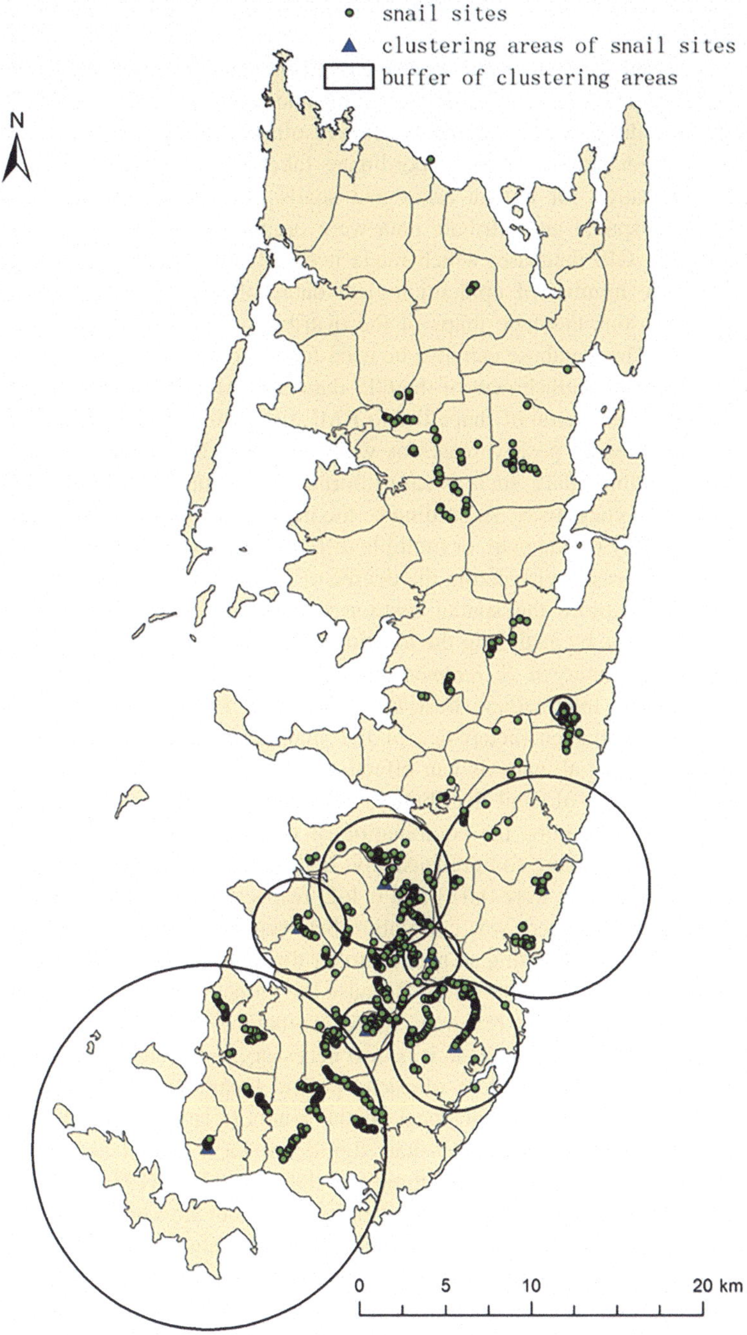

Fig. 8.10 Buffer of clustering areas of snail sites in 50 shehias of Pemba Island

epidemiological monitoring and the development of early warning systems (EWS) (Simonsen et al. 2016).

An on-site trial with GE as the operative platform achieved good results. Indeed, already after a few weeks of simple training and some practice, the NTD staff mastered all the different steps from recording results to displaying them. The spatial data were first generated using the coordinates taken from the GPS device adding attribute information for human cases and snails previously recorded in Excel. Secondly, both spatial and attribute data were organically integrated via ArcGIS establishing the GIS database, which made it possible to effectively manage and utilize the large number of epidemiological data collected. ArcGIS permits the provision of various thematic maps of the distribution of both human cases and positive snails. This database can also be used for information query and statistical analysis. After the establishment of the GIS database, ArcGIS provided the link to GE through the conversion of shapefiles to KML files and thus clearly visualize the geographic position of positive cases as well as ponds and streams with infected snails, while attribute data, such as case information or snail distribution information, were only a click away. According to the needs, a certain type of data can be selected for display on the map, or multiple different types of data can be displayed at the same time, easily visualizing the degree of severity of the endemic situation. We could also explore the spatial relation between infected human cases and *B. globosus* habitats by analyzing the location of human cases and waterbodies.

The epidemic range of *S. haematobium* is consistent with the geographical distribution of the intermediate host *B. globosus*. It is of great significance to understand the spatial characteristics of the snail distribution for the control of schistosomiasis. Spatial analysis can effectively detect the spatial autocorrelation and spatial clustering of snail distribution. Spatial autocorrelation is measured by global and local indicators, the combination of the two indicators can reflect the spatial characteristics more comprehensively. In this study, the global Moran's I and the general G indicators were 0.013 and 0.462, respectively, with statistical significance, indicating that there was correlation between the distribution of snail in a certain area and its surrounding area, and further there was a high value aggregation area according to general G indicator. Because the global indicator only reveals the general properties of the research space, sometimes it will cover up the non-stationary of the local state, so it is also necessary to use the local indicator to reflect the spatial autocorrelation of each specific location. In this study, LISA method was used to analyze the local spatial autocorrelation of snail sites. The results showed that 161 snail sites had significant local spatial autocorrelation, which could be further divided into three correlation models: H-H, L-L, and L-H. Although the snail distribution was positively correlated in the whole, there was also negative correlation between the high value and the low value in the local snail sites, which objectively revealed the uneven distribution of snail. The spatial heterogeneity of snail distribution suggests that the environment of snail distribution is relatively complex, and the distribution of snail in adjacent areas may be quite different. Therefore, it is necessary to take control measures according to local conditions.

The spatial clustering of diseases or vectors is one of the important characteristics of the spatial distribution of diseases. In this study, the location and the number of live snails in 50 shehias of Pemba Island were analyzed by the method of spatial scan statistics. It was found that there were 8 clustering areas of snails, all of which were high value clustering areas. The results of spatial clustering analysis were consistent with the results of local spatial autocorrelation analysis, but there were some differences between two analyzing results. The reason is that the principles of the two methods are different. LISA method calculates the relevant indicators of each spatial location and judges clustering by hypothesis test, while spatial scan statistics method is based on likelihood ratio, and detects clustering area by establishing dynamic circular window.

There are some limits in this study. For snail classification, we just conducted preliminary identification using shell morphology, while it is advisable to apply molecular diagnosis based on DNA detection for further identification. Secondly, the location of households with infected cases is not a proxy of where transmission takes place. In the next step, the proximity of a household to the nearest water body could be used to identify clusters of infected cases which might point towards a transmission source.

It should be borne in mind that, in contrast to complete eradication of a disease, elimination means controlling it so well that it no longer is of public health importance. However, it should also be remembered that keeping a disease eliminated in the face of ongoing transmission can be a tall order in the absence of strong economic development. A surveillance tool like GE that can connect all information needed and also play the role of EWS would be most useful for the rapid deployment of an effective response. As such, GE would not only be useful on Pemba but also in other countries on the African continent and elsewhere.

References

Allan F, Dunn AM, Emery AM, Stothard JR, Johnston DA, Kane RA, Khamis AN, Mohammed KA, Rollinson D (2013) Use of sentinel snails for the detection of *Schistosoma haematobium* transmission on Zanzibar and observations on transmission patterns. Acta Trop 128:234–240

Barbosa DS, Belo V, Rangel ME, Werneck GL (2014) Spatial analysis for identification of priority areas for surveillance and control in a visceral leishmaniasis endemic area in Brazil. Acta Trop 131:56–62

Bergquist R, Rinaldi L (2010) Health research based on geospatial tools: a timely approach in a changing environment. J Helminthol 84:1–11

Brooker S, Alexander N, Geiger S, Moyeed RA, Stander J, Fleming F, Hotez PJ, Correa-Oliveira R, Bethony J (2006) Contrasting patterns in the small-scale heterogeneity of human helminth infections in urban and rural environments in Brazil. Int J Parasitol 36:1143–1151

Cheng G, Li D, Zhuang D, Wang Y (2016) The influence of natural factors on the spatio-temporal distribution of *Oncomelania hupensis*. Acta Trop 164:194–207

Clements AC, Lwambo NJ, Blair L, Nyandindi U, Kaatano G, Kinung'hi S, Webster JP, Fenwick A, Brooker S (2006) Bayesian spatial analysis and disease mapping: tools to enhance

planning and implementation of a schistosomiasis control programme in Tanzania. Trop Med Int Health 11:490–503
Clennon JA, King CH, Muchiri EM, Kariuki HC, Ouma JH, Mungai P, Kitron U (2004) Spatial patterns of urinary schistosomiasis infection in a highly endemic area of coastal Kenya. Am J Trop Med Hyg 70:443–448
Elliott P, Wartenberg D (2004) Spatial epidemiology: current approaches and future challenges. Environ Health Perspect 112:998–1006
Escamilla V, Emch M, Dandalo L, Miller WC, Martinson F, Hoffman I (2014) Sampling at community level by using satellite imagery and geographical analysis. Bull World Health Organ 92:690–694
Gao FH, Abe EM, Li SZ, Zhang LJ, He JC, Zhang SQ, Wang TP, Zhou XN, Gao J (2014) Fine scale Spatial-temporal cluster analysis for the infection risk of Schistosomiasis japonica using space-time scan statistics. Parasit Vectors 7:578
Goodman M, LaKind JS, Fagliano JA, Lash TL, Wiemels JL, Winn DM, Patel C, Van Eenwyk J, Kohler BA, Schisterman EF, Albert P, Mattison DR (2014) Cancer cluster investigations: review of the past and proposals for the future. Int J Environ Res Public Health 11:1479–1499
Guidi A, Andolina C, Makame Ame S, Albonico M, Cioli D, Juma Haji H (2010) Praziquantel efficacy and long-term appraisal of schistosomiasis control in Pemba Island. Trop Med Int Health 15:614–618
Hundessa SH, Williams G, Li S, Guo J, Chen L, Zhang W, Guo Y (2016) Spatial and space-time distribution of *Plasmodium vivax* and *Plasmodium falciparum* malaria in China, 2005–2014. Malar J15:595
Kirby RS, Delmelle E, Eberth JM (2017) Advances in spatial epidemiology and geographic information systems. Ann Epidemiol 27:1–9
Knopp S, Mohammed KA, Ali SM, Khamis IS, Ame SM, Albonico M, Gouvras A, Fenwick A, Savioli L, Colley DG, Utzinger J, Person B, Rollinson D (2012) Study and implementation of urogenital schistosomiasis elimination in Zanzibar (Unguja and Pemba islands) using an integrated multidisciplinary approach. BMC Public Health 12:930
Knopp S, Person B, Ame SM, Mohammed KA, Ali SM, Khamis IS, Rabone M, Allan F, Gouvras A, Blair L, Fenwick A, Utzinger J, Rollinson D (2013) Elimination of schistosomiasis transmission in Zanzibar: baseline findings before the onset of a randomized intervention trial. PLoS Negl Trop Dis 7:e2474
Knopp S, Person B, Ame SM, Ali SM, Muhsin J, Juma S, Khamis IS, Rabone M, Blair L, Fenwick A, Mohammed KA, Rollinson D (2016) Praziquantel coverage in schools and communities targeted for the elimination of urogenital schistosomiasis in Zanzibar: a cross-sectional survey. Parasit Vectors 9:5
Lai WT, Chen CH, Hung H, Chen RB, Shete S, Wu CC (2018) Recognizing spatial and temporal clustering patterns of dengue outbreaks in Taiwan. BMC Infect Dis 18:256
Lefer TB, Anderson MR, Fornari A, Lambert A, Fletcher J, Baquero M (2008) Using Google Earth as an innovative tool for community mapping. Public Health Rep 123:474–480
Lisle RJ (2006) Google Earth: a new geological resource. Geol Today 22:29–32
Lwambo NJ, Savioli L, Kisumku UM, Alawi KS, Bundy DA (1997) The relationship between prevalence of *Schistosoma haematobium* infection and different morbidity indicators during the course of a control programme on Pemba Island. Trans R Soc Trop Med Hyg 91:643–646
Manyangadze T, Chimbari MJ, Gebreslasie M, Mukaratirwa S (2015) Application of geo-spatial technology in schistosomiasis modelling in Africa: a review. Geospat Health 10:99–110
Manyangadze T, Chimbari MJ, Gebreslasie M, Mukaratirwa S (2016) Risk factors and micro-geographical heterogeneity of *Schistosoma haematobium* in Ndumo area, uMkhanyakude district, Kwa Zulu-Natal, South Africa. Acta Trop 159:176–184
Olsen A, Kinung'hi S, Magnussen P (2015) *Schistosoma mansoni* infection along the coast of Lake Victoria in Mwanza region, Tanzania. Am J Trop Med Hyg 92:1240–1244

Pennance T, Person B, Muhsin MA, Khamis AN, Muhsin J, Khamis IS, Mohammed KA, Kabole F, Rollinson D, Knopp S (2016) Urogenital schistosomiasis transmission on Unguja Island, Zanzibar; characterisation of persistent hot-spots. Parasit Vectors 9:646

Savioli L, Mott KE (1989) Urinary schistosomiasis on Pemba Island: low-cost diagnosis for control in a primary health care setting. Parasitol Today 5:333–337

Savioli L, Dixon H, Kisumku UM, Mott KE (1989) Control of morbidity due to *Schistosoma haematobium* on Pemba island; selective population chemotherapy of schoolchildren with haematuria to identify high-risk localities. Trans R Soc Trop Med Hyg 83:805–810

Simonsen I, Gog JR, Olson D, Viboud C (2016) Infectious disease surveillance in the big data era: towards faster and locally relevant systems. J Infect Dis 214:S380–S385

Simoonga C, Kazembe LN, Kristensen TK, Olsen A, Appleton CC, Mubita P, Mubila L (2008) The epidemiology and small-scale spatial heterogeneity of urinary schistosomiasis in Lusaka province, Zambia. Geospat Health 3:57–67

Simoonga C, Utzinger J, Brooker S, Vounatsou P, Appleton CC, Stensgaard AS, Olsen A, Kristensen TK (2009) Remote sensing, geographical information system and spatial analysis for schistosomiasis epidemiology and ecology in Africa. Parasitology 136:1683–1693

Smith CM, Comber SC, Fry H, Bull M, Leach S, Hayward AC (2015) Spatial methods for infectious disease outbreak investigations: systematic literature review. Euro Surveill 20

Sun LP, Liang YS, Wu HH, Tian ZX, Dai JR, Yang K, Hong QB, Zhou XN, Yang GJ (2014) A Google Earth-based surveillance system for schistosomiasis japonica implemented in the lower reaches of the Yangtze River, China. Parasit Vectors 4:223

Sun J, Huang H, Xiao GX, Feng GS, Yu SC, Xue YT, Wan X, Yang GH, Sun X (2015) Spatial distribution of liver cancer incidence in Shenqiu County, Henan Province, China: a spatial analysis. Biomed Environ Sci 28:214–218

Tieu HV, Laeyendecker O, Nandi V, Rose R, Fernandez R, Lynch B, Hoover DR, Frye V, Koblin BA (2018) Prevalence and mapping of hepatitis C infections among men who have sex with men in New York City. PLoS One 13:e0200269

Wang WC, Zhan T, Zhu YF (2014) A manufacture method of schistosomiasis epidemic maps based on Google Earth. Chin J Schist Cont 26:96–98

Wang XY, He J, Yang K, Liang S (2016) Applications of spatial technology in schistosomiasis control programme in the People's Republic of China. Adv Parasitol 92:143–163

WHO (2013) Schistosomiasis: progress report 2001–2011, strategic plan 2012–2020. World Health Organization, Geneva, 22pp

WHO (2018) World Health Organization fact-sheets. http://www.who.int/news-room/fact-sheets/detail/schistosomiasis

Xiao BZ, Meng JX, Wen JY, Dong LG, Xuan Z, Bing L, Xiao JM (2016a) Establishment of a monitoring database of historical *Oncomelania hupensis* environments at grass – root level for schistosomiasis prevention based on Google Earth. Chin J Schisto Cont 28:306–309

Xiao WS, Jian B, Shun XC, Zu WT, Xiao RZ, Yan YC, Bo L (2016b) Establishment of surveillance and forecast platform of schistosomiasis in key water regions in Hubei Province. Chin J Schist Cont 28:125–127

Yang K, Sun LP, Huang YX, Yang GJ, Wu F, Hang DR, Li W, Zhang JF, Liang YS, Zhou XN (2012) A real-time platform for monitoring schistosomiasis transmission supported by Google Earth and a web-based geographical information system. Geospat Health 6:195–203

Chapter 9
Bulinus Snails Control by China-made Niclosamide in Zanzibar: Experiences and Lessons

Yun-Tian Xing, Jian-Rong Dai, Kun Yang, Tao Jiang, Cheng-Gong Jiang, Saleh Juma Mohammed, and Fatma Kabole

Abstract *Schistosomiasis haematobium* is one of the most important, but often neglected, tropical diseases transmitted by snails of *Bulinus* in Zanzibar. Snail control by mollusciciding is the cornerstone in preventing the spread of schistosomiasis. In the past thirty years of schistosomiasis elimination in China, suspension concentrate of metaldehyde and niclosamide (MSCN) as a molluscicide has played a crucial role. However, there is no report about MSCN against *Bulinus*. Therefore, it is necessary to evaluate the effects of MSCN against *Bulinus* both in laboratory and field. The effects of MSCN were tested both in laboratory and field setting. The LC_{50} of MSCN for *Bulinus* at 24 h and 48 h was 0.0758 mg/L, while the LC_{50} wettable powder of niclosamide (WPN) at 24 h and 48 h was 0.400 mg/L. The lowest concentration of MSCN that killed all *Bulinus* after 24 h and 48 h was 0.200 mg/L and the highest concentration permitting snail survivals were 0.013 mg/L and 0.006 mg/L, respectively. In the field setting, one month after MSCN treatment, no living snails were found in any of the four treated ponds. MSCN had better

Y.-T. Xing · J.-R. Dai (✉) · K. Yang
Jiangsu Institute of Parasitic Diseases, Key Laboratory of National Health and Family Planning Commission on Parasitic Disease Control and Prevention, Jiangsu Provincial Key Laboratory on Parasites and Vector Control Technology, Wuxi, China

China Aid of Schistosomiasis Control Project Pemba, Zanzibar, United Republic of Tanzania

T. Jiang
China Aid of Schistosomiasis Control Project Pemba, Zanzibar, United Republic of Tanzania

Danyang Center for Disease Control and Prevention, Danyang, China

C.-G. Jiang
China Aid of Schistosomiasis Control Project Pemba, Zanzibar, United Republic of Tanzania

Jurong Center for Disease Control and Prevention, Jurong, China

S. J. Mohammed · F. Kabole
China Aid of Schistosomiasis Control Project Pemba, Zanzibar, United Republic of Tanzania

Ministry of Health, Zanzibar, United Republic of Tanzania

K. Yang, H. Mehlhorn (eds.), *Sino-African Cooperation for Schistosomiasis Control in Zanzibar*, Parasitology Research Monographs 15,
https://doi.org/10.1007/978-3-030-72165-7_9

molluscicidal effects for adult snails compared with WPN. The observation of the short-term effect (one month) and the long-term (three months) application showed that MSCN had a high molluscicidal effect against *Bulinus* and is therefore a powerful weapon in the control of *Schistosomiasis haematobium*.

Keywords Schistosomiasis · Snail control · Niclosamide · Zanzibar · *Bulinus*

Abbreviations

LC50	50% lethal concentration
MDA	Mass drug administration
SCN	Suspension concentrate of metaldehyde and niclosamide
WHO	World Health Organization
WPN	Wettable powder of niclosamide

9.1 Background

Schistosomiasis, a snail-transmitted parasitic disease, is one of the world's major tropical diseases affecting more than 250 million people in tropical and subtropical regions around the world (Khonde Kumbu et al. 2016; Lewis and Tucker 2014). It is endemic in 51 out of the 54 countries on the Africa. More than 200,000 people die of schistosomiasis in sub-Saharan Africa annually (Galappaththi-Arachchige et al. 2018; Sacolo et al. 2018; Woodall and Kramer 2018). In most parts of Africa human schistosomiasis is mainly caused by *Schistosoma haematobium* (*S. haematobium*) and *Schistosoma mansoni* (*S. mansoni*) (Gower et al. 2017; Hegertun et al. 2013). Schistosomiasis is an acute and chronic disease caused by parasitic worms. Transmission occurs when people suffering from schistosomiasis contaminate freshwater sources with their excreta that contain the parasite's eggs, which subsequently hatch in the water. The larval forms of the parasite—released by freshwater snails—penetrate human skin during contact with infested water. In the body, the larvae develop into adult worms that can cause progressive damage to organs. Snails are a crucial part of the life cycle of schistosomes and are necessary for this parasite's multiplication and transmission; without the presence of intermediate snail hosts, transmission is not possible (Diakite et al. 2017; Giacomet et al. 2010; Mohammed et al. 2016).

Snail control, mainly by mollusciciding, was the cornerstone of schistosomiasis control before the strategy for morbidity control and has contributed to many successful control outcomes (Civitello et al. 2018; Micati et al. 2018; Yasin et al. 2018). Niclosamide is the only recommended molluscicide for snail control by the World Health Organization (WHO) (King et al. 2015). First public mention of the discovery of niclosamide as a molluscicide was made at the Sixth International

Congress on Tropical Medicine and Malaria held in Lisbon in 1958 (Andrews et al. 1982); it was found to have high molluscicidal efficacy and low toxicity to animals and there was no evidence of poisoning caused by faulty operation or by accidental ingestion. Yet, there has been no large-scale and continuous worldwide application of niclosamide due to concerns about environmental pollution and its high cost (Hassan et al. 2008; McCullough et al. 1980; Mosta-Fa et al. 2005; Xia et al. 2014). Up until 1992, with the aid of a loan from the World Bank, a large quantity of niclosamide (an average of 2000–3000 tons annually) was used in China, which made a great contribution to the control of schistosomiasis. Furthermore, over the past 20 years, there have been no reports about the accumulation of niclosamide in the environment or about any negative ecological impact except economic loss caused by acute death of freshwater fish due to the accidental draining of niclosamide-treated water into pisciculture waters (Dai et al. 2014; Gray et al. 2014; Liang et al. 2007; Yang et al. 2012).

Initially, wettable powder of niclosamide (WPN) was applied against *Oncomelania* snails in China, which had several shortcomings (Dai et al. 2014; Yang et al. 2012) including that it needed to be dispersed in water with continuous stirring before use. WPN was also found to absorb water during storage, which prevented dispersal and resulted in uneven distribution and the appearance of sediments. To try to overcome these problems, MSCN, a novel formulation of niclosamide, was developed. The formulation has good dispersion and suspension and can be mixed with water in any ratio (Bao et al. 2011; Dai et al. 2008, 2015; Xing et al. 2013).

S. haematobium is endemic in Zanzibar, a semi-autonomous archipelago of the United Republic of Tanzania in East Africa, 25–50 km off the coast of the mainland. The archipelago consists of many small islands and two large populated ones: Unguja (the main island, referred to informally as Zanzibar) and Pemba Island are endemic for *S. haematobium* (French et al. 2007; Pennance et al. 2016; Rollinson et al. 2005; Stothard et al. 2002a; Webster et al. 2015). The estimated total resident population of Pemba was 511,576 in 2011 (the 2011 Zanzibar population report (office finance, economy and development planning Zanzibar)), with 10% of the population infected with *S. haematobium*. In Zanzibar, *Bulinus spp.* are the intermediate snail hosts for *S. haematobium*. Thus, control of this species is an important measure in the eradication of *S. haematobium* in this region. Hence this study involves MSCN in the control of *Bulinus spp.* in both the laboratory and in ponds in Zanzibar.

9.2 Methods

9.2.1 Laboratory Tests

50% WPN (batch no.: 1512031) and MSCN (batch no.:201604061) were provided by the Nanjing AI Jin Chemical Co. Ltd (Nanjing China); 70% WPN (batch no.:

PF90000118) was provided by Bayer (Leverkusen, Germany). We used Uhai drinking water (Bakhresa Food Products LTD, Dar es Salaam, Tanzania) for all laboratory experiments. 50% WPN, 70% WPN and MSCN were weighed and solutions of 0.400, 0.200, 0.100, 0.050, 0.025, 0.012, and 0.006 mg/L were prepared in water for the subsequent experiments.

Adult *Bulinus* were collected by individual picking with forceps, from Vitongoji, Pemba Island, Zanzibar and reared at 25 °C in plastic basins at a density of 200/L. The average size of the snails was 0.5–0.8 cm. *Bulinus* were randomly divided into groups after 48 h.

Five hundred milliliter flasks were filled with the above niclosamide solutions, and ten active adult snails were added to the flasks, which were covered with gauze to prevent their escape. Snails in flasks of water served as controls. After being immersed for 24 h and 48 h at 25 °C, the snails were washed with water and fed for a further 48 h. All data were double entered into Microsoft Excel 2003 (Microsoft Corporation; Redmond, WA, USA).

9.2.2 Filed Methods of Application

All field tests were performed in South Pemba Island in Chakechake and Uwandani. The average annual temperature in this region ranged between 23 °C and 32 °C during the experiments. There are two annual wet seasons in Pemba Island: the Masika rains from the south, which occur from March to June, and the Vuli rains from the north-east, which occur from October to November. Uwandani is densely covered by freshwater ponds and streams. There are four permanent perennial ponds and up to 17 ponds at the height of the rainy season. In Chakechake, 2840 people and 489 households were at risk of *S. haematobium* because of swimming in infested ponds and being exposed to the parasite during work. The prevalence of *S. haematobium* infection was found to be approximately 8.9%. The four permanent perennial ponds containing natural *Bulinus* populations were selected as test sites. Pond areas were computed by GPS positioning and Google earth. The handheld GPS instrument (Garmin Ltd., Kansas, USA) model used was eTrex 30x. The GPS trajectory was formed by walking around the edge of ponds and inputting the coordinates into Google earth. Pond areas were then automatically calculated from Google earth (Fig. 9.1, Table 9.1). The areas of the four ponds were found to be 40,993, 54,142, 40,010, and 9230 m^2, respectively.

Snail densities were assessed at each site. Two trained staff members collected snails for 15 min from different survey points setup at 50 m intervals. All snails were deposited into a basin and identified as either living or dead and the living *Bulinus* density was then calculated (Table 9.2). Other organisms were returned and evenly distributed to their original collection sites. MSCN was applied at 2 g/m^2. A motor spraying machine (Nanjing AI Jin Chemical Co., Ltd., Nanjing, China) was employed for application. The motor spraying machine consisted of a single cylinder engine (163 cc) and a closed impeller pump. The pump's lift and suction range were

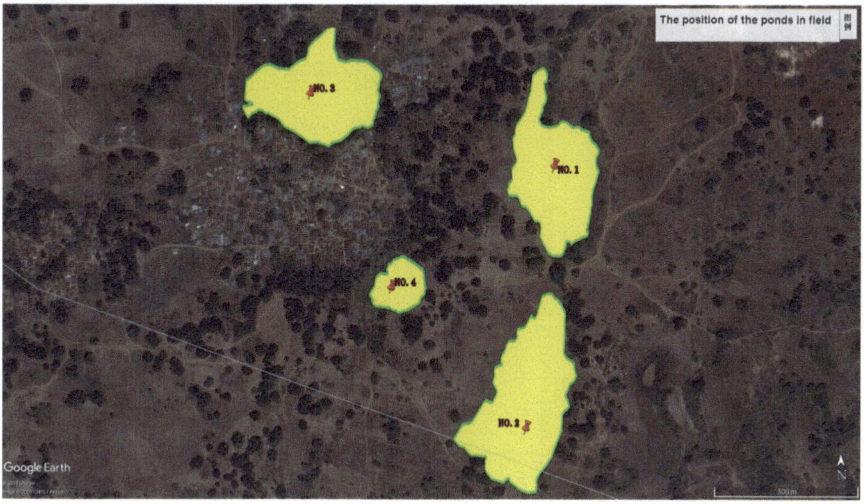

Fig. 9.1 Geographical position of the ponds in the field. The yellow squares represent four ponds treated with MSCN

Table 9.1 Geographic information of the four ponds

No	*S*	*E*	Area (m^2)
1	5.19846	39.83013	40,993
2	5.19955	39.82972	54,142
3	5.19601	39.82673	40,010
4	5.19842	39.82730	9230

33 and 8 m, respectively. MSCN was applied to an area, which was 15 m from the edge of the pond and which was covered with aquatic plants on the shore. Snail teams applying MSCN consisted of seven trained staff who were familiar with the procedure of snail control (Fig. 9.2). One was responsible for operating the molluscicide machine and adding the niclosamide solution and three were responsible for the arrangement and movement of the long hoses. The others were responsible for spraying the solution. The spraying dose was 2 g/m^2, and the spray area consisted of both sides along the water level line, including 10 m of beach (possible survival region of small snail) and 10 m of pond (the aquatic plant growth region). All the areas were carefully sprayed and included low-lying areas and branched area where the snails bred (Fig. 9.3). One and three months after MSCN application the living *Bulinus* density was monitored again. If there were still living snails, the whole pond was completely treated one more time according to the process mentioned above.

Table 9.2 Molluscicidal effect of MSCN in field

No	Snail investigation before application					Snail investigation 1 month later after application					Snail investigation 3 month later after application				
	No. of survey point	No. of snails	No. of living snails	Density of living snails	Natural mortality	No. of survey point	No. of snails	No. of living snails	Density of living snails	Decline rate of living snails	No. of survey point	No. of snails	No. of living snails	Density of living snails	Decline rate Of living snails
1	10	642	602	60.2	6.23	19	272	0	0	100	10	8	1	0.05	99.91
2	10	381	377	37.7	1.05	17	348	0	0	100	17	135	0	0	100
3	15	17	17	1.13	0	17	0	0	0	100	17	0	0	0	100
4	7	3	3	0.43	0	9	0	0	0	100	9	0	0	0	100

Fig. 9.2 Snail team was applying WPN with the motor spraying machine in the field

9.3 Results

9.3.1 Laboratory Tests

The results of the laboratory tests are set out in Table 9.3. The lowest concentration of MSCN that killed all *Bulinus* by immersion at 24 h and 48 h was 0.200 mg/L, and the highest concentrations permitting survival of the snails at these two time points were 0.013 mg/L and 0.006 mg/L, respectively. The LC_{50} of MSCN for *Bulinus* at 24 h and 48 h was 0.0758 mg/L. The LC_{50} of WPN at 24 h and 48 h was 0.400 mg/L, and the highest concentration permitting survival of the snails at these two time points was 0.100 mg/L. The LC_{50} of 50% WPN for *Bulinus* at 24 h and 48 h was 0.2144 mg/L. The lethal concentration of 70% WPN at 24 h and 48 h was found to be 0.400 mg/L, and the highest concentration permitting survival of the snails at these two time points was 0.100 mg/L. The LC_{50} of 50% WPN for *Bulinus* at 24 h and 48 h was 0.2462 mg/L and 0.2001 mg/L, respectively. The lowest concentration of WPN for killing snails was higher than MSCN, and the highest survival concentration of WPN was lower than that of MSCN. The LC_{50} for adult snails at 24 h at 48 h was only half that of WPN, suggesting that MSCN had better molluscicidal effects on adult snails compared with WPN.

Fig 9.3 Application of MSCN in the field. Snail team was applying MSCN with the motor spraying machine in the field

Table 9.3 Effect of three formulations of niclosamide against *Bulinus* in lab

Formulations	Time	Effective concentration (mg/l)							LC_{50}
		0.400	0.200	0.100	0.050	0.025	0.013	0.006	
MSCN	24	100	100	70	10	10	0	0	0.0758
	48	100	100	70	10	0	10	0	0.0758
50% WPN	24	100	40	0	0	0	0	0	0.2144
	48	100	40	0	0	0	0	0	0.2144
70% WPN	24	100	20	0	0	0	0	0	0.2462
	48	100	50	0	0	0	0	0	0.2001
Control	24	0	–	–	–	–	–	–	–
	48	0	–	–	–	–	–	–	–

9.3.2 Field Tests

In order to investigate the molluscicidal effect of MSCN, four permanent perennial ponds were treated at 2 g/m^2. One thousand and forty-three snails were found in the four ponds prior to treatment, of which 44 were dead (a natural death rate of 4.22%) and an average density of live snails at each point was 23.79. Six hundred and twenty dead snails were found in the four ponds one month after treatment and the density was reduced to 0.143 snails three months after treatment, resulting in one living snail in one pond. The pond in which the living snail was found was treated again. One month later, no living snails were found.

The results of the field tests are set out in Table 9.2. Analytical data one month after spraying indicate that no living snails were found and the MSCN induced decline of living snails in all four ponds was nearly 100%. The results of the survey were similar three months later.

9.4 The Key Point in Local Schistosomiasis Control

Schistosomiasis remains a global health problem with an estimated 250 million people in 78 countries infected, of whom 90% live in Africa (Hotez 2018; Inobaya et al. 2018; Savioli and Daumerie 2018). Thirty years ago, schistosomiasis patients in China accounted for more than 50% of the population in the world's endemic areas. After years of prevention and control, China now accounts for less than 1% of the population in the world's endemic areas. Unfortunately, the decline of schistosomiasis infection in China has been counterbalanced by increasing rates in Africa (Bergquist et al. 2017; Li et al. 2017, 2018; Song et al. 2017). In the thirty years since the eradication of schistosomiasis in China began, the technologies used for this process have advanced greatly. Therefore, at the request of the WHO, China's Ministry of Health, the WHO, and Zanzibar's Ministry of Health signed a cooperation agreement on schistosomiasis control. Large scale treatment for schistosomiasis had been carried out over many years in Zanzibar, but transmission persisted (Knopp et al. 2013; Stothard et al. 2002b). The schistosomiasis control project began to explore the feasibility of China's participation in the control schistosomiasis in Africa, with the goal: (1) To master the transmission pattern of schistosomiasis in Pemba Island and to formulate the control strategies and measures for schistosomiasis based on the local conditions; (2) to implement the control strategy and measures for schistosomiasis, eliminate schistosomiasis in three pilot sites on Pemba Island (achieving a human infection rate of less than 1%); and (3) to build practical and effective control methods for *S. haematobium* and related technologies for schistosomiasis control using standard operating procedures.

Humans become infected with *S. haematobium* when they wade, swim, or bathe in waters inhabited by compatible intermediate snail hosts, which were previously contaminated by human urine containing parasite eggs. The adult schistosomes live as paired male and female worms in the perivesical venous plexus of their human hosts. Eggs produced by female worms get trapped in the tissues and can lead to inflammatory and obstructive diseases in the urogenital system. Eggs that pass through the bladder wall are excreted in urine. If urination occurs in freshwater, the parasite eggs hatch in the water and release miracidia (a free swimming larval stage), which subsequently infect the specific intermediate host snails (Kagan 1955; Wright 1956). In Zanzibar, the only intermediate host snail for *S. haematobium* is *Bulinus spp*.

In the snail, asexual reproduction takes place and 4–6 weeks after infection, the snail starts to shed cercariae, the larvae that infect humans, into the water. Importantly, a single snail can shed thousands of cercariae over its life-time. On locating a human host, cercariae penetrate the skin and after migrating via the lungs to the liver, develop into adult worms within 9–10 weeks. These worms pair up, and on average, live and produce eggs for 3–5 years. Many eggs leave the body via the urine but many more become trapped in the tissues causing an inflammatory reaction that leads to morbidity.

Mass drug administration (MDA) was the main strategy to control schistosomiasis morbidity in endemic areas. Although the strategy reduced the rate of disease and sickness, reinfection remained a problem. In addition, cercariae and infected snails were still present after MDA and thus new infections occurred. Hence, we believe a better control of schistosomiasis can be achieved by cutting off the transmission through elimination of the intermediate host (snail) prior to MDA. Although the use of WPN is cheap and results in little environmental pollution we believe the short half-life of niclosamide in the environment and the lack of accumulation effect makes it superior in the control of *Bulinus*. Furthermore, our research shows that MSCN had superior molluscicidal effects against *Bulinus* compared with WPN. Snail control with MSCN is an important means to reduce the transmission of schistosomiasis, and a key point in local schistosomiasis control. Since 1950 the national schistosomiasis control program was initiated, snail control-based integrated control strategy has been involved in the National Schistosomiasis Control Program and played a critical role in the control of the disease. Since 1992 World Bank Loan Project for Schistosomiasis Control (WBLP) initiated in 1992, as the only recommended molluscicide drug by the World Health Organization (WHO), niclosamide has been widely used for killing snails in areas of China where the parasite is endemic, which has contributed significantly to the interruption and control of transmission of Schistosoma japonicum in China. It is reported that the molluscicidal efficacy of niclosamide is affected by the field environment and climate conditions. In addition, the molluscicidal effect of the same batch and dose of drug varies in different endemic foci. It is therefore considered that niclosamide with the uniform dose may achieve various molluscicidal efficacies among different endemic foci in the country. On the one hand, the expected molluscicidal effect is not achieved in some endemic areas, which leads to snail breeding or spread and thus is not beneficial for the prevention and control of the disease. On the other hand, excessive drugs are used in some endemic areas, leading to wasting of resources and aggravation of environmental burden. However, the causes for the inconsistence of molluscicidal efficacy remain unknown. The present study was conducted to quantitatively assess the effects of environmental temperature and immersion time on sensitivity of Oncomelania snails to niclosamide and determined the sensitivity of snails collected at different seasons and from various areas to niclosamide in laboratory, so as to establish the basic database of niclosamide sensitivity in snails in Chinese mainland. Such a study would provide theoretical evidence for selection of appropriate drug dose for snail control in different various endemic foci. Based on the current study, a standardized protocol of mollusciciding which meets the requirement of snail control in the field would be developed, so as to achieve the maximal molluscicidal efficacy and reduce the consumption of the drug. On the one hand, the cost of snail control will be saved, and the resource collocation will be effectively optimized, so as to facilitate the progress and development of schistosomiasis control. On the other hand, the environmental pollution and environmental burden would be reduced to the maximal degree, so as to the sustained development of mollusciciding. A large number of applications in China in the past ten years confirmed that MSCN was more efficient against *Oncomelania* (the intermediate hosts of S. *japonicum*) and was less toxic to fish than WPN (Li et al. 2012). MSCN was developed to overcome the problems associated with WPN formulation including clumping during transport and difficulty in dispersion. To date though, field application of MSCN in Africa against *Bulinus* has not been reported. This research studied the effects of MSCN on the control of *Bulinus*. The trials carried out on Pemba Island with MSCN are very encouraging. In the laboratory, immersion with 0.200 mg/L of MSCN and higher concentrations of WPN achieved 100% snail mortality. MSCN was therefore found to be more effective against *Bulinus* than WPN. The observation of the short-term (one month) and the long-term (three months) effects after application showed that MSCN had a high molluscicidal effect against *Bulinus* and is likely to be a powerful weapon in the control of *S. haematobium*.

References

Andrews P, Thyssen J, Lorke D (1982) The biology and toxicology of molluscicides, Bayluscide. Pharmacol Ther 19(2):245–295

Bao ZP et al (2011) Evaluation on molluscicidal effect of suspension concentrate of metaldehyde and niclosamide. Chin J Schistosomiasis Control 23(1):48–53

Bergquist R, Zhou XN, Rollinson D, Reinhard-Rupp J, Klohe K (2017) Elimination of schistosomiasis: the tools required. Infect Dis Poverty 6(1):158. https://doi.org/10.1186/s40249-017-0370-7

Civitello DJ, Fatima H, Johnson LR, Nisbet RM, Rohr JR (2018) Bioenergetic theory predicts infection dynamics of human schistosomes in intermediate host snails across ecological gradients. Ecol Lett. https://doi.org/10.1111/ele.12937

Dai JR, Wang W, Liang YS, Li HJ, Guan XH, Zhu YC (2008) A novel molluscicidal formulation of niclosamide. Parasitol Res 103(2):405–412. https://doi.org/10.1007/s00436-008-0988-2

Dai J, Li Y, Wang W, Xing Y, Qu G, Liang Y (2014) Sensitivity of *Oncomelania hupensis* to niclosamide: a nation-wide survey in China. Int J Environ Res Public Health 11(3):3086–3095. https://doi.org/10.3390/ijerph110303086

Dai JR, Li YZ, Wang W, Xing YT, Qu GL, Liang YS (2015) Resistance to niclosamide in *Oncomelania hupensis*, the intermediate host of *Schistosoma japonicum*: should we be worried? Parasitology 142(2):332–340. https://doi.org/10.1017/S0031182014000870

Diakite NR, Winkler MS, Coulibaly JT, Guindo-Coulibaly N, Utzinger J, N'Goran EK (2017) Dynamics of freshwater snails and *Schistosoma* infection prevalence in schoolchildren during the construction and operation of a multipurpose dam in central Côte d'Ivoire. Infect Dis Poverty 6(1):93. https://doi.org/10.1186/s40249-017-0305-3

French MD, Rollinson D, Basanez MG, Mgeni AF, Khamis IS, Stothard JR (2007) School-based control of urinary schistosomiasis on Zanzibar, Tanzania: monitoring micro-haematuria with reagent strips as a rapid urological assessment. J Pediatr Urol 3(5):364–368. https://doi.org/10.1016/j.jpurol.2007.01.198

Galappaththi-Arachchige HN et al (2018) Evaluating diagnostic indicators of urogenital *Schistosoma haematobium* infection in young women: a cross sectional study in rural South Africa. PLoS One 13(2):e0191459. https://doi.org/10.1371/journal.pone.0191459

Giacomet V, Manfredini V, Mameli C, Marinacci F, Zuccotti GV (2010) A dangerous freshwater contact. BMJ Case Rep. https://doi.org/10.1136/bcr.01.2010.2620

Gower CM, Vince L, Webster JP (2017) Should we be treating animal schistosomiasis in Africa? The need for a One Health economic evaluation of schistosomiasis control in people and their livestock. Trans R Soc Trop Med Hyg 111(6):244–247. https://doi.org/10.1093/trstmh/trx047

Gray DJ et al (2014) A multi-component integrated approach for the elimination of schistosomiasis in the People's Republic of China: design and baseline results of a 4-year cluster-randomised intervention trial. Int J Parasitol 44(9):659–668. https://doi.org/10.1016/j.ijpara.2014.05.005

Hassan AA, Shoukary NM, Ismail NM (2008) Efficacy of temperature, and two commonly used molluscicides and fertilizers on *Fasciola gigantica* eggs. J Egypt Soc Parasitol 38(2):621–634

Hegertun IE et al (2013) *S. haematobium* as a common cause of genital morbidity in girls: a cross-sectional study of children in South Africa. PLoS Negl Trop Dis 7(3):e2104. https://doi.org/10.1371/journal.pntd.0002104

Hotez PJ (2018) The global fight to develop antipoverty vaccines in the antivax era. Hum Vaccines Immunother. https://doi.org/10.1080/21645515.2018.1430542

Inobaya MT et al (2018) Mass drug administration and the sustainable control of schistosomiasis: community health workers are vital for global elimination efforts. Int J Infect Dis 66:14–21. https://doi.org/10.1016/j.ijid.2017.10.023

Kagan IG (1955) Studies on the serology of schistosomiasis. I. The in vitro activity of cercariae and miracidia in serum of experimental, natural, and immunized hosts. Exp Parasitol 4(4):361–376

Khonde Kumbu R, Mbanzulu Makola K, Bin L (2016) Prevalence of *Schistosoma mansoni* infection in four health areas of Kisantu health zone, Democratic Republic of the Congo. Adv Med 2016:6596095. https://doi.org/10.1155/2016/6596095

King CH, Sutherland LJ, Bertsch D (2015) Systematic review and meta-analysis of the impact of chemical-based mollusciciding for control of *Schistosoma mansoni* and *S. haematobium* transmission. PLoS Negl Trop Dis 9(12):e0004290. https://doi.org/10.1371/journal.pntd.0004290

Knopp S et al (2013) Elimination of schistosomiasis transmission in Zanzibar: baseline findings before the onset of a randomized intervention trial. PLoS Negl Trop Dis 7(10):e2474

Lewis FA, Tucker MS (2014) Schistosomiasis. Adv Exp Med Biol 766:47–75. https://doi.org/10.1007/978-1-4939-0915-5_3

Li BG et al (2012) Molluscicidal effect of suspension concentrate of metaldehyde and niclosamide in fields of mountainous areas in Yunnan Province. Chin J Schistosomiasis Control 24(5):564–566

Li FY, Tan HZ, Ren GH, Jiang Q, Wang HL (2017) Research of prevalence of schistosomiasis in Hunan province, 1984–2015. Zhonghua liu xing bing xue za zhi = Zhonghua liuxingbingxue zazhi 38(3):350–353. https://doi.org/10.3760/cma.j.issn.0254-6450.2017.03.015

Li S et al (2018) The spatial-temporal trend analysis of schistosomiasis from 1997 to 2010 in Anhui Province, Eastern China. Am J Trop Med Hyg. https://doi.org/10.4269/ajtmh.17-0475

Liang S et al (2007) Environmental effects on parasitic disease transmission exemplified by schistosomiasis in western China. Proc Natl Acad Sci U S A 104(17):7110–7115. https://doi.org/10.1073/pnas.0701878104

McCullough FS, Gayral P, Duncan J, Christie JD (1980) Molluscicides in schistosomiasis control. Bull World Health Organ 58(5):681–689

Micati DJ, Hime GR, McLaughlin EA, Abud HE, Loveland KL (2018) Differential expression profiles of conserved snail transcription factors in the mouse testis. Andrology. https://doi.org/10.1111/andr.12465

Mohammed NA, Madsen H, Ahmed AA (2016) Types of trematodes infecting freshwater snails found in irrigation canals in the East Nile locality, Khartoum, Sudan. Infect Dis Poverty 5:16. https://doi.org/10.1186/s40249-016-0108-y

Mosta-Fa BB, el-Deeb FA, Ismail NM, el-Said KM (2005) Impact of certain plants and synthetic molluscicides on some fresh water snails and fish. J Egypt Soc Parasitol 35(3):989–1007

Pennance T et al (2016) Urogenital schistosomiasis transmission on Unguja Island, Zanzibar: characterisation of persistent hot-spots. Parasit Vectors 9(1):646. https://doi.org/10.1186/s13071-016-1847-0

Rollinson D, Klinger EV, Mgeni AF, Khamis IS, Stothard JR (2005) Urinary schistosomiasis on Zanzibar: application of two novel assays for the detection of excreted albumin and haemoglobin in urine. J Helminthol 79(3):199–206

Sacolo H, Chimbari M, Kalinda C (2018) Knowledge, attitudes and practices on schistosomiasis in sub-Saharan Africa: a systematic review. BMC Infect Dis 18(1):46. https://doi.org/10.1186/s12879-017-2923-6

Savioli L, Daumerie D (2018) WHO leadership is essential for the elimination of NTDs. Lancet 390 (10114):2765. https://doi.org/10.1016/S0140-6736(17)33300-7

Song LG, Zheng XY, Lin DT, Wang GX, Wu ZD (2017) Parasitology should not be abandoned: data from outpatient parasitological testing in Guangdong, China. Infect Dis Poverty 6(1):119. https://doi.org/10.1186/s40249-017-0332-0

Stothard JR, Mgeni AF, Khamis S, Seto E, Ramsan M, Rollinson D (2002a) Urinary schistosomiasis in schoolchildren on Zanzibar Island (Unguja), Tanzania: a parasitological survey supplemented with questionnaires. Trans R Soc Trop Med Hyg 96(5):507–514

Stothard JR et al (2002b) New insights into the transmission biology of urinary schistosomiasis in Zanzibar. Trans R Soc Trop Med Hyg 96(5):470–475

Webster BL et al (2015) Development of novel multiplex microsatellite polymerase chain reactions to enable high-throughput population genetic studies of *Schistosoma haematobium*. Parasit Vectors 8:432. https://doi.org/10.1186/s13071-015-1044-6

Woodall PA, Kramer MR (2018) Schistosomiasis and infertility in East Africa. Am J Trop Med Hyg. https://doi.org/10.4269/ajtmh.17-0280

Wright CA (1956) Some factors affecting the susceptibility of snails to infection with miracidia of *Schistosoma* spp. Bulletin de la Societe de pathologie exotique et de ses filiales 49 (6):1211–1220

Xia ZG et al (2014) Molluscicidal effects of 3 molluscicides by spraying and poudrage methods in dry season. Chin J Schistosomiasis Control 26(4):461–2, 465

Xing YT et al (2013) Preparation and molluscicidal effect of 5% niclosamide ethanolamine granules. Chin J Schistosomiasis Control 25(5):473–476

Yang GJ et al (2012) Optimizing molluscicide treatment strategies in different control stages of schistosomiasis in the People's Republic of China. Parasit Vectors 5:260. https://doi.org/10.1186/1756-3305-5-260

Yasin MG et al (2018) Trematode infections in farm animals and their vector snails in Saint Martin's Island, the southeastern offshore area of Bangladesh in the Bay of Bengal. J Vet Med Sci. https://doi.org/10.1292/jvms.17-0308

Chapter 10
Human Chemotherapy by China-Made Praziquantel in Zanzibar: Efficacy and Safety

Xinyao Wang and Salum Abubakar

Abstract Schistosomiasis is a global neglected tropical parasitic disease of great medical and economic significance. Since praziquantel was developed in 1970s, it has replaced other antischistosomal agents to become the only drug of choice for the treatment of human schistosomiases. Currently, mass drug administration with praziquantel remains the global strategy for schistosomiasis control. In the roadmap on the neglected tropical diseases (NTD) the World Health Organization (WHO) aims at attaining at least 75% coverage of preventive chemotherapy in pre-school and school-age children by 2020. A randomized controlled trial was performed to compare the effectiveness and safety of praziquantel in treating *Schistosoma haematobium* in Africa using two different sources for the drug, Merck Limited Partnership (KgaA), Germany and Nanjing Pharmaceutical Factory (NPF), China. More than 6000 participants testing positive for *S. haematobium* infection were enrolled from three villages (shehias) situated in the northern, middle and southern part of Pemba Island, Zanzibar. Applying criteria of inclusion and exclusion, resulted in a study population of 152 people (84 males, 68 females). A randomized controlled trial was conducted assigning participants to either praziquantel from NPF or Merck KGaA. After 1 month, the cure rate of *S. haematobium* and adverse events were compared. The ratio of male to female, the ratio of light/high infection intensity, and the average value of age were calculated between the two drug manufacturers. Chi-squared test and T-test were used for consistency analysis. Out

X. Wang (✉)
Jiangsu Institute of Parasitic Diseases, Wuxi, Jiangsu, China

Key Laboratory on Technology for Parasitic Disease Prevention and Control, Ministry of Health, Wuxi, Jiangsu, China

Jiangsu Provincial Key Laboratory on the Molecular Biology of Parasites, Jiangsu Institute of Parasitic Diseases, Wuxi, Jiangsu, China

College of Medicine, Jiangnan University, Wuxi, Jiangsu, China
e-mail: wangxinyao@jipd.com

S. Abubakar
Ministry of Health of Zanzibar, Zanzibar, United Republic of Tanzania

K. Yang, H. Mehlhorn (eds.), *Sino-African Cooperation for Schistosomiasis Control in Zanzibar*, Parasitology Research Monographs 15,
https://doi.org/10.1007/978-3-030-72165-7_10

of the total of 73 cases receiving praziquantel from NPF, the cure rate achieved was 97.3% (73/75), while the 74 cases receiving the drug from Merck KgaA reached a similar cure rate (96.1% or 74/77). There was no significant difference between the two outcomes ($\chi^2 = 0.003$, $P = 0.956$). Among the 75 patients treat, only one (a 16-years old female student), who had received the drug made in China had slight adverse reactions manifested as dizziness, headache and abdominal pain. The efficacy of China-made praziquantel does not differ significantly from praziquantel made by Merck KGaA in Germany.

Keywords China-made praziquantel · Schistosoma haematobium · Randomized controlled trial

10.1 Background

Schistosomiasis remains as an important public health issue in tropical and subtropical regions of the world, and this neglected tropical disease ranks second only to malaria in terms of its medical and socioeconomic significance among all parasitic diseases. Currently, this tropical parasitic disease is estimated to affect over 200 million people globally, with further 800 million people at risk of infections (Collins et al. 2012; van Dam et al. 2015). Human schistosomiasis syn. Bilharziasis is a parasitic disease caused by trematode flukes of the genus Schistosoma. The trematode digenetic parasites in the family Schistosomatidae (phylum Platyhelminthes) infect a wide range of mammals, and six species of the genus *Schistosoma* are of major medical importance, including *Schistosoma mansoni*, *S. japonicum*, *S. haematobium*, *S. intercalatum* and *S. mekongi* and *S. malayensis* (Barsoum et al. 2013; Lewis and Tucker 2001). *S. mansoni* is widely prevalent in Africa, South America and Caribbean countries, and *S. japonicum* is mainly distributed in China and Indonesia, while the distribution of *S. mekongi* is limited in Cambodia and Laos, and *S. intercalatum* is only found in parts of Central and West Africa. In addition, *S. haematobium* is distributed through African countries and in some regions of the Middle East. However, *S. haematobium* has been recently detected in Europe and even outbreaks of schistosomiassi haematobia occur in Corsica, France, which has received global attention. Upon infection, *S. haematobium* may cause urinary schistosomiasis in humans, presenting a range of complications, and *S. haematobium* infections have been linked with bladder cancer (Antoni et al. 2017; Khurana et al. 2005).

Praziquantel, 2-(cyclohexylcarbonyl)-1,2,3,6,7, 11b-hexahydro-4H-pyrazino [2, 1-a] isoquinolin-4-one with a molecular formula of $C_{19}H_{24}N_2O_2$, is an anthelmintic drug that was developed by E. Merck and Bayer A at early 1970s. As a broad-spectrum anthelmintic, this agent is widely active against a broad range of parasitic helminths, including schitsosomiasis, clonorchiasis, opisthorchiasis, tapeworm infections, cysticercosis and echinococciasis (Bobyreva et al. 2016). Although several hypotheses have been proposed for explaining the mechanisms of praziquantel actions, such as calcium ion channels, tegument damages, and multi-drug resistant proteins MRP1 and P- glycoprotein, the exact mechanisms remain. Praziquantel is in the anthelmintic class of medications. It is highly effective and remains the drug of choice for the treatment of all forms of schistosomiasis infection (Cioli and Picamattoccia 2003) Although safe and generally without serious side effects, praziquantel can still cause poor coordination, abdominal pain, vomiting, headache and allergic reactions. While it may be used in women during pregnancy for the second trimester, it is not recommended for the mother when breastfeeding (Xu et al. 2014).

The global schistosomiasis elimination strategy proposed by the World Health Organization (WHO) involves a large-scale treatment for affected populations through periodic, targeted treatment of school-children with praziquantel (Organization 2014). In 2008, Merck KgaA, Darmstadt, Germany started a program donating 20 million tablets (600 mg) annually, which is still ongoing (Hotez and Fenwick 2009). However, in 2012, Merck-KGaA committed to increase its donation to 250 million tablets of praziquantel per year until schistosomiasis is eliminated. To bolster this donation, additional amounts of praziquantel and resources for implementation were provided by other partners (Tchuem Tchuente et al. 2017). Many companies around the world produce praziquantel, such as E-merck, Bayer, Miles, Korea Daewoo, etc. The largest output is Bayer, with an annual output of more than 50 tons. In recent years, due to various reasons, the output of these companies has been reduced to varying degrees, and some have also stopped production. In previous years, the total annual output of praziquantel around the world was about 200 tons. Whilst there is now growing access to praziquantel for schistosomiasis control in sub-Saharan Africa, it is not at the level of projected requirements to reach all people at risk and requiring treatments (Stothard et al. 2013). Analysis of data reported on treatment coverage for schistosomiasis show that utilization of available praziquantel by NTD programmes is not yet optimal in many countries (Organization 2015, 2016).

Praziquantel was successfully synthesized in China in 1977 and began to be used in clinical trials, with officially put on the market in 1982. For more than 30 years, China's annual production capacity and output of praziquantel have steadily increased. In the 1980s, China's annual production capacity of praziquantel was not below 20 tons, and the annual output was only more than 10 tons. By the 1990s, the annual production capacity reached more than 30 tons, with an annual output of more than 20 tons. At the end of the twentieth century, the annual production capacity reached nearly 50 tons and the annual output was about 35 tons. For many years, Nanjing Pharmaceutical Factory and Shanghai No. 6 Pharmaceutical Factory have been the main producers of praziquantel, with annual output and exports accounting for more than 80% of the country's share. Since the twenty-first century, many enterprises in Jiangsu, Zhejiang, Anhui and other places had successively invested in the production of praziquantel. In the past, Zhejiang Hisun Pharmaceutical had also cooperated with foreign countries to build a 40-ton annual production line of praziquantel. China's annual production capacity had increased to more than 150 tons, with an annual output of more than 80 tons, thus becoming the world's major producer of praziquantel. For a long time, sales of praziquantel in the domestic market have been relatively stable, usually 10–20 tons. In years of severe floods in areas with frequent schistosomiasis infections in southern China, sales of praziquantel have sharply increased and even doubled. From 1993 to 1995, the annual export volume was only about ten tons. Since then, China's export of praziquantel has continued to grow. From 1996 to 1998, the annual export volume reached about 20 tons; at the end of the twentieth century, the export volume increased significantly, reaching more than N3o tons; at the beginning of the twenty-first century, China's praziquantel export volume again There has been a substantial increase, exceeding 40 tons, and the export volume now exceeds 60 tons (Zhang 2013).

The increased use of the drug is attributable to many factors, including improved availability of donated praziquantel, essentially from Merck, which has led to some countries implementing large-scale schistosomiasis control programmes and scale-up of praziquantel treatment (Knopp et al. 2016). In 2012, WHO's roadmap on neglected tropical diseases set out a comprehensive plan for the control, elimination and eradication of 17 neglected tropical diseases by 2020, and the global target of WHO in the roadmap on neglected tropical diseases (NTD) is to attain at least 75% coverage of preventive chemotherapy in pre-school and school-age children by 2020 (Molyneux et al. 2017). Praziquantel has been used to treat schistosomiasis for a long time in China and the experience demonstrates that preventive chemotherapy (i.e., large-scale treatment without individual diagnosis) with high coverage significantly impacts infection indices and even reduces transmission (Wu et al. 1993).

Schistosomiasis was once highly prevalent in China. In 1950s, over 11 million Chinese people were infected with this endemic disease, which was named "God of Plague" by Chairman Zedong Mao, the founder of the People's Republic of China. Since middle 1950s, a strong political will was committed to schistosomiasis control in China, and the national schistosomiasis control programme was initiated. Following the concerted efforts for decades, snail control-based integrated control strategy, praziquantel chemotherapy-based integrated control strategy and the integrated strategy with emphasis on management of the source of *S. japonicum* infections were sequentially implemented in China, which resulted in great successes in the national schitsosomiasis control programme in the country. In 2015, the transmission of schistosomiasis was effectively controlled in China, which encourages the goal set for global elimination of schistosomiasis in this wormy world. In the context of China-Africa Forum, the Chinese government committed to transfer its successful experiences gained from public health concerns to African counties, including parasitic diseases.

In 2014, WHO signed a tripartite memorandum of understanding (MOU) with China and Zanzibar, paving the way for the launch of a pilot schistosomiasis elimination plan in Zanzibar. The main goal is to control and eliminate schistosomiasis based on the experience of China's National Schistosomiasis Control Program. The project on Pemba Island is an emergency project for schistosomiasis control in China. The early stage of the project is to carry out the prevention and control of schistosomiasis in three villages (shihias) named Mtangani, Kiuyu and Wingwi through snail control, population survey and treatment. (LV 2017 May 11;

Michuzi 2014 May 24; Organization 2014 May 21; Xu et al. 2016). A randomized trial study of the efficacy of two different sources for the drug, Merck Limited Partnership (KgaA), Germany and Nanjing Pharmaceutical Factory (NPF), China has been carried out in the early stage. Based on this, this study evaluated the effectiveness and safety of China-made praziquantel in the treatment of schistosomiasis.

10.2 Methods

10.2.1 Ethics Statement

In the preliminary randomized controlled study, all subjects have signed the informed consent form for this study. For all underage participants, parents or guardians provided informed consent on their behalf. The praziquantel made in China used has been certified by the Zanzibar Food and Drug Administration (ZFDB). The preliminary research was also approved by the local government of Pemba Island. In addition, the Zanzibar Ethics Review Board approved all studies described here (ZAMREC/002/MAY/014). The previous results have been published in the PLoS article https://journals.plos.org/plosntds/article?id=10.1371/journal.pntd.0007238.

10.2.2 The Efficacy of China-Made Praziquantel

Zanzibar is composed of two sister islands, Unguja and Pemba, situated off the eastern coast of Tanzania mainland between latitudes 4° and 5° South (Zanzibar). Zanzibar is endemic for schistosomiasis haematobium. The intermediate host is *Bulinus globosus* (Goatly and Jordan 1965; Knopp et al. 2013). The previous study was carried out on Pemba Island, and this study used a randomized controlled trial (Garcia et al. 2014). Three shehias, Mtangani, Kiuyu and Wingwi, situated in northern, central and southern Pemba, respectively, were selected as study area. Three pilot areas proceeded schistosomiasis control in the early stages of assistance. A series of meetings was held at the shehias and their schools to explain the objectives, procedures, and potential risks of the study. After more than a year to complete the study, the study eventually recruited more than 6000 participants. Applying the inclusion and exclusion criteria (see below), the final study population was 152 (84 males and 68 females). People were enrolled randomly into to two groups: group A received praziquantel (batch number: 20170303) from Nanjing Pharmaceutical Factory Co Ltd. (NPF), Nanjing donated from China and group B received batch number: M50812 from Merck Limited Partnership (Merck KgaA), Darmstadt, Germany donated from WHO for the treatment of schistosomiasis in Africa.

The dose is 40 mg/kg of drug body weight given in a single oral dose according to the manufacturer's instructions. Merck (Merck) KgaA Praziquantel is larger than NPF. However, the medication is taken at 40 mg/kg body weight. One day therapy for treatment. When administering, take it until the patient takes it. The staff of the NTD Office of the Ministry of Health of Zanzibar will instruct the drug distributors after confirming the allocation of treatments based on the random sequence generated by the computer. The NTD staff and participants was unmasked to the treatment assignment, but the laboratory technicians were masked to samples examination throughout the study.

10.2.3 The Safety of China-Made Praziquantel

The NTD staff recorded the exact time of drug ingestion. Participants were observed for 2 hours after taking the drug to see adverse events and were continually observed at home by the assistance of local public health community center(PHCU)staff. They followed up on the medications every day to see adverse events. If anybody was vomiting within 2 hours of drug ingestion, a second dose was given (Fig. 10.1). People without relevant symptoms were followed up at 1–2 hours after chemotherapy, at the second day, 2 weeks, and 1 month after chemotherapy, to observe the side effects of the nervous system, digestive system, and cardiovascular system after chemotherapy, and whether the occurrence of allergic reactions and their side effects have delayed effects (Table 10.1).

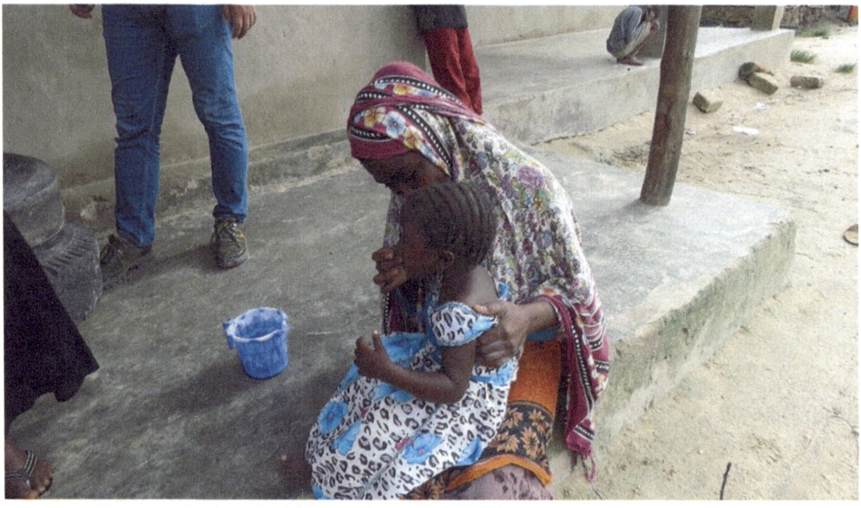

Fig. 10.1 Participant taking medication

Table 10.1 The efficacy of praziquantel for treatment of Schistosomiasis haematobium in Zanzibar

Controlled number: JIPD-401					
A. Basic information					
1.Name:	Gender: M/ F	Birth date: Year	Month		
2. Project ID:		National ID:			
3. Name of householder:			Tel:		
4. Shehia:	Village:		School:		
5. GPS of household: South	°	East	°		
6. Occupation: *1= Farman; 2= Fishman;3=Student; 4= Workman;5=Businessman; 6= Others*					
7. Education: *1=Unschooled; 2= Primary school; 3=Secondary school; 4= High School; 5=College*					
B. Diagnose and treatment					
1. The diagnose:					
Date:	; Visible haematuria:	*1=Yes; 2=No*;	Egg amount:		
Date:	; Visible haematuria:	*1=Yes; 2=No*;	Egg amount:		
Date:	; Visible haematuria:	*1=Yes; 2=No*;	Egg amount:		
Date:	; Visible haematuria:	*1=Yes; 2=No*;	Egg amount:		
Date:	; Visible haematuria:	*1=Yes; 2=No*;	Egg amount:		
2. The treatment:					
Date:	; Praziquantel type:	*1=China, 2 = WHO;*	Weight:	Kg; Tables:	
Date:	; Praziquantel type:	*1=China, 2 = WHO;*	Weight:	Kg; Tables:	
Date:	; Praziquantel type:	*1=China, 2 = WHO;*	Weight:	Kg; Tables:	
Date:	; Praziquantel type:	*1=China, 2 = WHO;*	Weight:	Kg; Tables:	
Date:	; Praziquantel type:	*1=China, 2 = WHO;*	Weight:	Kg; Tables:	
C. Feeling after treatment *1 = seious,2 = mide;3 = No*					
1. Date:					
Dizziness: Vomiting	Headache: Body weakness:	Nausea:	Anorexia:	Abdominal pain:	Diarrhoea:
Cough:	Itchy skin:	Skin rash:			
2. Date:					
Dizziness: Vomiting	Headache: Body weakness:	Nausea:	Anorexia:	Abdominal pain:	Diarrhoea:
Cough:	Itchy skin:	Skin rash:			
3. Date:					
Dizziness: Vomiting	Headache: Body weakness:	Nausea:	Anorexia:	Abdominal pain:	Diarrhoea:
Cough:	Itchy skin:	Skin rash:			

10.2.4 Procedures

Inclusion criteria were the following: people permanently living on the island including residents and students of ages between 5 and 60 years who had been shown to be infected by *S. haematobium* as confirmed by urine test (worm eggs detected), and agreed to receive treatment and a 1-month follow-up visit; the resident stayed on Pemba Island for a month after taking the medicine. Exclusion criteria

Fig. 10.2 Collection of urine and questionnaire

were the following: people refusing chemotherapy of praziquantel; women pregnant or lactating at the time of the study; people with serious adverse drug reactions; patients with severe heart, liver or kidney problems; those with a history of mental illness; and patients who for various reasons did not take medication on time; patients who disappeared less than a month after taking the medicine.

Urine was collected and tested in 2017. Every participant provided a fresh urine sample used to detect the presence of *S. haematobium*. The NTD staff subjected all potential participants to physical examination and eligibility check (Fig. 10.2). We used a 10 ml syringe to extract 10 ml and filtered the urine, repeatedly shaking the cup to include eggs that that has a tendency to collect at the bottom. The filter device was prepared according to WHO recommendations with the filter taken out and placed on a glass slide before microscopy (Fig. 10.3). Only participants found positive and otherwise meeting the criteria were included in the study.

Within 1 month after taking the medicine, follow-up was conducted with urine were collected and tested in the same way again. Signs and symptom indicating adverse effects, inter-current illness or abdominal pain was recorded for the preceding month. Quality control measure was used for inter-observation variability, and technician retested a random selection of 10% of slides in the laboratory. At end of the study, all participants who still excreted *S. haematobium* eggs (i.e., who had not been cured) were again treated with praziquantel.

Fig. 10.3 Testing of collected urine in a laboratory

10.2.5 Outcome

The outcome was a comparison of the cure rates after either having received praziquantel from NPF or Merck KgaA and the rate of severe adverse events.

10.2.6 Statistical Analysis

Serial report forms were used to collect the data from the participants. Epi Info (Centers for Disease Control and Prevention), SPSS 20.0 (International Business Machines Corporation) and PASS 16 (NCSS) were used for data entry and analysis.

The sample size was calculated by cross-sectional survey, 95% confidence interval (CI) ($Z\alpha/2 = 1.96$), 5% margin of error and design effect of 2.5. According to previous study, the infection rate of schistosomiasis in Pemba was about 6.0% and following the function of $N = Z_{1-\alpha/2}^{2}(1-p)/\varepsilon^{2}p$ (Lv and Feng 2016). The outcome of power was 0.99975, which met the requirements. For the estimation of sample size in the early stage, we set the accuracy to 10% of the expected incidence rate to estimate the sample size.

In this study, *S. heamatobium* infection intensity was determined as light (1–49 egg/10 ml urine) and heavy (≥50 eggs/10 ml urine) (Organization 2002). The ratio of male to female, the ratio of light infection intensity to heavy, and the average value of age was firstly calculated between the NPF and Merck KgaA praziquantel groups, Chi-squared test and T-test were used to analysis the consistency.

A two-sided of Chi-squared test was used to compare the cure rate of positive and adverse reactions between different groups. The size of the test or α level was set to 5%, with 95% CI, with exact binomial estimates when necessary. The number of eggs and logarithmic transformation was in a skewed state, so the median value was used to analysis the change of infection intensity between the NPF and Merck KgaA praziquantel groups after praziquantel treatment.

10.3 Results

10.3.1 Baseline Characteristics

A total of 6740 people were enrolled from three shehias in Pemba Island Zanzibar, Urine samples were collected and tested in April 2017 (Fig. 10.4). According to the criteria of inclusion and exclusion, 152 people (84 males, 68 females) were included consisting of 45 participants from Mtangani, 76 participants from Kiuyu and 31 participants from Wingwi. Participants were randomly assigned to study groups, 75 people received NPF praziquantel (males: 37, females: 38) and 77 Merck KgaA praziquantel (males: 47, females: 30) (Table 10.2). There was no significant difference of age in the two groups ($t = -0.424$, $P = 0.672$) with a median of 14 for both Merck KGaA and NPF. There was no significant difference of the number of eggs in the two groups ($t = -1.130$, $P = 0.261$) with the median value in Merck KGaA and NPF 8 and 9, respectively. There was good consistency with respect to age and gender for *S. haematobium* infection and intensity between the two groups.

10.3.2 Efficacy of China-Made Prazqiuantel

Seventy three of the 75 cases receiving the NPF praziquantel were all negative translating into a cure rate of 97.3%. With regard to the Merck KgaA praziquantel, treatment the cure rate was 96.1% (74/77). There was no significant difference in the cure rate between the two groups ($\chi2 = 0.003$, $P = 0.956$).

In Mtangani, 20 were people tested negative out of the 21 who had received the NPF praziquantel (cure rate = 95.2%), while 21 of the 24 receiving the Merck KgaA praziquantel were cured translating into a rate = 87.5%. There was no statistically significant difference ($\chi^2 = 0.038$, $P = 0.845$). In Wingwi, 16 patients received praziquantel from NPF and 15 were negative (cure rate = 93.8%). With regard to Merck KgaA praziquantel, all 15 treated patients were negative, so the cure rate was

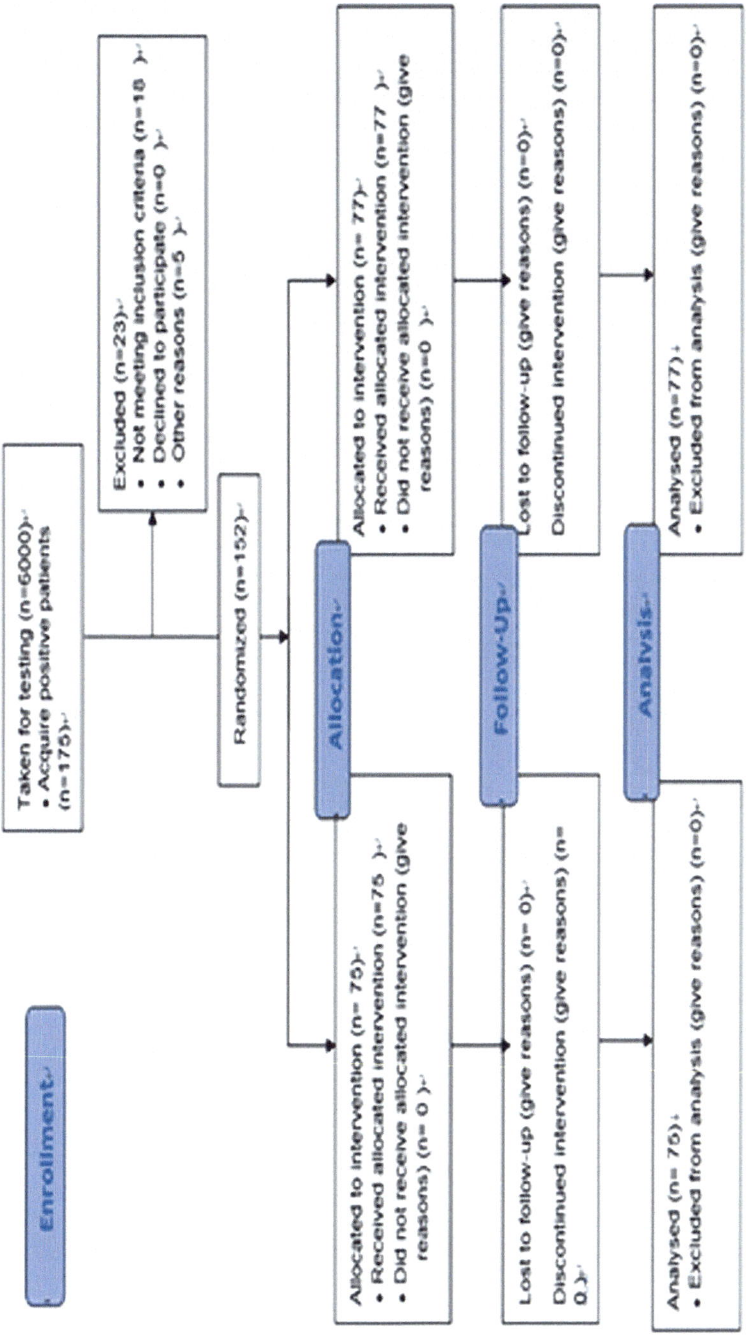

Fig. 10.4 Flow diagram

Table 10.2 The characteristics of participants in three Shehias in Pemba island, Zanzibar

Shehia	Drug type	Participant no. (female pro- portion %)	Number of males	Males (% pro- portion)	Mean age (SD)	Median age (interquartile range)	Light infection* (% proportion)	Heavy infection** % proportion)
Mtangani	NPF	21 (9.4%)	9	43%	13.4 (2.4)	13.0 (12.0–16.0)	17 (81%)	4 (19%)
	Merck KgaA	24 (14.6%)	15	63%	14.2 (2.4)	13.0 (13.0–17.0)	22 (92%)	2 (8%)
Wingwi	NPF	16 (5.3%)	5	31%	31.0 (19.3)	27.0 (14.0–46.8)	11 (69%)	5 (31%)
	Merck KgaA	15 (6.4%)	6	40%	28.0 (19.6)	27.0 (12.5–45.0)	11 (73%)	4 (27%)
Kiyuyu	NPF	38 (23.6%)	23	61%	17.1 (12.1)	13.5 (10.0–17.0)	31 (82%)	7 (18%)
	Merck KgaA	38 (26.7%)	26	68%	16.3 (12.0)	14.0 (11.0–19.8)	33 (87%)	5 (13%)
Total	NPF	75 (37.5%)	37	49%	18.9 (13.7)	14.0 (11.0–17.0)	59 (79%)	16 (21%)
	Merck KgaA	77 (47.6%)	47	61%	18.0 (14.3)	14 (5–70)	66 (86%)	11 (14%)

*1–49 eggs per milliliter of urine; **≥50 eggs per milliliter of urine

Table 10.3 Results of urine sample testing before and after treatment of NPF and Merck KgaA praziquantel with respect to gender

Drug	Gender	Positive before treatment (no.)	Negative after treatment (no.)	Negative rate after treatment (% proportion)	Results of chi-square test (p-value)
NPF praziquantel	Male	37	37	100.00%	0.869
	Female	38	36	94.74%	
Merck KgaA praziquantel	Male	47	45	95.74%	0.977
	Female	30	29	96.67%	

Table 10.4 Results of urine sample testing before and after treatment of NPF and Merck KgaA praziquantel with respect to the degree of infection

Degree of infection	Drug	Positive before treatment (no.)	Negative after treatment (no.)	Negative rate after treatment (% proportion)	Results of chi-square test (p-value)
Light degree	NPF praziquantel	28	27	96.43%	0.986
	Merck KgaA praziquantel	66	64	96.97%	
Heavy degree	NPF praziquantel	9	8	88.89%	0.973
	Merck KgaA praziquantel	11	10	90.91%	

100.0%. Also here, there was no statistical significant differences between the two groups ($\chi^2 = 0.016$, $P = 0.900$). In Kiuyu, finally, there were 38 people in both groups and they were all cured, so the cure rate was again 100.0% in two groups.

As seen in Tables 10.3 and 10.4, there was no statistical difference with respect to gender (male: $\chi^2 = 0.027$, $P = 0.869$; female: $\chi^2 = 0.001$, $P = 0.977$) or the degree of infection (light: $\chi^2 = 0.0003$, $P = 0.986$; heavy: $\chi^2 = 0.001$, $P = 0.973$).

One month after treatment, only two cases were found in the NPF group; the number of eggs was one and six, respectively, while three cases appeared in the Merck KGaA group with 3, 12 and 20 eggs, respectively. The final egg reduction rate was thus 99.8% (4285/4292) in the NPF group and 98.7% (2620/2655) in the Merck KGaA group.

10.3.3 Safety of China-Made Praziquantel

The side effects included dizziness, headache, nausea, anorexia, abdominal pain, diarrhea, vomiting, body weakness, cough, itchy skin and skin rash. Among the 152 patients, only one 16-year old female student experienced slight adverse reactions, manifested as dizziness and headache of receiving the NPF product. No other adverse effects occurred.

10.4 Discussion

Praziquantel is a heterocyclic isoquinolinazine derivative synthesized in 1972 by the two Germany firms E. Merck and Bayer Pharmaceuticals (Xu et al. 2009). Outside Germany, the drug was first synthesized by Chinese company in 1977, which was subsequently used in clinical practices in the country in 1978. Due to a high curative effect, easy administration, low cost and a high safety, praziquantel iswidely used for clinical treatment on a large scale of various stages of schistosomiasis in China (Guo 2006; Xu 1988). Since 1980s, praziquantel has been widely used in the morbidity control, transmission control and transmission interruption, as well as prevention of schistosomiasis japonica in China, and synchronous chemotherapy with praziquantel for humans and livestock is the most important component in the praziquantel-based integrated schistosomiasis control strategy. Since 1992 when the World Bank Loan Project for Chinese Schistosomiasis Control Programme began in China, mass drug administration with praziquantel was recommended in endemic foci with a > 15% prevalence of *S. japonicum* human infections, expanded chemotherapy in endemic foci with 10–15% prevalence rates of infections, and selective chemotherapy in endemic foci with <5% prevalence of infections. Praziquantel chemotherapy, the cornerstone of the integrated control strategy during the stage moving from morbidity control to transmission control, has been proved to be effective to reduce the prevalence and intensity of *S. japonicum* infections in China. More importantly, there is no evidence of resistance to *S. japonicum* detected in main endemic foci of China, although this agent has been extensively used for chemotherapy in both humans and livestock in China for more than 30 years. This implicates that praziquantel remains a strong weapon for the fight against the God of Plague in China during the stage moving towards the progress of elimination of schitsosomiasis. However, the efficacy and safety of China-made praziquantel for the treatment of *S. haematobium* infections have never been reported in African countries.

Although *S. haematobium* is widespread in Africa, no large amounts of China-made praziquantel made by NPF have been used on a large scale. Most praziquantel used in Africa comes is from Merck KgaA through the WHO donation mentioned above (Gryseels et al. 2006; Olds et al. 1996; Webbe and James 1977). However, the actual demand over the last 5 years was about 400 million tablets, which above what can be met by the donation. It is thus important to find another source of effective praziquantel that is not expensive (Fenwick 2006; Hotez and Fenwick 2009; Threats 2011).

This study investigated the feasibility of China-made praziquantel to complement what is currently available. Non-inferiority margins limit the allowable range of clinically acceptable medicines and comparator drugs. From the perspective of clinical efficacy, and against the background of previous test results, it must be repeatedly demonstrated that a drug is superior to these margins. Sometimes, a cost-benefit analysis may be necessary. However, this study did not investigate the efficacy of praziquantel per se, but used traditional statistical methods to compare the differences in efficacy between the two products to prove the possibility of using Chinese praziquantel for the treatment and prevention of schistosomiasis haematobium. For non-inferiority margins, relevant studies will be conducted again in the future.

This study may be limited in that the study of testing only relatively more than 6000 people from three shehias. However, this was done to serve as a reflection of the overall situation on Pemba Island. The studies proved to be a strong indication there was no difference between the NPF praziquantel and that made by Merck KgaA in efficacy. In the results, the cure rates of different shehias were compared, and then the overall cure rates were compared, just to observe whether there were statistical differences in different shehias, and then make an overall comparison. The drug made by Merck KgaA is a 600 mg per tablets, the price is about 2.39 US dollars. The China-made praziquantel is 200 mg per tablets, the price is about 0.25 US dollars. With the same curative effect, praziquantel made in China is cheap, which is conducive to large-scale application in Africa to meet the needs of schistosomiasis control in the region (Frye 2012; Kusel and Hagan 1999; Reich et al. 1998; Xiao 2005).

Although praziquantel is effective in the treatment of schistosomiasis, side effects have been reported clinically and in the field (Ming-Gang 2005; Stelma et al. 1995). Since praziquantel is made from a mixture of left-handed and right-handed isomers during the production process, animal experiments have confirmed that lev-of-praziquantel is the main insecticidal component, while dextro-praziquantel has little insecticidal effect, which is main factor for causing side reactions. We recorded only one adverse effect, which could be due to differences in drug manufacture (Midzi et al. 2008). However, this occurrence was to minor, and a large-scale trail would be needed to rule this out or put it on a better footing. The fact that the common side effects, mainly reactions from the neuromuscular, digestive, or cardiovascular system sometimes due to allergy, are generally light and disappear in a short time, makes a comparison difficult without reverting to large-scale trails. Even severe reactions may occur in individual cases, such as syncope, ataxia, angina pectoris and severe arrhythmia, might be easier to record but they are very rare (Chen and Wen 2007). Finally, what makes the study of side effects difficult is that they may be related to the dose, route, purity, time, dosage form, physical properties of the drug, and individual differences in the drug administration (Julien 1995; Lu and Xu 1998; Xu et al. 2009). Through studying of other researches, China-made praziquantel had also good effectiveness to treat schistosomiasis haematobium (Wu 1994; Wu and Qiu 1990). Through the previous study, it was found that domestic praziquantel only had one mild adverse reaction, and its safety is worthy of affirmation. It can be used on a large scale in Africa, thus contributing to the elimination of schistosomiasis by Chinese.

In conclusion, this study indicates that China-made praziquantel has good efficacy and minor side effects. China-made praziquantel can be used to treat *S. haematobium* infections in Africa on a large scale.

Acknowledgments We especially express our gratitude to all patients who participated in the study. We also thank caregivers from all the NTD staff of Pemba and health authorities from Ministry of Public Health for their great support in facilitating this work. We are also grateful to provide praziquantel from Nanjing pharmaceutical factory co., ltd and Merck KgaA. We also thank Dr. Robert for the modification to the original article https://doi.org/10.1371/journal.pntd.0007238 on which this chapter is based.

References

Antoni S, Ferlay J, Soerjomataram I, Znaor A, Jemal A, Bray F (2017) Bladder cancer incidence and mortality: a global overview and recent trends. Eur Urol 71(1):96

Barsoum RS, Esmat G, Elbaz T (2013) Human schistosomiasis: clinical perspective: review. J Adv Res 4(5):433–444

Bobyreva NS, Korneeva YA, Degteva GN (2016) Analysis of parasitological situation in nenets autonomous district, vol 95, pp 157–162

Chen W, Wen LY (2007) The side-effects of praziquantel against schistosomiasis japonica. Int J Med Parasit Dis 34(4):44–46

Cioli D, Picamattoccia L (2003) Praziquantel. Parasitol Res 90(1):S3–S9

Collins C, Jing X, Tang S (2012) Schistosomiasis control and the health system in P.R. China. Infect Dis Poverty 1(1):1–8

Fenwick A (2006) New initiatives against Africa's worms ☆. Trans R Soc Trop Med Hygiene 100 (3):200–207

Frye JE (2012) International drug price indicator guide. Management Sciences for Health, Arlington, VA

Garcia HH et al (2014) Efficacy of combined antiparasitic therapy with praziquantel and albendazole for neurocysticercosis: a double-blind, randomised controlled trial. Lancet Infect Diseases 14(8):687–695

Goatly KD, Jordan P (1965) Schistosomiasis in Zanzibar and Pemba. East Afr Med J 42:1–9

Gryseels B, Polman K, Clerinx J, Kestens L (2006) Human schistosomiasis. Lancet 368 (9541):1106–1118. https://doi.org/10.1016/S0140-6736(06)69440-3

Guo JG (2006) The history and status of comprehensive management of schistosomiasis in China. Chin J Prev Med 40(4):225–228

Hotez PJ, Fenwick A (2009) Schistosomiasis in Africa: an emerging tragedy in our new global health decade. PLoS Negl Trop Diseases 3(9):e485

Julien RM (1995) A primer of drug action: a concise, nontechnical guide to the actions, uses, and side effects of psychoactive drugs, 7th edn. W H Freeman/Times Books/Henry Holt & Co

Khurana S, Dubey ML, Malla N (2005) Association of parasitic infections and cancers. Ind J Med Microbiol 23(2):74

Knopp S et al (2013) Elimination of schistosomiasis transmission in Zanzibar: baseline findings before the onset of a randomized intervention trial. PLoS Negl Trop Dis 7(10):e2474. https://doi.org/10.1371/journal.pntd.0002474

Knopp S et al (2016) Praziquantel coverage in schools and communities targeted for the elimination of urogenital schistosomiasis in Zanzibar: a cross-sectional survey. Parasites Vectors 9(1):5

Kusel J, Hagan P (1999) Praziquantel--its use, cost and possible development of resistance. Parasitol Today 15(9):352

Lewis FA, Tucker MS (2001) Schistosomiasis. Adv Exp Med Biol 766(6):47

Lu Y, Xu M (1998) Experimental observation of transdermal delivery of different dosage forms of praziquantel. Chin J Schisto Control 4:202

Lv YQ, Feng GS (2016) Common sample size estimation methods in medical research. Chron Pathematol J 4:359–361

Lv S (2017 May 11) China-Africa collaboration in schistosomiasis control. http://eliminateschisto.org/repository/media/1510590334-5a09c77e8441d-853731.pdf

Michuzi MI (2014 May 24) Ministry of health Zanzibar signs mou with China to implement a programme to elimanate schistosomiasis in Geneva. https://issamichuzi.blogspot.com/2014/05/ministry-of-health-zanzibar-signs-mou.html

Midzi N et al (2008) Efficacy and side effects of praziquantel treatment against *Schistosoma haematobium* infection among primary school children in Zimbabwe. Trans R Soc Trop Med Hygiene 102(8):759–766

Ming-Gang C (2005) Use of praziquantel for clinical treatment and morbidity control of schistosomiasis japonica in China: a review of 30 years' experience. Acta Trop 96(2):168–176. https://doi.org/10.1016/j.actatropica.2005.07.011

Molyneux DH, Savioli L, Engels D (2017) Neglected tropical diseases: progress towards addressing the chronic pandemic. Lancet 389(10066):312–325. https://doi.org/10.1016/S0140-6736(16)30171-4

Olds GR et al (1996) Immunity and morbidity in schistosomiasis japonicum infection. Am J Trop Med Hygiene 55(5 Suppl):121–126

Organization TWH (2014 May 21) WHO signs memorandum of understanding with China and Zanzibar for collaboration on schistosomiasis elimination in Zanzibar. http://www.who.int/neglected_diseases/schistosomiasis_china_zanzibar/en/

Organization WH (2002) Prevention and control of schistosomiasis and soil-transmitted helminthiasis: report of a WHO expert committee

Organization WH (2014) Schistosomiasis fact sheet from WHO providing key facts and information on transmission, epidemiology, symptoms, diagnosis, prevention and control, WHO response

Organization WH (2015) Schistosomiasis: number of people treated worldwide in 2013. Releve epidemiologique hebdomadaire 90(5):25–32

Organization WH (2016) Schistosomiasis: number of people treated worldwide in 2014. Releve epidemiologique hebdomadaire 91(5):53–60

Reich MR, Govindaraj R, Dumbaugh K, Yang B, Brinkmann A, Elsaharty S (1998) International strategies for tropical disease treatments: experiences with praziquantel. Harvard Initiative for Global Health—HIGH

Stelma FF et al (1995) Efficacy and side effects of praziquantel in an epidemic focus of *Schistosoma mansoni*. Am J Trop Med Hygiene 53(2):167

Stothard JR, Sousa-Figueiredo JC, Navaratnam AM (2013) Advocacy, policies and practicalities of preventive chemotherapy campaigns for African children with schistosomiasis. Exp Rev Anti-infective Ther 11(7):733–752. https://doi.org/10.1586/14787210.2013.811931

Tchuem Tchuente LA, Rollinson D, Stothard JR, Molyneux D (2017) Moving from control to elimination of schistosomiasis in sub-Saharan Africa: time to change and adapt strategies. Infect Dis Poverty 6(1):42. https://doi.org/10.1186/s40249-017-0256-8

Threats IMFM (2011) The causes and impacts of neglected tropical and zoonotic diseases: opportunities for integrated intervention strategies. Phys Fluids (1994-present) 23 (6):2656–2657

van Dam GJ et al (2015) An ultra-sensitive assay targeting the circulating anodic antigen for the diagnosis of *Schistosoma japonicum* in a low-endemic area, People's Republic of China. Acta Trop 141(Pt B):190–197

Webbe G, James C (1977) A comparison of the susceptibility to praziquantel of *Schistosoma haematobium*, *S. japonicum*, *S. mansoni*, *S. intercalatum* and *S. mattheei* in hamsters. Z Parasitenkunde 52(2):169–177

Xiao SH (2005) Development of antischistosomal drugs in China, with particular consideration to praziquantel and the artemisinins. Acta Trop 96(2):153–167

Xu J et al (2016) China-Africa and China-Asia collaboration on schistosomiasis control: a SWOT analysis. Adv Parasitol 92:435

Xu LL et al (2014) Efficacy and safety of praziquantel, tribendimidine and mebendazole in patients with co-infection of clonorchis sinensis and other helminths. PLoS Negl Trop Diseases 8(8): e3046

Wu WX (1994) Clinical observation of praziquantel with different doses in treating 1,627 cases of schistosomiasis haematobium. Chin J Parasitol Parasit Diseases 12(4):288–290

Wu WX, Qiu YZ (1990) The clinical analysis of schistosomiasis in Egypt treating with praziquantel: 1508 cases. Chin J Schisto Control 2:61–62

Wu ZW, Bu KM, Liu QL et al (1993) Study on the effect of praziquantel chemotherapy to reduce the prevalence and spread of schistosomiasis in Dongting lake. Chin J Schisto Control 6:325–329

Xu ZY (1988) Diagnosis and treatment of common parasitic diseases in China (6) summary of domestic praziquantel treatment of worms. Chin J Med 11:25–28

Xu J, Guo JG, Wu XH et al (2009) Efficacy and adverse effects of film coated praziquantel for treatment of schistosomiasis japonica. Chin J Prev Med 43(8):718–722

Zhang L (2013) Praziquantel market analysis. Shanghai Med Inform Res 1:41+6

Further Reading

Abdel-Wahab MF et al (1994) Is *Schistosoma mansoni* replacing *Schistosoma haematobium* in Fayoum? Am J Trop Med Hygiene 49(6):697–700

Elfakahany AF, Abdalla KF, Elhady HM, Abd elAziz SM, Afifi LM (1993) The effect of praziquantel treatment on the liver functions, worm burden, and granuloma size using two drug regimen in murine *Schistosoma mansoni* infection. J Egypt Soc Parasitol 23(3):877–886

Fakahany AFE, Fetouh K, Hadi HE, Aziz SAE, Afifi L (1993) Effect of praziquantel treatment on the liver functions, worm burden, and granuloma size using murine *Schistosoma mansoni* infection. J Egypt Soc Parasitol 23(3):877–886

King C et al (1989) Dose-finding study for praziquantel therapy of *Schistosoma haematobium* in Coast Province, Kenya. Am J Trop Med Hygiene 40(5):507–513

Mansoury ST (1997) Effect of two trematodicidal drugs on the morphology and tegumentary ultrastructure of *Schistosoma mansoni*. J Egypt Soc Parasitol 27(1):233–241

Markoski MM et al (2006) Praziquantel and albendazole damaging action on in vitro developing Mesocestoides corti (Platyhelminthes: Cestoda). Parasitol Int 55(1):51–61

Mohamed-Ali Q et al (1991) Ultrasonographical investigation of periportal fibrosis in children with *Schistosoma mansoni* infection: reversibility of morbidity seven months after treatment with praziquantel. Am J Trop Med Hygiene 44(4):444–451

Sacko M et al (2009) The effect of single dose versus two doses of praziquantel on *Schistosoma haematobium* infection and pathology among school-aged children in Mali. Parasitology 136 (13):1851–1857

Wang X-Y, He J, Juma S, Kabole F, Guo J-g, Dai J-R et al (2019) Efficacy of China-made praziquantel for treatment of Schistosomiasis haematobium in Africa: a randomized controlled trial. PLoS Negl Trop Dis 13(4):e0007238. https://doi.org/10.1371/journal.pntd.0007238

Zanzibar (n.d.) http://zanzibar.go.tz

Chapter 11
A Package of Health Education Materials: Effectiveness for Schistosomiasis Control in Zanzibar

Jian He, Da-Cheng Xu, and Racheal Nassiwa

Abstract This chapter starts by the evaluation of community's and schoolchildren's knowledge, attitude, practices (KAPs) on urogenital schistosomiasis caused by *Schistosoma haematobium* in Pemba, Zanzibar. After the baseline survey, applications of health education were also introduced. At the end, there is a study of the effectives of schistosomiasis health education sessions in primary and secondary schools.

Keywords Knowledge-attitude-practise · Health education · Behavior intervention · Zanzibar

Abbreviations

KAPs	Knowledge, Attitude and Practices
NTD	Neglected tropical disease
YLDs	Years lived with disability
WHO	World Health Organization
HIV	Human immunodeficiency virus
AIDS	Acquired immune deficiency syndrome
CASE	China Aid Schistosomiasis Elimination
PZQ	Praziquantel
MoH	Ministry of Health

J. He (✉)
Global Health Center, Jiangsu Institute of Parasitic Diseases, Wuxi, Jiangsu, China
e-mail: hejian@jipd.com

D.-C. Xu
Jintan Center for Disease Control and Prevention, Changzhou, Jiangsu, China

R. Nassiwa
Jiangnan University, Wuxi, Jiangsu, China

K. Yang, H. Mehlhorn (eds.), *Sino-African Cooperation for Schistosomiasis Control in Zanzibar*, Parasitology Research Monographs 15,
https://doi.org/10.1007/978-3-030-72165-7_11

Health education and behavior interventions have a beneficial impact on schistosomiasis control, they are recognized as not only key components of schistosomiasis control, but also good supplements of disease control strategy (Chen et al. 2016; Price et al. 2015). The experience of Schistosomiasis control for China also showed that health education and behavior interventions had a beneficial impact on disease control (Shi et al. 2015; Zhang et al. 2016). A series of trials and evaluation of health education and behavior interventions were carried out in Zanzibar, and most of them were school-based (Celone et al. 2016; Stothard et al. 2016; Stothard et al. 2006). The community was always considered as an important part of health education and behavior interventions targets (Israel et al. 1998; Kloos 1995). To improve people's schistosomiasis control awareness of people in Pemba, a series of methods were implemented by the project.

11.1 Baseline Survey of Knowledge, Attitude and Practice

At the beginning of the project, a knowledge, attitude, and practices (KAPs) survey towards schistosomiasis haematobium, along with urine sample collection among residents and schoolchildren, was implanted, it would help find schistosomiasis-related KAPs of the targeted population, provide the basis for future health education and behavior interventions strategy.

11.1.1 Methods

11.1.1.1 Study Area and Study Design

The United Republic of Tanzania was established in 1964 by Tanganyika and Zanzibar. Unguja and Pemba are two main islands of Zanzibar archipelago in the Indian Ocean, which are located to the east of mainland Tanzania. Pemba island has two provinces, which are further split into small areas called shehias, according to population and housing census in 2012, the number of shehias was 121 (Knopp et al. 2016). The area of the island is 1014 km^2, it is broken up into many small valleys with vegetation-choked watercourses and ponds; larger ponds and lakes are found in the eastern-central region of the island (Savioli and Mott 1989). It has two rainy seasons occurs every year, a long rainy season usually occurs between March–June while a short one between October–December (Ali et al. 2006). The prevalence of schistosomiasis was 6% in adults and 10% in school children respectively in 2012 (Knopp et al. 2016; Knopp et al. 2013).

Before the survey, main water sources including streams, rivers and ponds were investigated in the island, so, in this study, 29 shehias which are near the water sources were selected as study area, cause these areas could be the most endemic areas of the island (Fig. 11.1). The map shapefile map was freely downloaded from

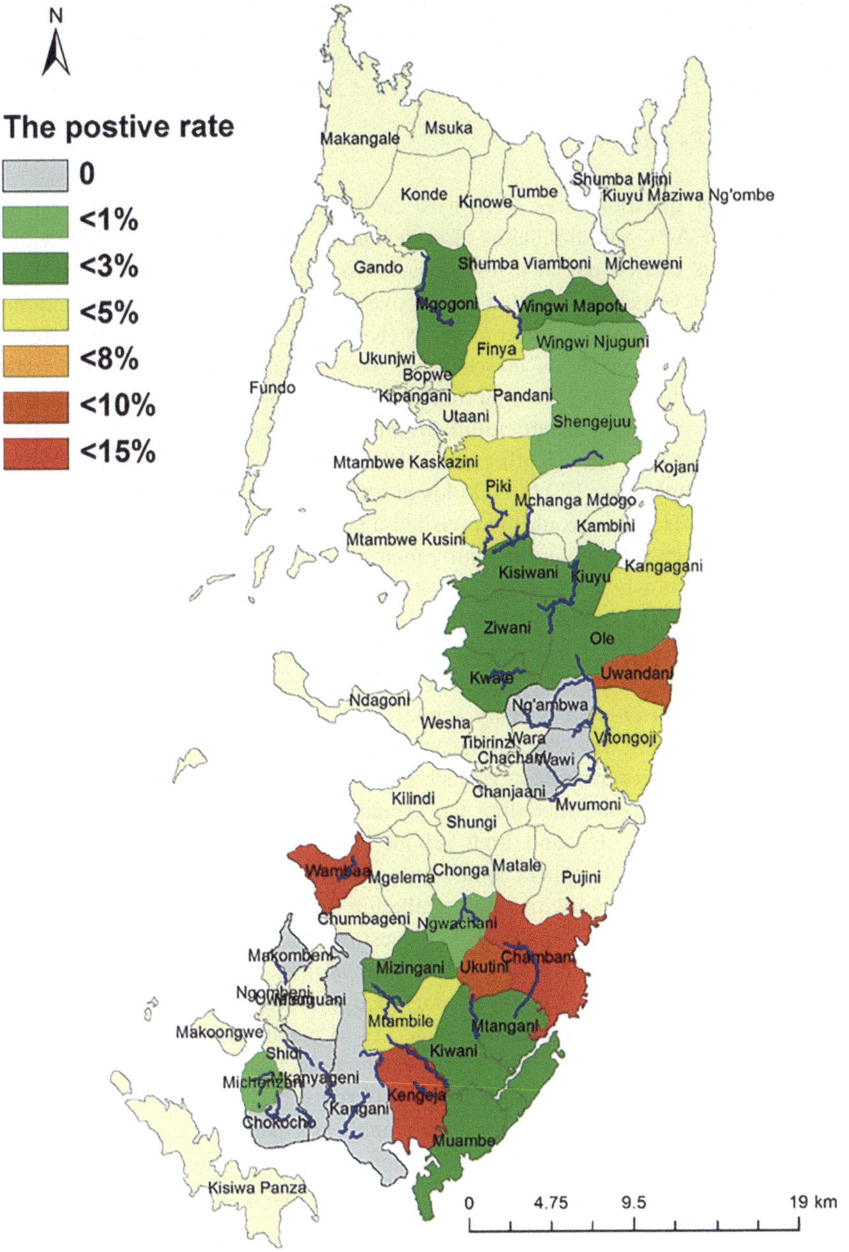

Fig. 11.1 The distribution of study region and prevalence of urogenital schistosomiasis in Pemba island, Zanzibar

https://www.cdc.gov. A cross-sectional survey was conducted among the residents and schoolchildren, including KAPs survey and urine sample collection.

11.1.1.2 KAPs Survey

Three types of questionnaires were used in the survey: Family information for households, KAPs questionnaires for residents and students separately. A pre-survey was carried out in a few households before the survey, the feedback from the interviewers was collected to improve the quiz for an easier understanding of them. Validated bilingual questionnaires (Kiswahili and English) were designed for the households, residents and students respectively. Inclusion criteria was: (1) People who lived near the waterbodies in the villages, (2) The livelihood of them were mostly centred on agriculture, livestock keeping and housework (washing and looking after kids etc.) (Koffi et al. 2018). Other similar studies were taken as reference for simple size, 120 households in each shehia were investigated (Knopp et al. 2012). Households would be revisited again if they weren't available in the previous day. In each family, one (family members < 4) or two (family members ≥ 4) family members whose age over 12 were questioned about KAPs separately, the investigator used a random number table (range from 1 to 10, ship the random number if it was bigger than the number of family members) to choose the interviewees. In this part, if the interviewees were still students, they would be excluded. Five shehias didn't contain primary school, and in the rest shehias, schools had largest pupil members was selected in each place, so 21 schools were taken into the study, and one classroom students of each school were selected, which were chosen by random number table, including the primary school ($6 \leq$ age ≤ 12), middle school ($12 \leq$ age ≤ 15) and high school ($15 \leq$ age ≤ 18). The household questionnaire was used to collect some socioeconomic information including demographic information, family income, sanitation facilities, family's past medical history of schistosomiasis, etc. The KAPs questionnaire contained medical records, including past medical history and history of treatment, knowledge about schistosomiasis, its host and transmission, and knowledge sources. In attitude and practice part, attitude toward disease prevention, access to praziquantel (PZQ) and doctor, urinating in ponds or rivers and other water contact activities like swimming, washing and passing were questioned. The difference between the KAPs questionnaire of residents and students was the personal information, the students need to write their village and household name to find their family information. A unique code for the interviewees contained shehia, village, house and individual information.

Trained research assistants and staff from the Ministry of Health (MoH) NTD office collected information from residents via face-to-face interviews. For students, the interview was in the classroom, students accompanied by their teachers, the NTD staffs collected the data in the classroom.

11.1.1.3 Parasitology

The urine samples collection was implemented after the KAPs survey, based on other similar reference, 70 residents who received KAPs survey before in each shehia and 30 students received KAPs survey in each school were selected by random number table (Knopp et al. 2012). Urine samples were collected from the participants in 100 ml clean, labelled, and screw-capped plastic containers. The samples were usually collected between 9–11 a.m. and stored in suitable cool boxes (4–6 °C), then, they were delivered to the laboratory in the afternoon. Visible hematuresis was checked by experienced laboratory testers, diagnose of schistosomiasis haematobium were checked by filtration method under the microscope, and for the positive urine samples, the number of eggs was accounted (Wynn et al. 1993). Ten percent of the samples were selected randomly and re-examined by another staff to confirm the results.

The family information, KAPs and urine test results were checked again for the association study, records which had wrong or missing interviewee code would be excluded. In this part, we aimed to get an interviewee' family information, KAPs about the disease and urine test result, which can be used to discuss the relationship between them.

11.1.1.4 Data Analysis

Data from the questionnaire and laboratory examinations were entered into Epidata version 3.1 (Atlanta, USA). Microsoft Office Excel 2016 (Microsoft, Redmond, WA) and SPSS Version 24.0 (IBM Corp, Armonk, NY) were used for data management and analysis. Descriptive analysis was used to display the KAPs in residents and students, and Chi-square test was used to compare the difference of percentage between them, while binary logistic regression was used to find risk KAPs factors of the disease.

The quality control of our data rooted in standardization of methods and documentation: interviewers were ensured comprehending the meaning of the questionnaire through training, the questionnaire data and lab examination result were traceable, double data entry by NTD staffs ensured accuracy of data.

11.1.1.5 Ethical Consideration

The study was carried out according to the guidelines laid down in the Declaration of Helsinki and all procedures involving human subjects were approved by the MoH of Zanzibar (reference number: ZAMREC/002/MAY/014). Before the interview, information about the objectives of the study and the role of the participants were given to the participants or their parents and guardians. When students doing the interview, they were accompanied by their teachers and headmasters, teachers and students were all told about the interview, with the permit of the teacher, the interview would carry on. Moreover, both residents and students were informed that their participation was voluntary and that they could withdraw from the study at any stage without any consequences. After the urine test, in the next day, the infected residents and students were treated with a single dose of 40 mg/kg body weight praziquantel tablet by staffs of NTD office.

11.1.2 Results

11.1.2.1 Demographic Characteristics at Household and Individual Level

The mean prevalence of schistosomiasis was 2.4% in these 29 shehias. The range of the prevalence was 0–13.5%. Three thousand two hundred twelve households were face-to-face interviewed in the study to get their family information, 64.7% (2078/3212) of the families had a low per capita income every month (< TSh 70,000; US$1 = TSh 2255). Only 49.2% (1580/3212) of the families had a permanent toilet, which meant less than half of study families had a fixed place to release themselves. There were 3526 valid residents KAPs questionnaires and valid 1590 students KAPs questionnaires were collected. The average age of residents and students was 42.36±15.54 and 13.5±2.64 respectively. 37.8% (601/1590) of students were under 12 years old. The proportion of female was 64.6% (2278/3526) and 56.0% (890/1590) among residents and students respectively. Due to some missing visits, 1945 (95.8%) residents' and 550 (87.3%) students' urine test results were recorded, there were 40 of residents who received urine test ($n = 1945$) and 19 of students who received urine test ($n = 550$) infected with *S. haematobium*. The disease prevalence of students (3.4%) was higher than residents (2.0%, $p = 0.02$). The prevalence in male students (6.0%) was higher than the female (1.6%, $p = 0.04$).

11.1.2.2 Past Medical Record

Form the KAPs questionnaires, the reported prevalence of schistosomiasis in residents ($n = 3526$) and students ($n = 1590$) were 19.9% and 9.3%, respectively. In the last 2 years, the reported history of PZQ treatment were all more than 90% in both residents (95.5%) and students (98.7%).

11.1.2.3 Knowledge about Schistosomiasis

The results of knowledge were shown in Table 11.1. 75.4% of residents (2657/3526) and 56.4% of students (897/1590) heard about schistosomiasis. Most residents (2762/3526) and students (1405/1590) knew the infected way was contacting with

Table 11.1 Knowledge, attitude and practices of schistosomiasis among residents and students in Pemba, Zanzibar

	Residents number (%)	Students number (%)	Total number (%)
Have you heard about schistosomiasis? (yes)	2657 (75.4)	897 (56.4)	3552 (69.5)
Do you know snail? (yes)	1306 (37.0)	507 (31.9)	1813 (35.4)
How about the infected way?			
Contact with sick people	5 (0.1)	25 (1.6)	30 (0.6)
Eating undercooked fish	15 (0.4)	16 (1.0)	31 (0.6)
Contact with infected water	2762 (78.3)	1405 (88.4)	4167 (81.5)
Don't know	744 (21.1)	144 (9.1)	888 (17.4)
What are the sources of your schistosomiasis–related knowledge?			
Doctor	1058 (30.0)	261 (16.4)	1319 (25.8)
School	1037 (29.4)	1240 (78.9)	2277 (44.5)
Never heard	417 (11.8)	135 (8.5)	552 (10.8)
Do you think schistosomiasis could be prevented?	2488 (70.6)	1278 (80.4)	3766 (73.6)
Do you follow doctor's advice?	3476 (98.6)	1562 (98.2)	5038 (98.5)
Do you follow praziquantel treatment?	3491 (99.0)	1552 (97.6)	5043 (98.6)
Do you pee in ponds?	1601 (45.4)	1064 (66.9)	2665 (52.1)
How do you always contact ponds or rivers?			
Swimming	818 (23.2)	660 (41.5)	1478 (28.9)
Washing	1213 (34.5)	594 (37.3)	1807 (35.3)
Passing across	1433 (40.6)	473 (29.7)	1906 (37.3)
No contact	654 (18.5)	285 (17.9)	939 (18.4)

infected water. However, only 37.0% of residents and 31.9% of students knew that *Bulinus* snails were the intermediate hosts of *S. haematobium*. 21.1% of residents and 9.1% of students still didn't know the route of infection. 30.0% of residents received schistosomiasis knowledge from doctors and 78.9% of students the school.

11.1.2.4 Attitude and Practice Towards Schistosomiasis

70.6% of residents and 80.4% of students thought the schistosomiasis could be prevented. Nearly all residents (98.6%) and students (98.2%) were willing to follow the doctor's advice. 99.0% of residents and 97.6% of students consented to receive PZQ treatment. 45.4% of residents and 66.9% of students still had the act of urinating in ponds or rivers. Among three water contact behaviors (swimming, washing and passing across), the main behaviors of contacting water were passing across (40.6%) and swimming (41.5%) for residents and students respectively.

11.1.2.5 Association of Schistosomiasis Knowledge with Demographic and Socioeconomic Factors

One thousand seven hundred forty-four (89.6%) of residents and 523 (95.1%) students were included in the association study, the results were showed in Table 11.2. For residents, gender, education level and age were important factors, and male had more knowledge of schistosomiasis, the percentage of male and female knowing *Bulinus* snails as intermediate hosts were 51.4% and 29.5%, respectively. 73.2% of uneducated residents (people neither have been received primary school education nor higher education) had knowledge of schistosomiasis compared with 85.6% of educated residents (people at least have been received primary school education). The percentage of residents under and over 40 years old knowing *Bulinus* snails were 29.6% and 43.0%, respectively. For students, the age and average income of family were two factors of schistosomiasis knowledge. The percentage of students over 12 years who knew schistosomiasis, *Bulinus* snail, and the infected way were 92.3%, 38.4% and 96.6%, respectively. The percentage of students in low-income family knowing schistosomiasis, *Bulinus* snail, and the infected way were 82.9%, 20.9% and 87.1%.

11.1.2.6 Association of Attitude and Practices with Demographic and Socioeconomic Factors

For residents, the attitude of schistosomiasis could be prevented differed in gender, age, past medical history and education level. Male (80.6%), residents whose age was under 40 (78.5%), residents hadn't infected before (78.1%), educated residents (83.1%) more likely had the positive attitude.

The percentage of urinating in ponds was higher among residents under 40 (60.5%) and with infection history (58.1%). The male would like swimming more than women (23.8% vs. 15.2%), while female did washing more than male (47.8% vs. 39.6%). Resident who had a schistosomiasis history (55.0%) more likely contacted with water bodies.

For students, elder students (90.3%) and students in rich families (96.5%) more likely had the attitude of schistosomiasis could be prevented. Boys (88.3%) and young students (85.1%) were more likely urinating in ponds. The percentage of students in poor families were more likely going swimming (47.3%), washing (68.1%) and passing across water bodies (71.9%).

11.1.2.7 Logistic Regression of Factors Associated with KAP

The result of the binary logistic regression of factors associated with KAPs variables among residents showed that age was the most important factor. Residents over 40 years old had more knowledge about *Bulinus* snail (Exp(B) = 1.9, 95%CI,

Table 11.2 Association of knowledge, attitude, practices about schistosomiasis with demographic characteristics and socioeconomic factors among residents and students in Pemba, Zanzibar

		Heard of schistosomiasis		Know *bulinus* snail		Know infected way		Schistosomiasis could be prevented		Pee in ponds		Swimming		Washing		Passing across	
		Number (%)	OR	N (%)	OR	N (%)	OR	N (%)	OR	N (%)	OR	N (%)	OR	N (%)	OR	N (%)	OR
Residents																	
Gender	Male	448 (85.3)	1.6**	270 (51.4)	2.5**	469 (89.3)	2.1**	423 (80.6)	1.46**	284 (54.1)	0.9	125 (23.8)	1.7**	208 (39.6)	0.7**	280 (53.3)	1.1
	Female	961 (78.8)		359 (29.5)		975 (80.0)		908 (74.0)		703 (57.7)		185 (15.2)		583 (47.8)		635 (58.1)	
Age	≥40	664 (79.0)	0.8*	361 (43.0)	1.8**	677 (80.6)	0.7*	615 (73.2)	0.7**	440 (52.4)	0.7**	185 (22.0)	1.7**	353 (42.0)	0.8**	436 (51.9)	1.0
	<40	745 (82.4)		268 (29.6)		767 (84.8)		710 (78.5)		547 (60.5)		125 (13.8)		438 (48.5)		479 (53.0)	
Past medical history	Not infected	1072 (78.8)	0.4**	483 (35.5)	0.8	1118 (82.2)	0.7*	1062 (78.1)	1.5**	790 (58.1)	1.3*	242 (17.8)	1.0	636 (46.7)	1.3*	748 (55.0)	1.6**
	Infected before	333 (91.2)		145 (39.7)		315 (86.3)		257 (70.4)		190 (52.1)		66 (18.1)		150 (41.1)		158 (43.3)	
Education	Unschooled	495 (73.2)	0.5**	222 (32.8)	0.8*	492 (72.8)	0.3**	438 (64.8)	0.4**	375 (55.5)	0.9	124 (18.3)	1.0	295 (43.6)	0.9	336 (49.7)	0.8*
	Educated	914 (85.6)		407 (38.1)		952 (89.1)		887 (83.1)		612 (57.3)		186 (17.4)		496 (46.4)		579 (54.2)	
Average income	<70,000TSh	911 (81.4)	1.1	413 (36.9)	1.1	931 (83.2)	1.1	862 (77.0)	1.2	615 (55.0)	0.8*	211 (18.9)	1.2	495 (44.2)	0.9	588 (52.5)	1.0
	≥70,000TSh	498 (79.7)		216 (34.6)		513 (82.1)		463 (74.1)		372 (59.5)		99 (15.8)		296 (47.4)		327 (52.3)	
Students																	
Gender	Boys	197 (88.3)	1.2	77 (34.5)	1.3	200 (89.7)	0.7	197 (88.3)	1.1	197 (88.3)	2.6**	116 (52.0)	2.6**	111 (49.8)	0.7*	142 (63.7)	1.2
	Girls	260 (86.7)		86 (28.7)		277 (92.3)		261 (87.0)		224 (74.4)		88 (29.3)		179 (59.7)		178 (59.3)	
Age	≤12	135 (77.6)	0.3**	29 (16.7)	0.3**	140 (80.5)	0.1**	143 (82.2)	0.5**	148 (85.1)	1.6*	74 (42.5)	1.2	91 (52.3)	0.8	101 (58.0)	0.8
	>12	322 (92.3)		134 (38.4)		337 (96.6)		315 (90.3)		273 (78.2)		130 (37.2)		199 (57.0)		219 (62.8)	
Past medical history	Not infected	429 (87.2)	0.3	152 (30.9)	0.7	447 (90.9)	0.9	434 (88.2)	1.8	395 (80.3)	0.7	191 (38.8)	1.0	274 (55.7)	0.9	305 (62.0)	1.4
	Infected before	25 (96.2)		10 (38.5)		26 (100.0)		21 (80.8)		22 (84.6)		10 (38.5)		15 (57.7)		14 (53.8)	
Average income	<70,000TSh	218 (82.9)	0.4**	55 (20.9)	0.4**	229 (87.1)	0.3**	207 (78.7)	0.1**	197 (74.9)	0.1**	81 (30.8)	0.5**	113 (43.0)	0.4**	133 (50.6)	0.4**
	≥70,000TSh	239 (91.9)		108 (41.5)		248 (95.4)		251 (96.5)		224 (86.2)		123 (47.3)		117 (68.1)		187 (71.9)	

$^{*}p < 0.05$, $^{**}p < 0.01$

1.5–2.4, $p < 0.01$), but they were more likely to swim (Exp(B) = 1.6, 95%CI, 1.2–2.1, $p < 0.01$). Residents under 40 years old were more likely to urinate in ponds (Exp(B) = 0.7, 95%CI, 0.6–0.8, $p < 0.01$), however, less of them thought that schistosomiasis could be prevented (Exp(B) = 0.6, 95%CI, 0.4–0.7, $p < 0.01$).

For students, family income and gender were important factors. Students who did not know *Bulinus* snails (Exp(B) = 0.5, 95%CI, 0.3–0.7, $p < 0.01$) and schistosomiasis could be prevented (Exp(B) = 0.4, 0.3–0.6, $p < 0.01$) were more likely living in a relative poor family (average income < 70,000 Tsh/month). Male students were more likely going swimming (Exp(B) = 2.2, 95%CI, 1.5–3.3, $p < 0.01$) and peeing in ponds (Exp(B) = 2.1, 95%CI, 1.2–3.5, $p < 0.01$).

11.1.2.8 Association of the Prevalence of Schistosomiasis with KAPs

The residents have a high frequency of passing water bodies had a higher prevalence of schistosomiasis (2.7% vs. 1.3%, $x^2 = 4.2$, $p < 0.05$). The students who had more knowledge had lower prevalence of schistosomiasis (2.5% vs. 8.7%, $x^2 = 5.4$, $p < 0.05$).

11.1.3 Discussion

Although PZQ administration has a clear effect of reducing the infection prevalence, limitations like reinfection, poor drug coverage and infrequent monitoring and evaluation, call for integrated methods to enhance control efforts and reduce prevalence and intensity of infection further (Guidi et al. 2010; Ross et al. 2014). If an endemic area moves towards an elimination goal, implement of health education and behavior interventions will be needed (Evan Secor 2014). This study was a survey of KAPs among local people, which is associated with health education and behavior interventions strategies (Chitsulo et al. 2000; Sturrock 1995). Compared to other studies of KAPs, health education or behavior interventions in Zanzibar, this study expands the interviewee population, including schoolchildren and residents in the community (Person et al. 2016; Stothard et al. 2006). Besides the KAPs information, urine samples and socioeconomic information were also collected. So the potential factors could affect people's KAPs and KAPs influence on the disease prevalence could also be discussed.

In the study, the disease prevalence of students was higher than residents, it seemed that this population had a higher risk of schistosomiasis infection, which indicated that schoolchildren may be the research focus in the disease control (Hatz et al. 1998; Magnussen et al. 1997; Nduka et al. 1995; Traoré et al. 1998). School boys were more likely infected by the disease, this finding was consistent with another study in Zanzibar (Person et al. 2016). People's knowledge of schistosomiasis was still insufficient after a series schistosomiasis control program in Pemba (Savioli et al. 1989; Savioli et al. 1990), more than 25% of residents and nearly half of the students did not have knowledge about schistosomiasis, although they had heard about the disease. Furthermore, few residents and students knew the intermediate host-*Bulinus* snails, which may let them ignore the risk of contacting water bodies containing infected snails. Male residents seemed to have a better knowledge of the disease than female, the result is consistent with a similar study in Africa, and strengthening the health education for females and uneducated peoples women will be urgent (Dawaki et al. 2015; Gazzinelli et al. 2006; Mwai et al. 2016; Sady et al. 2015). Middle and high school students had more knowledge, and current educational methods may need to be changed for younger students (Stothard et al. 2006, 2016). As with other studies, we also thought that more acceptable health education materials with local culture should be designed and provided (Stothard et al. 2016).

Education level, as well as gender, age, past medical history were all factors which had an impact on attitude towards the disease in residents group, and this indicated that school played an important role in health-promoting (Freudenthal et al. 2006; Montresor et al. 1999; Suzuki et al. 2005). So, community-based health education and behavior change should be established, which would help unschooled people get more engagement in health education. The practice of different gender among students was consistent with other studies, female student did more washing while male more likely went swimming, the health education focus on practice should be changed due to different gender (Person et al. 2016).

MDA is considered highly cost-effective in schistosomiasis and soil-transmitted helminthiasis (Lo et al. 2015). High-risk behaviors were common among both residents and students, some alterable water contact activities like urinating in the open waterbody were quite common in both residents and students, swimming which had a high rate in students, these activities can be changed by health education and behavior interventions. Unfortunately, passing across and washing in the open waterbody were hard to change due to the local poor living environment, these two activities were major water contact activities among residents. We observed that the road and low-lying areas were easily flooded in the rain season, some roads would become a temporary stream or river where people needed to pass or walk along, this would add the risk of infection. Females always did the most of domestic washing in open water bodies, due to the insufficient supply of safe water source, and could be infected by the disease (Downs et al. 2011).

The logistic regression results showed some important factors had an impact on KAPs. More efforts in knowledge education are needed for young adults and residents with infection history, which may raise their awareness of the disease. Disseminating the harms of swimming to elderly residents and change the habit of peeing in the open waterbody among young residents. What's more, we need to pay more attention to risk practice like washing and passing the waterbody, which could be an important part of behavior change in the future.

The study showed that practice and knowledge could impact the disease prevalence among residents and students. So, the behavior interventions including the habits changing of passing the river among residents and improvement of the schistosomiasis knowledge education among students will help to reduce the incidence of the disease.

This study showed that KAPs were at a low level among residents and students, more health education and behavior interventions should be involved based on the local living environment and their living habits. The protection equipment can be provided, like rubber boots, gloves and anti-cercariae cream, which encircling the skin and block the process of cercariae drilling into the shin. Health education is still needed in the school, explore an easy-understanding health promotion materials especially for primary school students.

11.1.4 Conclusion

In this study, we conducted a baseline survey of KAPs in Pemba island, Zanzibar, and found that both residents and students had insufficient knowledge on schistosomiasis, and risk behaviors were common in these two population. Influence factors are quite different between residents and students. We showed that age, gender, income level, medical history had different influence on KAPs among residents and students respectively. A more flexible, localized, different population based health promotion plan for different population is needed to eliminate the disease in the island.

11.2 Health Education Methods

After the baseline survey, the project implemented some health education trails to fill the schistosomiasis knowledge gaps and change some unsafe behavior for both schoolchildren and residents.

Before implementing the health education activities, the project prepared some health education materials, including health education paper materials, stationery, daily use goods, etc. Paper materials were widely used in health education activities due to its low cost and high flexibility, brochures and posters were the main forms of them. The English and Swahili version were both prepared, which could help young children and low education level residents to understand the knowledge easier. Stationery include schoolbags, pens, pencil bags, notebooks etc., pictures and schistosomiasis knowledge were printed in these goods. Swahili was the first language choice due to the targeted population are students. Daily use goods which had schistosomiasis control knowledge printed were intended to improve the life quality of people in Pemba, some of them could help people to change the dangerous behavior of getting infected by *schistosoma*. The project prepared bags, buckets, slippers and battery-free torches etc. These goods were very popular among the people in Pemba since they are very practical in people's life.

11.2.1 School Children

According to the baseline survey results, students had a higher risk of get infected by *schistosoma* than residents, and their insufficient schistosomiasis knowledge demanded more health education activities. The project made some health education activities for the students according to the local situation, which can help students to understand the knowledge easier.

11.2.1.1 Health Education Class

Schistosomiasis health education classes are the most common way to improve the knowledge of students. The classes were usually implemented by the specialist and school teachers to the students. Education videos, posters, brochures and snail samples were usually used in the class. Education videos in Swahili was usually played at the beginning of the class. It can attract students attention quickly through the vivid pictures and video, and a brief impression of what schistosomiasis is, the life cycle of *schistosoma* were showed to them. After the video play, more specific knowledge were classified by the specialists, and the teachers helped the students to understand the knowledge easier by explaining the complex part of the class. After the theory teaching, example explaining like the snails sample teaching could let students to learn teaching contents through operation, which can enhance students' awareness of *schistosoma* infection. At the end of the class, health education stationery were given to the students, which was also considered as a souvenir for them.

Health education class was the main health education activity for students, due to its low degree of difficulty to conduct, and the class content could be flexible according to students' feedback (Fig. 11.2).

11.2.1.2 Health Promotion Competition

Two kinds of health promotion competition have been carried out, one is schistosomiasis knowledge composition competition, the other is schistosomiasis control poster competition.

The schistosomiasis knowledge composition was held by the project, Ministry of Health and Ministry of Education in Zanzibar. At first, systematic health education courses were carried out in all primary and secondary schools in the pilot and demonstration area. In order to further deepen local students' cognition and understanding of schistosomiasis control knowledge, then the first "schistosomiasis control cup" composition competition was implemented in these schools. This movement was warmly welcomed by local schools and teachers and students. Students actively participated in composition writing. In the end, 510 competition compositions were received, including 195 English composition and 315 Swahili

Fig. 11.2 Students showing the health education pens in the health education class

ones. Six winning compositions and three excellent organization schools were chosen.

The schistosomiasis control poster competition was intended to improve the schistosomiasis knowledge among the students. The project and Ministry of Health in Zanzibar conducted the competition, Ministry of Education, and other related departments like District Administrator, Zanzibar Water Authority, Ministry of Environment were also engaged in this activity. Shehias where the prevalence of schistosomiasis was more than 1% were taken into the competition. The procedure of poster competition was initial meeting, poster collection, multi-department posters review and reward meeting. The following is the competition protocol:

11.3 Schistosomiasis Poster Competition Protocol

To enhance the schistosomiasis health promotion among schools in Pemba island, China Aid Project of Schistosomiasis Control, Ministry of Health/NTD office decide to conduct a poster competition to assess the knowledge or understanding of schistosomiasis haematobium. Primary school students will be enrolled in this competition, yet in some shehias where schistosomiasis is an endemic.

Table 11.3 List of competition shehias and schools

S.no.	Shehia	School
1	MTANGANI	MTANGANI, CHWAKA
2	WINGWI MTEMANI	MTEMANI
3	WINGWI MJANANZA	MJANANZA
4	KIUYU	MJINI KIUYU, KIUYU
5	KANGAGANI	KANGAGANI
6	OLE	OLE
7	ZIWANI	ZIWANI, PEMBENI
8	KISIWANI	KISIWANI
9	PIKI	PIKI
10	FINYA	FINYA
11	MGOGONI	MGOGONI
12	CHAMBANI	CHAMBANI
13	UKUTINI	UKUTINI
14	VITONGOJI	VITONGOJI
15	WAMBAA	WAMBAA
16	KIWANI	JUNDAMITI, TASINI
17	UWANDANI	UWANDANI
18	KWALE	KWALE
19	KIZIMBANI	KIZIMBANI, IKRA
20	MTAMBILE	MTAMBILE
21	KENGEJA	KENGEJA
22	MWAMBE	MWAMBE
23	MGELEMA	MGELEMA
24	PUJINI	KUMVINI, DODO
25	PANDANI	PANDANI
26	KILINDI	KILINDI
27	MGAGADU	MIZINGANI
28	CHUMBAGENI	CHUMBAGENI
29	SHUMBA VYAMBONI	SHUMBA VYAMBONI
30	KINYASINI	ISTIQAMA, KINYASINI
31	MJINI OLE	KANGAGANI
32	BOPWE	BOPWE
33	MBUZINI	MBUZINI
34	MAKANGALE	MAKANGALE
35	CHANJAANI	CHANJAANI

11.3.1 Participants

The list of shehias and schools are showed in the Table 11.3. Two posters are supposed to make by students in each school, and the maximum number authors of one poster should be no more than 2. The age of the students should be more than 10 years.

11.3.2 Poster Requirements

11.3.2.1 Subject and Contents

- The knowledge of schistosomiasis control knowledge.
- The health impact of the disease.
- The way of disease transmission patterns and prevention.

The subject and contents are supposed to be clear and easy to understand.

11.3.2.2 Making Requirements

Handwritten text and hand-drawn illustrations are used in layout making, clipart images are not allowed to use. Typos and wrongly written should not be appeared in the poster. Both English and Swahili can be used. The size of poster is A3. Color pencils (illustration) and black gel ink pen (text) are used for making posters. Lower right corner of the reverse side of the poster should write: name of shehia, name of school, class, author names and teacher names. The poster should be given to NTD office before the cut-off date of Poster making.

11.3.3 Rewards

Number of the first prize: five works; Number of the second prize: ten works; Number of the third prize: 20 works. We will also set ten organization awards. The award-winning works will get certificate and prizes. Every participant will get souvenir.

11.3.4 Funding

China Aid Project of Schistosomiasis Control will provide financial support and equipment for this competition.

11.3.5 Competition Timetable

September 2019–8 October 2019: Preparation and starting the competition.

10 October 2019–1 November 2019: Poster making.

1 November 2019–28 November 2019: Evaluation of poster.

29 November 2019: Awarding Ceremony.

11.3.6 Contact Information

Name: **********.

Name: **********.

After the competition, books of schistosomiasis control collection was spread to schools in Pemba. These books have some advantages of health education: the posters were combined by the pictures and text, they were more attractive to young students, besides, due to the texts were written by the students, which could be easier for students to understand the schistosomiasis knowledge (Figs. 11.3 and 11.4).

11.3.7 Residents

To fill the gap of insufficient knowledge and risk behavior in the community, the project conducted some health education activities to help the community people to gain more knowledge and change risk behaviors. Community health promotion conference was the main type to deliver the knowledge to the community. Behavior intervention activities were mainly learned from Chinese successful schistosomiasis,

Fig. 11.3 Giving the rewards to the award winning students of schistosomiasis control poster competition

Fig. 11.4 A poster work of schistosomiasis control poster competition

slogans were painted in the communities and risk areas, a new type of voice alarm machine was also used.

11.3.7.1 Community Health Promotion Conference

Health education conference, the conduct method was almost the same as the class in school. In the conference, schistosomiasis knowledge was taught by the specialist and community doctor, and residents were organized by the community leaders. At the end of the class, health education daily use goods would be given to the attendants. Sometimes, to expand the number of attendants, the project would hold the class with China Medical Team and health department in Zanzibar. When the residents attend this kind of health education conference, they will receive the basic healthcare like blood pressure measuring and health consultation, this kind of basic healthcare could attract more residents. With the help of local health department, it was easier to arrange the activity (Fig. 11.5).

11.3.7.2 Community Slogan Painting

In China, using slogans is a good way to advocate schistosomiasis control knowledge to residents, due to the obvious painting and easy-remembered words of them (Wang et al. 2009). Before the project initiating this activity, a survey of residents'

Fig. 11.5 Advocate schistosomiasis control knowledge to community leaders in the community health promotion conference

willing to accept this activity was implemented. By holding the community meeting, the project explained the reason and advantages of painting slogans in the community. With the approval of the community and residents, places where can paint the slogans were chosen. To expand the influence of slogans, they will be painted in the entrance of community, places near the mosque, schools et al. The language of slogans was Swahili (Fig. 11.6).

11.3.7.3 Risk Area Alarming

To mark the risk areas of *schistosoma* infection, Alarming board and voice alarm machine were placed near the ponds where found infected snails before. Alarming board is a kind of passive way to inform people the dangerous waterbody. On the board, it contained the knowledge of schistosomiasis and pointed out the waterbody was dangerous. Voice alarm machine was a solar powered machine, it contained solar panels, sensor and voice announcer. When people were close to the risk waterbody, it would automatically broadcast warnings to them, which telling them the waterbody had *schistosoma* in it, people would get infected when contacting them directly. By inform the risk area, it can change the residents' risk behaviors (Figs. 11.7 and 11.8).

Fig. 11.6 A slogan of schistosomiasis control in the community

11.4 The Evaluation of the Effect of Health Education

This evaluation carried out schistosomiasis health education sessions for students in primary and secondary schools in three Shehias of Pemba Island, Zanzibar, as summarized in Xu et al. (2019). Through the pre- and post-health education, the effects of health education sessions and its feasibility were evaluated in order to find the weak links and, provide the basis for correcting the content and mode of health education.

11.4.1 Methods

11.4.1.1 Location and Population

All students in all primary and secondary schools in the Shehias of Mtangani (south), Kiuyu (central), and Wingwi (north), in Pemba island, Zanzibar. This schistosomiasis health education session covered five primary and three secondary schools in Mtangani, Kiuyu and Wingwi, Pemba Island, Zanzibar. Primary schools consist of grades 1–6, and secondary schools consist of forms 1–4.

Fig. 11.7 An alarming board in the risk waterbody

11.4.1.2 Health Education Sessions

Different from other forms of health education in the form of games, this schistosomiasis health education sessions are similar to traditional health education classes, but rich in content and form. Compared to the simple knowledge in the textbook, the sessions content is more systematic, it includes the history of *S. haematobia,* the medium, the way of infection, prevention and treatment, and other knowledge of schistosomiasis prevention and control, and the form is more rich, combining video broadcasting, case explanations, questioning and answering and so on. All of this is to enrich the live teaching mode and create a learning atmosphere. The sessions were conducted at selected schools, covering all students in the school. The number of students for class is kept within 100 people. The number of classes is based on the total number of students. The class time is within 1 hour. The lectures were carried out by the local NTD staff in Swahili, before the lecture, 30–50 students were randomly selected to finish a test of schistosomiasis prevention and control. After the lecture, 30–50 students were randomly selected to accept the same test a week later. The test questions were completely consistent before and after. Through the comparison of before and after, evaluation of the effect of health education was conducted.

The effect indicators include the awareness rate and the behavioral correctness rate of schistosomiasis. The awareness rate refers to the degree of understanding of

Fig. 11.8 A voice alarm machine in the risk area

schistosomiasis in the surveyed subjects. The calculation formula is the correct knowledge of schistosomiasis/the total knowledge of schistosomiasis*100%; the behavioral correctness rate refers to the correct proportion of schistosomiasis behavior in the survey. The calculation formula is the correct number of behaviors of the respondent/the total number of behaviors of the respondent*100%.

11.4.1.3 Statistical Analysis

Using EpiData3.1 to establish a database, students' test paper was double entered and consistency tested, using SPSS19.0 statistical software for data collation and statistical analysis. The rate was compared using a chi-square test, $P < 0.05$ was considered statistically significant.

11.4.2 Result

11.4.2.1 Population Samples before and after Health Sessions

A total of 712 questionnaires were distributed in this study. Three hundred fifty-six students were tested before health education and another 356 students were tested after health education sessions, and 712 valid questionnaires were returned. The sampled students were grades 3–6 in primary schools and grades 1–3 in secondary schools. There were 163 male students (45.79%) and 193 female students (54.21%) before health education, 182 male students (51.12%) and 174 female students (48.88%) after health education. The average age was 14.65 and 14.83 years before and after the health education. The knowledge and behavior of schistosomiasis prevention and control before and after the health education are shown in Table 11.4.

The awareness rate and behavioral correctness rate before health education sessions.

Table 11.5 shows that the awareness rates of schistosomiasis among primary school and secondary school were 64.62% and 68.75%, and the behavioral correctness rate were 58.24% and 59.77%, respectively. The awareness rates of schistosomiasis among male and female was 65.64% and 67.49%, and the was 63.96% and 54.79%. Only difference of behavioral correctness rate between male and female was statistically significant ($\chi 2 = 11.895$, $P = 0.0006$).

11.4.2.2 The Awareness Rate and Behavioral Correctness Rate after Health Education

Between primary school and secondary school, the awareness rate of schistosomiasis was 79.74% and 84.70%, with the statistically significant difference, and the behavioral correctness rate was 78.30% and 81.32%. Between male and female students, the awareness rates of schistosomiasis among male and female were 82.69% and 81.61%, and the behavioral correctness rate was 82.83% and 76.58%. The difference of the awareness rate between primary school and secondary school ($\chi 2 = 11.607$, $P = 0.0007$), and the behavioral correctness rate between male and female students ($\chi 2 = 8.23$, $P = 0.0041$) were statistically significant (Table 11.6).

Table 11.4 Number of students for questionnaire before and after health education in Pemba island, Zanzibar

No	Contents	Before health education ($n = 356$)	After health education ($n = 356$)
1	Do you know schistosomiasis?		
	Yes	204	309
	No	144	46
2	What disease is schistosomiasis?		
	Respiratory disease	10	5
	Mental disease	60	52
	Intestinal infection	21	39
	Parasitic disease	260	260
3	What species cause *S. haematobium?*		
	Vescular blood flukes	196	280
	Intestinal worms	44	21
	Leechs worms	71	38
	Mosquito parasites	41	15
4	What's the intermediate host of S. *haematobium* in Zanzibar?		
	Fly	42	10
	Snail	265	335
	Mosquito	24	6
	Mended	22	5
5	Which of the following pictures is *Bulinus globosus*?		
		17	15
		109	195
		114	55
		110	90
6	Where does the *Bulinus globosus* live in?		
	Sea	15	10
	Land	29	16
	Pond or stream	293	323
7	What's the way that schistosomiasis enters the human body?		
	Nose	28	11

(continued)

Table 11.4 (continued)

No	Contents	Before health education ($n = 356$)	After health education ($n = 356$)
	Mouth	41	16
	Skin	273	327
8	What are the main symptom of S. *haematobium* infection?		
	High blood pressure (hypertension)	27	16
	Fever	17	12
	Hematuresis	298	311
	Dizzy	8	17
9	What's the way to prevent schistosomiasis?		
	Do not touch the contaminated water	208	291
	Vaccination	54	28
	Wash hands before meal or after pee	48	21
	Don't touch the infected patient	40	12
10	Can schistosomiasis patient be infected again after treatment?		
	Yes	182	231
	No	154	117
11	Do you think schistosomiasis can be treated?		
	Yes	278	297
	No	61	44
12	Do you pee in ponds or streams?		
	Yes	123	32
	No	172	317

Note: There are missing answers, so the total is not necessarily 356

11.4.2.3 Comparison of the Awareness and Behavioral Correctness Rate before and after Sessions

The awareness rates of schistosomiasis among all students were significant increasing from 66.64% to 82.16% after health education ($\chi2 = 179.279$, $p < 0.0001$). The behavioral correctness rates were also significant increasing from 58.99% to 79.78% ($\chi2 = 143.84$, $P < 0.0001$) (Table 11.7).

For the primary school, the awareness rates were significant increasing from 64.63% to 79.74% ($\chi2 = 82.0284$, $P < 0.0001$). The behavioral correctness rates were significant increasing from 58.24% to 78.30% ($\chi2 = 66.66$, $P < 0.0001$). For the secondary school, the awareness rate was significant increasing from 68.75% to 84.70% ($\chi2 = 98.238$, $P < 0.0001$), and the behavioral correctness rates were also significant increasing from 59.77% to 81.32% (566/696) ($\chi2 = 76.757$, $P < 0.0001$).

For the boy student, the awareness rates were significant increasing from 65.64% to 82.69% ($\chi2 = 110.652$, $P < 0.0001$) and the behavioral correctness rates were also significant increasing from 63.96% to 82.83% ($\chi2 = 62.561$, $P < 0.0001$). For

Table 11.5 The detail students' awareness and behavioral correctness rate of *S. haematobium* before health education in Pemba island, Zanzibar

Classification	Content	School		Chi-square test	*P* value	Gender		Chi-square test	*P* value
		Primary school (n = 182)	Secondary school (n = 174)			Male (n = 163)	Female (n = 193)		
The awareness rate	Do you know schistosomiasis?	73 (40.11%)	131 (75.29%)	5.2520	0.0219	98 (60.12%)	106 (54.92%)	0.999	0.3176
	What disease is schistosomiasis?	125 (68.68%)	135 (77.59%)			114 (69.94%)	146 (75.65%)		
	What species cause *S. haematobium*?	106 (58.24%)	90 (51.72%)			83 (50.92%)	113 (58.55%)		
	What's the intermediate host of *S. haematobium* in Zanzibar?	136 (74.73%)	129 (74.14%)			125 (76.69%)	140 (72.54%)		
	Which of the following pictures is *Bulinus globosus*?	62 (34.07%)	47 (27.01%)			43 (26.38%)	66 (34.20%)		
	Where does the *Bulinus globosus* live in?	146 (80.22%)	147 (84.48%)			137 (84.05%	156 (80.83%)		
	What's the way that schistosomiasis enters the human body?	144 (79.12%)	129 (74.14%)			124 (76.07%)	149 (77.20%)		
	What are the main symptom of *S. haematobium* infection?	149 (81.87%)	149 (85.63%)			132 (80.98%)	166 (86.01%)		
The behavioral correctness rate	What's the way to prevent schistosomiasis?	103 (56.59%)	105 (60.34%)	0.2830	0.5946	100 (61.35%)	108 (55.96%)	11.895	0.0006
	Can schistosomiasis patient be infected again after treatment?	97 (53.30%)	85 (48.85%)			97 (59.51%)	85 (44.04%)		
	Do you think schistosomiasis can be cured?	138 (75.82%)	140 (80.46%)			137 (84.05%)	141 (73.06%)		
	Do you pee in ponds or streams?	86 (47.25%)	86 (49.43%)			83 (50.92%)	89 (46.11%)		

Table 11.6 The detail students' awareness and behavioral correctness rate of *S. haematobium* after health education in Pemba island, Zanzibar

Classification	Content	School		Chi-square test	*P* value	Gender		Chi-square test	*P* value
		Primary school ($n = 182$)	Secondary school ($n = 174$)			Male ($n = 182$)	Female ($n = 174$)		
The awareness rate	Do you know schistosomiasis?	144 (79.12%)	165 (94.83%)	11.607	0.0007	164 (90.11%)	145 (83.33%)	0.498	0.4803
	What disease is schistosomiasis?	134 (73.63%)	126 (72.41%)			131 (71.98%)	129 (74.14%)		
	What species cause *S. haematobium*?	132 (72.53%)	148 (85.06%)			139 (76.37%)	141 (81.03%)		
	What's the intermediate host of *S. haematobium* in Zanzibar?	168 (92.31%)	167 (95.98%)			169 (92.86%)	166 (95.40%)		
	Which of the following pictures is *Bulinus globosus*?	93 (51.10%)	102 (58.62%)			110 (60.44%)	85 (48.85%)		
	Where does the *Bulinus globosus* live in?	162 (89.01%)	161 (92.53%)			162 (89.01%)	161 (92.53%)		
	What's the way that schistosomiasis enters the human body?	169 (92.86%)	158 (90.80%)			166 (91.21%)	161 (92.53%)		
	What are the main symptom of *S. haematobium* infection?	159 (87.36%)	152 (87.36%)			163 (89.56%)	148 (85.06%)		
The behavioral correctness rate	What's the way to prevent schistosomiasis?	147 (80.77%)	144 (82.76%)	1.835	0.1755	154 (84.62%)	137 (78.74%)	8.23	0.0041
	Can schistosomiasis patient	119 (65.38%)	112 (64.37%)			125 (68.68%)	106 (60.92%)		

(continued)

Table 11.6 (continued)

Classification	Content	School		Chi-square test	P value	Gender		Chi-square test	P value
		Primary school (n = 182)	Secondary school (n = 174)			Male (n = 182)	Female (n = 174)		
	be infected again after treatment?								
	Do you think schistosomiasis can be cured?	150 (82.42%)	147 (84.48%)			161 (88.46%)	136 (78.16%)		
	Do you pee in ponds or streams?	154 (84.62%)	163 (93.68%)			163 (89.56%)	154 (88.51%)		

the girl student, the awareness rates were significant increasing from 67.49% to 81.61% ($\chi2 = 75.493$, $P < 0.0001$); the behavioral correctness rates were significant increasing from 54.79% to 6.58% ($\chi2 = 75.541$, $P < 0.0001$).

11.4.3 Discussion

Schistosomiasis is an infectious disease affected by behavioral factors (Satayathum et al. 2006). The frequency of exposure to epidemic water is closely related to its educational level, occupation, and living habits, and is affected by local social and economic factors. The cognition of schistosomiasis is the basis, which affects people's beliefs and behaviors and is an important factor in the infection and transmission of schistosomiasis (Hu et al. 2018).

In African countries, students have a higher rate of schistosomiasis infection (Tsega et al. 2018), as well as in Zanzibar (Knopp et al. 2013). Due to the high proportion of local students, through health education, they can quickly improve their cognition and spread relevant knowledge to their peers and family members to increase the awareness rate of the entire population and have a good blocking effect on disease transmission (Bobbie et al. 2016; Guanghan et al. 2000).

In this study, we conducted the schistosomiasis health education sessions at eight schools in three Shehais in Pemba Islands, and conducted awareness and behavioral tests before and after health education. Through statistical analysis, we have more accurate understanding of cognition of schistosomiasis and health education mode among Pemba Island students.

First, despite its own efforts and the help of international organizations, the infection rate of schistosomiasis in Zanzibar (Pemba Island) has been declining, but the cognition level of schistosomiasis is still low. The pre-health education showed that the overall awareness rate and the behavioral correctness rates of students were only 66.64% and 58.99%, respectively, which is similar to previous studies in Tanzania, Cameroon, Senegal, Kenya and other countries (Folefac et al. 2018; Munisi et al. 2017; Odhiambo et al. 2014).

Second, the difference in awareness rate in pre-health education is mainly between primary and secondary schools ($\chi2 = 5.252$, $P = 0.0219$), which has no relationship with gender, but the difference is not obvious (secondary school students have higher awareness rate, but only about 68%); the difference in the behavioral correctness rate is mainly between boys and girls (the behavioral correctness rates of boys is high, but only about 63%), Not related to the type of school; It suggest that whether it is primary school or secondary school students, The overall cognitive level of students is poor, and the systematic health education training and learning should be strengthened, especially in the identification of intermediate host vesicular snails (Q5) and urinary behavior habits (Q10), the students' cognition is weaker and should be improved.

Third, through this systematic health education activity, the overall awareness rate of students has increased from 66.64% to 82.16%. The difference is significant

Table 11.7 The total students' awareness and behavioral correctness rate of *S. haematobium* before and after health education in Pemba island, Zanzibar

Classification	All			Primary school			Secondary school			Male			Female		
	Before ($n = 356$)	After ($n = 356$)	Chi-square test	Before ($n = 182$)	After ($n = 182$)	Chi-square test	Before ($n = 174$)	After ($n = 174$)	Chi-square test	Before ($n = 163$)	After ($n = 182$)	Chi-square test	Before ($n = 193$)	After ($n = 174$)	Chi-square test
The awareness rate	1898/2848	2340/2848	179.3	941/1456	1161/1456	82.028	957/1392	1179/1392	98.238	856/1304	1204/1456	110.652	1042/1544	1136/1392	75.493
The behavioral correctness rate	840/1424	1136/1424	143.8	424/728	570/728	66.66	416/696	566/696	76.757	417/652	603/728	62.561	423/772	533/696	75.541

($\chi2 = 179.279$, $P < 0.0001$), it increased of 23.29%. The awareness rate of primary school students increased from 64.63% to 79.74%, it increased of 23.38%, and the rate of secondary school students increased from 68.75% to 84.70%, it increased of 23.2%. There are similar situations in terms of behavioral correctness rate, and this suggested that the schistosomiasis health education sessions have a significant effect on the improvement of local students' cognition level of schistosomiasis, and the effect on primary and secondary school students is similar.

In view of the lack of corresponding health education materials and professional teachers in Zanzibar, it is not enough to rely on the health department's employees to carry out health education for local students. In is necessary to carry out schistosomiasis health education in Zanzibar in the future, and improve the local students' knowledge of schistosomiasis prevention and control. Next, Zanzibar health department should provide professional training for the local teachers with the education department, and they can master the knowledge of schistosomiasis. Schools should set up special health education classes, teaching the knowledge of schistosomiasis with full-time or part-time teachers, and then strengthen assessment. The health department can conduct related essay competitions, snail-survey activities and other activities every year, so that to deepen students' understanding of schistosomiasis. The health department should develop more publicity materials and products (Gazzinelli et al. 2016), through which these knowledge of schistosomiasis prevention and control will reach thousands of families. Students' health education should be constantly carried out and persevered, and the students' cognitive level should be improved through long-term indoctrination.

11.4.4 Conclusion

The KAP model is one of the modes that change human health-related behaviors. It is also a behavioral intervention theory, which divides human behavior changes into three successive stages: acquiring knowledge, generating beliefs and forming behaviors. Therefore, it takes a certain time to form an effect. Acquiring knowledge is the foundation, and health education is one of the effective ways. Children are ideal candidates because of their plasticity and high acceptability. At present, the cognitive level of schistosomiasis in students in Pemba, Zanzibar is low. Through our study (schistosomiasis health education sessions), the cognitive level of students (awareness rate and behavioral correctness rate) has been significantly changed. Therefore, this method can be recommended to be promoted throughout Zanzibar.

Ethics Approval and Consent to Participate The study was carried out according to the guidelines laid down in the Declaration of Helsinki and all procedures involving human subjects were approved by the MoH of Zanzibar (reference number: ZAMREC/002/MAY/014). Before the commencement of the study, we sought approval from local authorities, community leaders, school headmasters and teachers. We sought individual oral consent from community participants directly in face interview. The range of student's age was 6–22. For all students whose age was under 18, their oral consent was obtained from their parents or legal guardians and teachers prior to the interview, and for those whose age was above 18, the oral consents were obtained from them directly. The procedure for verbal consent was approved by the ethical committees, the reason was the high degree of illiteracy in the rural area that participants there were not familiar with dealing with forms. Verbal consents were documented and captured on the digital audio recorder in the researcher's notes. Participation was entirely voluntary, and no names were recorded. We took care to phrase the questions appropriately and respectfully.

References

Ali AI, Schulthess F, Ogol CK, Omwega CO (2006) Effect of nitrogen fertilizer level on infestations of lepidopterous stemborers and yields of maize in Zanzibar. Ann Soc entomol France 42:481–486

Bobbie P et al (2016) Community knowledge, perceptions, and practices associated with urogenital schistosomiasis among school-aged children in Zanzibar, United Republic of Tanzania. PLoS Negl Trop Diseases 10(7):e0004814

Celone M et al (2016) Increasing the reach: involving local muslim religious teachers in a behavioral intervention to eliminate urogenital schistosomiasis in Zanzibar. Acta Trop 163:142–148

Chen L, Zhong B, Xu J, Li R-Z, Cao C-L (2016) Health education as an important component in the national schistosomiasis control programme in the People's Republic of China. Adv Parasitol 92:307–339

Chitsulo L, Engels D, Montresor A, Savioli L (2000) The global status of schistosomiasis and its control. Acta Trop 77(1):41–51

Dawaki S et al (2015) The menace of schistosomiasis in Nigeria: knowledge, attitude, and practices regarding schistosomiasis among rural communities in Kano state. PLoS One 10(11):e0143667

Downs JA et al (2011) Urogenital schistosomiasis in women of reproductive age in Tanzania's Lake Victoria region. Am J Trop Med Hygiene 84(3):364–369

Evan Secor W (2014) Water-based interventions for schistosomiasis control. Pathogens Global Health 108(5):246–254

Folefac LN, Nde-Fon P, Verla VS, Tangye MN, Luma HN (2018) Knowledge, attitudes and practices regarding urinary schistosomiasis among adults in the Ekombe Bonji health area, Cameroon. Pan African Med J 29:1–9

Freudenthal S, Ahlberg BM, Mtweve S, Nyindo P, Poggensee G, Krantz I (2006) School-based prevention of schistosomiasis: initiating a participatory action research project in northern Tanzania. Acta Trop 100(1–2):79–87

Gazzinelli A, Velasquez-Melendez G, Crawford SB, LoVerde PT, Correa-Oliveira R, Kloos H (2006) Socioeconomic determinants of schistosomiasis in a poor rural area in Brazil. Acta Trop 99(2–3):260–271

Gazzinelli MF, Lobato L, Andrade G, Matoso LF, Diemert DJ, Gazzinelli A (2016) Improving the understanding of schistosomiasis among adolescents in endemic areas in Brazil: a comparison of educational methods. Patient Educ Counsel 99:1657–1652

Guanghan H, Dandan L, Shaoji Z, Xiaojun Z, Guojun C (2000) The role of health education for schistosomiasis control in heavy endemic area of Poyang Lake region, People's Republic of China. Southeast Asian J Trop Med Public Health 31(3):467–472

Guidi A, Andolina C, Makame Ame S, Albonico M, Cioli D, Juma Haji H (2010) Praziquantel efficacy and long-term appraisal of schistosomiasis control in Pemba Island. Trop Med Int Health 15(5):614–618

Hatz C et al (1998) Evolution of Schistosoma haematobium-related pathology over 24 months after treatment with praziquantel among school children in southeastern Tanzania. Am J Trop Med Hygiene 59(5):775–781

Hu GH, Xu J, Cao CL, Jin JN, Zhou XN (2018) Challenges and strategies of health education and health promotion in stage of schistosomiasis elimination. Chin J Schistosomiasis Control 30 (2):117–120

Israel BA, Schulz AJ, Parker EA, Becker AB (1998) Review of community-based research: assessing partnership approaches to improve public health. Ann Rev Public Health 19 (1):173–202

Kloos H (1995) Human behavior, health education and schistosomiasis control: a review. Soc Sci Med 40(11):1497–1511

Knopp S et al (2012) Study and implementation of urogenital schistosomiasis elimination in Zanzibar (Unguja and Pemba islands) using an integrated multidisciplinary approach. BMC Public Health 12(1):930

Knopp S et al (2013) Elimination of schistosomiasis transmission in Zanzibar: baseline findings before the onset of a randomized intervention trial. PLoS Negl Trop Dis 7(10):e2474. https://doi.org/10.1371/journal.pntd.0002474

Knopp S et al (2016) Praziquantel coverage in schools and communities targeted for the elimination of urogenital schistosomiasis in Zanzibar: a cross-sectional survey. Parasites Vectors 9(1):5

Koffi AJA, Doumbia M, Fokou G, Keita M, Koné B (2018) Community knowledge, attitudes and practices related to schistosomiasis and associated healthcare-seeking behaviours in northern Côte d'Ivoire and southern Mauritania. Infect Diseases Poverty 7(1):70

Lo NC et al (2015) Comparison of community-wide, integrated mass drug administration strategies for schistosomiasis and soil-transmitted helminthiasis: a cost-effectiveness modelling study. Lancet Global Health 3(10):e629–e638

Magnussen P, Muchiri E, Mungai P, Ndzovu M, Ouma J, Tosha S (1997) A school-based approach to the control of urinary schistosomiasis and intestinal helminth infections in children in Matuga, Kenya: impact of a two-year chemotherapy programme on prevalence and intensity of infections. Trop. Med Int Health 2(9):825–831

Montresor A, Gyorkos TW, Crompton DW, Bundy D, Savioli L, Organization WH (1999) Monitoring helminth control programmes: guidelines for monitoring the impact of control programmes aimed at reducing morbidity caused by soil-transmitted helminths and schistosomes, with particular reference to school-age children. World Health Organization, Geneva, pp 1–57

Munisi DZ, Buza J, Mpolya EA, Angelo T, Kinung'Hi SM (2017) Knowledge, attitude, and practices on intestinal schistosomiasis among primary schoolchildren in the Lake Victoria basin, Rorya District, North-Western Tanzania. BMC Public Health 17(1):1–9

Mwai J, Njenga S, Barasa M (2016) Knowledge, attitude and practices in relation to prevention and control of schistosomiasis infection in Mwea Kirinyaga county, Kenya. BMC Public Health 16 (1):819

Nduka F, Ajaero C, Nwoke B (1995) Urinary schistosomiasis among school children in an endemic community in South-Eastern Nigeria. Appl Parasitol 36(1):34–40

Odhiambo GO et al (2014) Low levels of awareness despite high prevalence of schistosomiasis among communities in Nyalenda informal settlement, Kisumu city, western Kenya. PLoS Negl Trop Diseases 8(4):e2784

Person B et al (2016) Community knowledge, perceptions, and practices associated with urogenital schistosomiasis among school-aged children in Zanzibar, United Republic of Tanzania. PLoS Negl Trop Diseases 10(7):e0004814

Price A, Verma A, Welfare W (2015) Are health education interventions effective for the control and prevention of urogenital schistosomiasis in sub-Saharan Africa? A systematic review. Trans R Soc Trop Med Hygiene 109(4):239–244

Ross AG et al (2014) Can mass drug administration lead to the sustainable control of schistosomiasis? J infect diseases 211(2):283–289

Sady H et al (2015) Knowledge, attitude, and practices towards schistosomiasis among rural population in Yemen. Parasites Vectors 8(1):436

Satayathum SA, Muchiri EM, Ouma JH, Whalen CC, King CH (2006) Factors affecting infection or reinfection with Schistosoma haematobium in coastal Kenya: survival analysis during a nine-year, school-based treatment program. Am J Trop Med Hygiene 75(1):83–92

Savioli L, Dixon H, Kisumku U, Mott K (1989) Control of morbidity due to Schistosoma haematobium on Pemba Island: programme organization and management. Trop Med Parasitol 40(2):189–194

Savioli L, Hatz C, Dixon H, Kisumku UM, Mott KE (1990) Control of morbidity due to Schistosoma haematobium on Pemba Island: egg excretion and hematuria as indicators of infection. Am J Trop Med Hygiene 43(3):289–295

Savioli L, Mott K (1989) Urinary schistosomiasis on Pemba Island: low-cost diagnosis for control in a primary health care setting. Parasitol Today 5(10):333–337

Shi Y, Liu C, Wang B (2015) Effect of health education intervention on schistosomiasis control in Mianyang City. Zhongguo xue xi chong bing fang zhi za zhi= Chin J Schistosomiasis Control 27(5):544–545

Stothard J, Khamis A, Khamis I, Neo C, Wei I, Rollinson D (2016) Health education and the control of urogenital schistosomiasis: assessing the impact of the Juma na Kichocho comic-strip medical booklet in Zanzibar. J Biosoc Sci 48(S1):S40–S55

Stothard J, Mook P, Mgeni A, Khamis I, Khamis A, Rollinson D (2006) Control of urinary schistosomiasis on Zanzibar (Unguja Island): a pilot evaluation of the educational impact of the Juma na Kichocho health booklet within primary schools. Mem Inst Oswaldo Cruz 101:119–124

Sturrock R (1995) Current concepts of snail control. Mem Inst Oswaldo Cruz 90(2):241–248

Suzuki C, Mizota T, Awazawa T, Yamamoto T, Makunike B, Rakue Y (2005) Effects of a school-based education program for schistosomiasis control. Southeast Asian J Trop Med Public Health 36(6):1388

Traoré M, Maude GH, Bradley DJ (1998) Schistosomiasis haematobia in Mali: prevalence rate in school-age children as index of endemicity in the community. Tropical Med Int Health 3 (3):214–221

Tsega T, Shewaye B, Desalegn T, Abraham A, Girmay T (2018) Prevalence of intestinal helminths and associated factors among school children of Medebay Zana wereda; North Western Tigray, Ethiopia 2017. BMC Res Notes 11(1):444

Wang JS, et al (2009) Effect of comprehensive control measures for schistosomiasis in marshland and lake regions. Chin J Schistosomiasis Control 21(1):39–42

Wynn T, Eltoum I, Cheever A, Lewis F, Gause W, Sher A (1993) Analysis of cytokine mRNA expression during primary granuloma formation induced by eggs of Schistosoma mansoni. J Immunol 151(3):1430–1440

Xu DC, Li W, Juma S, He MZ, Kabole F, Wang XY, Yang K (2019) The effectiveness of schistosomiasis health education session in primary and secondary schools in Pemba Island, Zanzibar. 23 October 2019, PREPRINT (Version 2) available at Research Square, https://doi.org/10.21203/rs.2.13577/v2

Zhang J et al (2016) Development and application of communication materials for participatory health education of schistosomiasis in fishermen and boatmen of Poyang Lake region. Zhongguo xue xi chong bing fang zhi za zhi= Chin J Schistosomiasis Control 28(1):58–61

Chapter 12
The Novel Strategy of China-Zanzibar Cooperation Project of Schistosomiasis Control: The Integrated Strategy

Xinyao Wang, Jianfeng Zhang, Yuntian Xing, and Fadhil M. Abdalla

Abstract Schistosomiasis is a serious global neglected tropical disease, and its prevalence is more serious in Africa. For a long time, international aid projects and local governments have mostly focused on mass drug administration (MDA) in the prevention and treatment of schistosomiasis in Africa. For the prevention and control of schistosomiasis, MDA cannot completely eliminate schistosomiasis in Africa. China has achieved great results in the prevention and control of schistosomiasis, and some provinces have eliminated schistosomiasis, the most important of which is the adoption of the integrated strategy. At this stage, the integrated strategy has not been implemented for schistosomiasis control in Africa, and most of them are based on chemotherapy. In 2017, China's aid to Zanzibar schistosomiasis project implemented the integrated strategy in cooperation with multiple departments in the local area. After 3 years of work, great results have been achieved in the local area. The schistosomiasis infection rate in some areas has dropped from 8.92% to 0.64%. It shows that the Chinese experience in schistosomiasis is applicable to the prevention and control of schistosomiasis in Africa. In order to eliminate schistosomiasis in

X. Wang (✉)
Jiangsu Institute of Parasitic Diseases, Wuxi, Jiangsu, China

Key Laboratory on Technology for Parasitic Disease Prevention and Control, Ministry of Health, Wuxi, Jiangsu, China

Jiangsu Provincial Key Laboratory on the Molecular Biology of Parasites, Wuxi, Jiangsu, China

College of Medicine, Jiangnan University, Wuxi, Jiangsu, China
e-mail: wangxinyao@jipd.com

J. Zhang · Y. Xing
Jiangsu Institute of Parasitic Diseases, Wuxi, Jiangsu, China

Key Laboratory on Technology for Parasitic Disease Prevention and Control, Ministry of Health, Wuxi, Jiangsu, China

Jiangsu Provincial Key Laboratory on the Molecular Biology of Parasites, Wuxi, Jiangsu, China

F. M. Abdalla
Ministry of Health of Zanzibar, Zanzibar, United Republic of Tanzania

K. Yang, H. Mehlhorn (eds.), *Sino-African Cooperation for Schistosomiasis Control in Zanzibar*, Parasitology Research Monographs 15,
https://doi.org/10.1007/978-3-030-72165-7_12

Africa as soon as possible, it is worth continuing to promote the integrated strategy in other regions of Africa.

Keywords China-Zanzibar cooperation project · The integrated strategy · Schistosoma haematobium · The Chinese experience in schistosomiasis

12.1 Introduction

In tropical and subtropical regions, schistosomiasis is also an important public health problem. The World Health Organization conservatively estimates at least 240 million people worldwide who are infected with Schistosoma spp. with severe disease manifestations in estimated 20 million people and an incidence of 280,000 deaths yearly. Schistosomiasis is caused by a snail-transmitted trematode that affects >200 million people in 76 countries, with >800 million at risk (Gryseels et al. 2006; Hiv/Aids 2010; Hotez et al. 2014; WHO 2010; Steinmann et al. 2006). It is among the top three highest burden infections among neglected tropical diseases, and seriously affects children (Hotez et al. 2014; Steinmann et al. 2006). Humans are infected when their skin is penetrated by cercariae of schistosomeswhich are released into freshwaters by infected intermediate host snails. The cercariae develop into adult worms in veins near the human gastrointestinal or urinary tract (Schistosomiasis japonica and S. haematobium, respectively). The adult worms produce eggs that pass through the intestine or bladder and are released with feces or urine, respectively, into the water to hatch into miracidia and infect new snails (Stelma et al. 1994). Infected humans can result in a loss of tissue function, stunted growth, learning deficits, and in some cases, increased risk of HIV, infertility, bladder cancer, liver failure, and death (Mohammed et al. 2007; King et al. 2005; Kjetland et al. 2006). While relatively effective drugs exist to clear infections, such as praziquantel, these drugs are not available to the majority of at-risk populations in Africa and offer no protection against reinfection, and humans are often re-exposed when they return to waterbodies with snails releasing schistosome cercariae (Carlton et al. 2011; Haggerty et al. 2020; Rollinson et al. 2013).

Schistosomiasis japonica is a zoonotic parasitic disease that seriously endangers the health of the people and an ancient infectious disease. It is endemic and natural epidemic, and more than 40 mammals except humans are the ultimate host of schistosomiasis. Through investigations during the early liberation period of China, 370 counties (cities) in 12 provinces, municipalities and autonomous regions in Hunan, Hubei, Jiangxi, Anhui, Jiangsu, Zhejiang, Fujian, Guangdong, Yunnan, Sichuan, Shanghai and Guangxi in southern China have been confirmed epidemic. The epidemic range is north to Baoying County, Jiangsu Province (north latitude 33°15′), south to Yulin County, Guangxi (north latitude 22°5′), east to Nanhui County, Shanghai (east longitude 121°51′), and Yunlong to Yunnan in the west County (east longitude 99°50′). The lowest altitude in the endemic area is zero

(Shanghai), and the highest is about 3000 m (Yunnan Province) (Committee 1986; Guo 2006a; Zhao 1996).

Zanzibar is composed of Unguja and Pemba islands, situated off the eastern coast of Tanzania mainland between latitudes 4° and 5° South (Zanzibar 2018). Schistosomiasis haematobium is one of endemic of Zanzibar. The intermediate host is Bulinus (Garcia et al. 2014; Knopp et al. 2013). In 2014, WHO signed a tripartite memorandum of understanding (MOU) with China and Zanzibar, paving the way for the start of a pilot schistosomiasis elimination programme in Zanzibar. The targets are to control and eliminated schistosomiasis based on the experience of schistosomiasis control in China. The early stage of the project was to carry out schistosomiasis prevention and control work in three villages (shihias), named Mtangani, Kiuyu and Wingwi, using snail control, population investigation, health education and treatment (Shan 2017; Michuzi 2014; WHO 2014; Xu et al. 2016).

12.2 The Integrated Strategy of Schistosomiasis in China

From the early days of the founding of the People's Republic of China to the early 1980s, the basic policy of schistosomiasis control in China was prevention-oriented, comprehensive measures, and the main measures were to eliminate snails in response to time and local conditions (Guo and Zheng 1999), which can be summarized as: (1) and large-scale implementation of snail eradication projects focusing on water conservancy and farmland infrastructure construction, as well as large-scale drug eradication; (2) large-scale investigation and treatment of patients and sick animals; (3) supplemented by safe water supply and feces Management, personal protection and publicity and education. In the early 1980s, with the advent of praziquantel, a highly effective and low-toxic therapeutic drug, and the development of immunological diagnostic technology, the rural economic system also began to gradually transition from a centralized system to a joint production contract system, and schistosomiasis control strategies also occurred. From 1980 to 1985, the country carried out research on the strategy of controlling schistosomiasis with chemotherapy as the mainstay, supplemented by the elimination of snails in susceptible areas. Focusing on the goal of disease control, some scholars in the country have done a lot of field countermeasure research work, laying the foundation for the chemotherapy control of schistosomiasis in the late 1980s. In 1985, the World Health Organization (WHO) schistosomiasis expert committee proposed a prevention and control strategy aimed at disease control, dividing the prevention and treatment of schistosomiasis into three stages: the first stage is disease control, reducing severe patients, especially acute and terminal patients, and patients with obvious symptoms; the second stage is to reduce the infection rate and degree of infection; the third stage is to control and block transmission. These three stages are complementary, the previous goal will create conditions for the next goal (WHO 1984, 1985). Controlling the source of infection through chemotherapy is also one of the effective prevention measures. Its purpose is to reduce transmission and improve patient symptoms, reduce morbidity and mortality by curing patients or reducing fecal ovulation.

Since 1980, some domestic experts have begun to discuss schistosomiasis control work from the health and economic aspects, and believe that when making decisions on schistosomiasis control, attention should be paid to the social and economic benefits of control or elimination. The use of limited resources, the least labor consumption, to achieve the best control effect (Guo et al. 1998). In the early 1990s, many domestic scholars analyzed different control strategies based on theories of health economics, and conducted a lot of research on the cost and benefit of implementing different control strategies in different regions (Guo 2006b). The World Bank loan of Chinese schistosomiasis control project is the largest health project invested by loans in the country so far. The goal is to reduce morbidity and implement the prevention and treatment strategies of universal chemotherapy in high-endemic areas and selective chemotherapy in medium- and low-endemic areas. After 3 years of cost and effect analysis, it is suggested that when the infection rate is greater than 44%, the cost of universal chemotherapy is lower than the cost of selective chemotherapy, and when the infection rate is less than 44%, the cost of selective chemotherapy is lower than the cost of universal chemotherapy (Guo et al. 1998). Since the founding of New China, the Chinese government and professional institutions have carried out a large number of schistosomiasis prevention propaganda work in the vast range of epidemic areas, mobilized and mobilized the people in the epidemic areas to participate in the campaign to eliminate Oncomelania snails, and achieved gratifying results. Since the 1980s, the country has begun to incorporate health education into the category of social medicine and has conducted a series of studies on it. Health education is not only a means of propaganda and agitation, but also emphasizes the role and evaluation of health education in the process of schistosomiasis control. After 1992, the country formulated a health education implementation plan, established a national schistosomiasis health education network, and organized health education workshops to incorporate schistosomiasis health education into the overall plan for schistosomiasis control. By the end of 1995, there were more than 40,000 full-time and part-time schistosomiasis prevention health education personnel nationwide, and 100,000 people participated in schistosomiasis prevention health education training courses. The training focused on changing the target population's exposure to infected water as the core content of health education intervention. Health education personnel used the mass media to conduct extensive publicity in the affected areas. During the project period, 2.318 million primary and secondary school textbooks were produced in various regions, and the coverage of schistosomiasis prevention and health education for students in the epidemic area was close to 100%. The central government and various popular provinces have produced more than 50 TV videos of different types and purposes, and produced a large number of broadcast materials, leaflets (books), posters, banned cards and slogans. Through systematic health education in a planned, purposeful, and targeted manner, the knowledge of schistosomiasis control among the people in the affected areas has been greatly improved, especially among the students in primary and secondary schools.

China has put forward the concept of comprehensive management since the 1950s. The comprehensive management method includes three aspects: (1) The source of infection is the investigation and treatment of patients and sick animals, and the treatment of feces and drinking water. Management; (2) Targeting the means of transmission is to check and eliminate snails in the epidemic area; (3) Targeting the susceptible population is to conduct publicity and education, personal protection, etc. With the improvement of diagnostic methods and the discovery of new therapeutic drugs, the focus of comprehensive management in different years has also changed greatly. In 1960, under the premise of backward diagnostic technology and large side effects of chemotherapy drugs, snail detection would become the key task of comprehensive management. Therefore, large-scale reclamation was carried out in conjunction with agricultural development. In the 1970s, large-scale drug killing was carried out. To eliminate snails, cultivation and mechanized farming became the main body of prevention and control work during the dry season of high-water beaches in the 1980s. After the 1990s, with the improvement of diagnostic technology and the emergence of praziquantel, a highly effective therapeutic drug, a strategy with disease control as the goal of prevention and treatment began to emerge. Patient investigation and treatment will become the focus of comprehensive management, so it will be expanded on a large scale. Chemotherapy for humans and livestock has become the main project and has received attention from all aspects.

According to the requirements of the country's schistosomiasis control goals, the country's main control strategy and the main comprehensive control measures that match it can be summarized as follows: Phase 1: The control goal is to eliminate schistosomiasis. The main prevention and control strategy is comprehensive management and comprehensive measures to eliminate Oncomelania snails, including: (1) Large-scale and large-scale implementation of snail control projects focusing on water conservancy and farmland infrastructure construction, and large-scale drug elimination (2) Large-scale investigation and treatment of patients and animals; (3) Supplemented by water pipes and fecal pipes, personal protection and publicity and education. After nearly 30 years of control work, great achievements have been made in the prevention and treatment of schistosomiasis. By the end of the 1980s, the area of oncomelania nationwide had dropped by 90%, and the number of patients had dropped by 80%. The four provinces, municipalities and autonomous regions of Guangdong, Guangxi, Fujian, and Shanghai have reached the standard for eradication of schistosomiasis (Guo and Zheng 2000). Phase 2: The goal of prevention and control is to control schistosomiasis and eliminate schistosomiasis wherever possible. The main prevention strategy is comprehensive treatment, comprehensive disease control based on chemotherapy, supplemented by comprehensive prevention strategies such as snail eradication in susceptible areas and health education. In the 1990s, a series of comprehensive intervention measures such as simultaneous chemotherapy for real people and livestock and snail eradication in susceptible areas, and for a long period of time, the examination and treatment of schistosomiasis was free. Through continuous intervention and treatment, the incidence of schistosomiasis in the country has shown a continuous and steady decline. By the end of 2000, the national schistosomiasis infection rate had dropped significantly.

There was a significant decrease in new-onset late-stage schistosomiasis. At this stage, Zhejiang Province announced that it has reached the standard for eradication of schistosomiasis (Yuan et al. 2000). Phase 3: Consolidation and maintenance of the control effect. The country has a large area of snails, the epidemic is severe, and the prevention and control are difficult. The most endemic areas of schistosomiasis in the country are mainly concentrated in lake areas and plateau mountainous areas. Because the water level cannot be controlled, the environment is complicated, the traffic is blocked, and the economy is underdeveloped, the prevention and control work are very difficult. In the past, some traditional effective schistosomiasis control measures were difficult to carry out and implement. Due to the reform of the social and economic system, large-scale mass movements in the past were difficult to carry out; due to consideration of ecological and environmental protection, methods such as flood diversion and chemical soaking, high reclamation, and blocking lakes to kill snails could not be implemented. At present, Oncomelania snails are mainly distributed in lake areas and mountainous areas. Due to environmental factors, it is very difficult to eliminate Oncomelania snails (Guo 2006b). Changes in the ecological environment have intensified the spread of schistosomiasis: the construction of the Three Gorges of the Yangtze River will benefit the country and the people, but it will also increase the risk of schistosomiasis transmission in the reservoir area; returning farmland to lakes, flooding through floodplains, and relocation to build towns will increase the area of snails The chances of recovery and infection of humans and animals have increased. In addition, frequent floods can spread the snails. Therefore, in the next few years, the area of Oncomelania snails in the country will rebound to a considerable extent (Zhou et al. 2004). Restricted by diagnostic technology and chemotherapeutic drugs, the prevention and treatment effects are difficult to consolidate. At present, the diagnosis methods of schistosomiasis are mainly divided into two categories: one is pathogenic diagnosis techniques, including modified Kato method, hatching method, and egg collection method; the other is immunological diagnosis methods, including intradermal test, blood antibody Detection and antigen detection. The sensitivity of pathogenic methods is low, and the detection rate in low-transmission areas is very low; the sensitivity of antibody detection is high, but the antibody disappears for a long time, and there is no curative effect evaluation value; the detection rate of circulating antigen is affected by the complex, so the detection rate is very low Therefore, these existing methods have deficiencies in either severe or mild epidemic areas (Guo and Guo 2005). Praziquantel is a safe and effective drug for the treatment of schistosomiasis. Existing indications indicate that the current dose of 40 mg/kg in severe epidemic areas may not be enough and may lead to the development of drug resistance. After 20 years of national chemotherapy, This issue should now be considered (Cao et al. 2003). Therefore, although the strategy of universal chemotherapy has played an extremely important role in disease control, the reinfection in severe endemic areas continues year after year, even as high as 40% (Guo et al. 2001). On the one hand, the compliance of residents from chemotherapy is decreasing year by year, on the other hand, the chemotherapy effect of praziquantel itself must be considered (Guo et al. 2005).

Through the phased implementation of comprehensive strategies and measures focusing on snail eradication or chemotherapy, although historical results have been achieved. However, a prevention strategy cannot be applied in all historical stages, nor can it be applied in various epidemic types. At the same time, prevention measures are sometimes time-effective, and may not be equally effective at different stages. Furthermore, the starting point and end point of schistosomiasis control work is to serve the people's health and serve the socialist economic construction. The fundamental purpose is to control schistosomiasis, not to eliminate the snail species. Therefore, the effectiveness indicators that reflect the goals of schistosomiasis control in the new period will mainly grasp three rates: (1) control and reduce the incidence of acute infection; (2) control and reduce the infection rate of humans and animals, of which the infection rate of livestock is particularly important; (3) Control and reduce the positive rate of Oncomelania snails. These three rates are indicators of schistosomiasis control that we should firmly grasp in the new era. To control and reduce these three rates, the focus of our work and the prevention and control measures we take should be changed accordingly. It is necessary to vigorously implement the use of tractors instead of cattle (replacement of cattle by machines) as the focus, focusing on the adjustment of the agricultural industry structure, combined with water reform, toilet improvement, health education, livestock breeding, simultaneous inspection and treatment of humans and animals, and susceptible areas or environmental transformation Such comprehensive measures take into account the maximization of economic and social benefits.

Governments at all levels and relevant departments in Jiangsu Province have carried out comprehensive management of schistosomiasis in light of local conditions. The agricultural sector conscientiously implements measures to replace cattle and livestock with machinery, strengthen management of livestock grazing, prohibit grazing in areas with snails, reduce the source of infection of livestock, and at the same time tilt the agricultural industry structure adjustment project to key areas for schistosomiasis control, and upgrade related construction projects and transformation of oncomelania. From 2012 to 2018, the province has implemented a total of 3870 cattle (sets) of replacement of cattle by machines, 36 culling cattle, 28,995,300 livestock in captivity, 2106.68 hm^2 and 896.13 hm^2 of fish farming and snail eradication. The water conservancy department listed the schistosomiasis control project as a key project of water conservancy construction, and carried out comprehensive treatment of schistosomiasis in combination with the construction of large-scale water conservancy projects such as the improvement of the Huaihe River into the river and the east route of the South-to-North Water Transfer. From 2012 to 2018, 2039 water conservancy and schistosomiasis control projects were implemented, 6336 micro water conservancy projects, 2321.51 km of small and medium-sized rivers, 223.51 km of hardened slope protection, 160 facilities for sinking snails, six rural safe drinking water projects, and water-saving irrigation projects 83 small watersheds were treated with 15.1 km^2; the forestry department planted 500.00 hm^2 of snail control and disease prevention forests, constructed 330.20 hm^2 of forest and agricultural ecological forests, and constructed 129.53 hm^2 of wetlands and nature reserves; the land and resources department carried out land remediation of 2.89 hm^2 and built field roads of 1662.83 km. Construction of irrigation, drainage (ditch) and pipelines are 4770.12 km, the land leveling area is 2.80 hm^2, and the area of relocated villages is 570.67 hm^2 (Li et al. 2019).

12.3 Schistosomiasis Prevention and Control Work in Africa

Within the past decade, significant progress has been made on large scale treatments through integrated control of schistosomiasis and other NTDs, due to a number of international organizations, donor foundations, bilateral institutions and non-governmental organizations that responded to the WHO's 2001 call for action (Savioli et al. 2009). Treatment with praziquantel (Doenhoff et al. 2009) is cost-effective and 'preventive chemotherapy' is currently the strategy of choice and endorsed by (WHO 2006, 2012b). With a support from international organizations, donor foundations, the pharmaceutical industry, and several not-for profit organizations, millions of children are regularly treated for schistosomiasis in Africa, through coordinated use of anthelminthic drugs (WHO 2015, 2016).

In the past, a key obstacle to implementation of preventive chemotherapy for control of schistosomiasis in Africa was the limited access to praziquantel, either purchased or donated (Hotez et al. 2010). With the expansion of activities of the Schistosomiasis Control Initiative, it was clear that the future need for large-scale quantities of praziquantel would grow in Africa (Fenwick et al. 2009). In 2007, Merck KGaA pledged to donate 200 million tablets of praziquantel over 10 years through WHO. However, in 2012, Merck-KGaA committed to increase its donation to 250 million tablets of praziquantel per year until schistosomiasis is eliminated. While there is now growing access to praziquantel for schistosomiasis control in Africa, it is not at the level of projected requirements to reach all people at risk and requiring treatment (Stothard et al. 2013). There reported on treatment coverage for schistosomiasis show that utilization of available praziquantel by NTD programme is not yet optimal in many countries (WHO 2015, 2016). In addition special attention is needed to develop new access plans and reporting frameworks to vulnerable demographic groups in high-risk areas, particularly pre-school aged children who are currently overlooked (Stothard et al. 2013; Stothard et al. 2011).

In 1975, the World Health Assembly (WHA) adopted the WHA 28.53 resolution calling for the preparation of guidelines and increased efforts in drug development, water projects and partner mobilization for schistosomiasis control (WHO 1975). In 1976, the WHA29.58 resolution urged countries with related diseases to consider the epidemiological aspects of schistosomiasis when planning and implementing water management schemes, and to undertake specific measures to prevent the spread of the disease into new areas and neighboring countries (WHO 1976). However, despite the existence of methods in the 1970s and 1980s, control of schistosomiasis was only sustained for a prolonged period in a few countries and almost no progress was made in African countries. In the 1990s, interest in the control of schistosomiasis in Africa waned, and disease control was overshadowed by other health priorities (Tchuem 2012).

Recent years have witnessed an increased interest in the control/elimination of NTDs, and the control of schistosomiasis has again become a priority on the agenda of many governments, donorsand international agencies. In 2001, all member states of WHO endorsed the WHA54.19 resolution on schistosomiasis, with the major objective of the regular treatment of at least 75% of all school-aged children at risk of morbidity by 2010 (WHO 2001). The resolution generated a greater political commitment and encouraged many countries to establish national action plans and programmes for schistosomiasis control. Ten years later, in January 2012, WHO published an NTD Roadmap that set targets for the period 2012–2020, and described the strategic approach to accelerate work to overcome the global impact of NTDs. This method identified preventive chemotherapy as a key strategy for tackling NTDs (WHO 2012b, 2012c). Partners and stakeholders (pharmaceutical companies, donors, endemic countries, Non-Governmental Organizations) endorsed the London declaration on NTDs, and committed to support the WHO roadmap and its 2020 targets for ten NTDs (WHO 2012a).

In 2012, WHO member states agreed with the WHA65.21 resolution on elimination of schistosomiasis, that called on all endemic countries to intensify control interventions and strengthen surveillance. Moving from control to elimination requires changes in the strategies of mass drug administration (MDA) with praziquantel to inclusion of complementary measures. Integrated intervention measures including public health education, access to clean water, improved sanitation, and snail control are highly recommended in this paradigm shift. A combination of these intervention measures can interrupt the lifecycle of the schistosomiasis (Mazigo 2019). Importantly, this resolution encouraged countries to embark on schistosomiasis elimination where possible (WHO 2012c). Finally, in 2013, the WHA66.12 resolution on NTDs encouraged countries to ensure continued country ownership of programmes for Neglected tropical diseases (NTD) prevention, control, elimination and eradication, devised plans for achieving and maintaining universal access to and coverage with interventions against NTDs, including provision of safe drinking-water, basic sanitation, health promotion and education (WHO 2013).

Clean water provision, sanitation and hygiene (WASH) are critical components in the prevention schistosomiasis scheduled for intensified control or elimination by 2020. For schistosomiasis, improved sanitation across the entire community to prevent contaminated faeces and urine from reaching surface water can reduce or eliminate transmission, by stopping worm eggs in faeces and urine from entering water–the snail habitat (Tchuem Tchuenté et al. 2017). Provision of safe water, sanitation and hygiene is one of the five key interventions within the global NTD roadmap. This requires a strong intersectoral collaboration. WASH providers must developed in the recent WASH strategy to accelerate and sustain progress on NTDs (Boisson et al. 2016).

It is worth noting that despite these recommendations, these measures have not been integrated into the MDA plan in the African environment. At this stage, the control of schistosomiasis in Africa is still dominated by MDA. In Africa, the control of schistosomiasis is a long-term project. The process from control to elimination will require policy changes, modifications, and strategic changes from MDA methods to integrated methods. In order to achieve the elimination goal, PZQ needs to be intensively treated to cover all age groups and hard-to-reach communities, implement complementary public health measures, such as improving the use of clean water and sanitation facilities, and take intervention measures to promote behavior change. New diagnostic tests are needed to help define elimination methods, which should also be part of the agenda. In general, political support and community involvement will be essential for the comprehensive approach to play a meaningful role. From control to elimination of schistosomiasis in Africa, a multisectoral approach should form the central pillar. (Mazigo 2019).

12.4 China Aids African Schistosomiasis Control Project

Africa is an endemic area of schistosomiasis. The prevention and control of schistosomiasis is a social system project. The success of schistosomiasis prevention and control in China is the cooperation of agriculture, water conservancy, health, education and other departments. With the bilateral assistance to Africa, it is possible to implement Chinese schistosomiasis control experience in Africa. However, Chinese most important schistosomiasis control experience is comprehensive management.

12.5 Multi-Sectoral Schistosomiasis Control Cooperation Meeting

Since Chinese aid to Zanzibar schistosomiasis prevention and control project started in March 2017, the work has expanded from the original three pilot areas to surrounding areas. However, most of the prevention and control work focuses on the investigation of blister snails and the elimination of snails with drugs, routine inspections and treatments, and health education. Relevant departments in Zanzibar are less involved. How to implement comprehensive management measures in the follow-up work to give China experience in schistosomiasis control Really effectively land in Africa, strengthen government leadership, and give full play to the role of departmental linkage. The third batch of project teams held the district government department's schistosomiasis control project promotion meeting from 15 to 23 March. Pemba Island is divided into North and South Pemba Province, Micheweni, Wete, Chakechake, Mkoani four regions. After thorough discussion with the non-NTD staff, the project team decided to hold district-level project promotion meetings in the four regions. The meeting invited the main leaders of the district and officials of the district finance, education, environment, agriculture, water conservancy, information, safety, health and other departments to participate. The project team introduced the development of the Zanzibar schistosomiasis project and the third batch of project teams since 2017. In the work plan, the participants had a lively discussion on schistosomiasis prevention and control knowledge, drug control effects, influencing factors, project background, and realization of specific goals, etc., thereby enhancing the understanding of schistosomiasis prevention and control in various departments. The project team held four project promotion meetings in four regions, with more than 120 representatives (JIPD 2018f). On 8 May, the project team held the Pemba Island community-level project promotion meeting and health education training class. A total of 46 staff from the President's Office under Pemba Island's four District participated in the training. Pemba Island's district-level president's office is subordinate to each district government, and staff members are elected by community residents to specifically coordinate the government affairs of the local community government, serve the people of each community, and be responsible to the district government. The staff is the coordinator and person in charge of the specific work of the grassroots shehia, and has a linking role in the grassroots work. The purpose of this conference is to smoothly promote the village-level health education for schistosomiasis through their training and their resources and capabilities. The project team pointed out that Chinese schistosomiasis prevention and control is a comprehensive prevention and control work. The government is the main body responsible for the control of schistosomiasis. All departments and units must participate together. The staff of the President's Office will play an important role in the elimination of Pemba schistosomiasis. Play an important role (JIPD 2018d). On 11 May, the project team held a coordination meeting for schistosomiasis prevention and control work departments on Pemba Island. The meeting invited officials from 16 departments of provincial finance, education, environment, agriculture, water conservancy, information, safety, health, etc. to participate. The Governor of South Pemba Province Hemes attended the opening ceremony. The project team introduced the successful experience of integrated control of schistosomiasis in China, clarified the purpose of this working meeting, and emphasized the importance of integrated control of schistosomiasis. Governor Hemed stated that he would do his best to ensure the smooth progress of the mulberry schistosomiasis control project and the greatest support to Chinese aid to the mulberry schistosomiasis control project. He hoped that the schistosomiasis in Pemba would be eliminated as soon as possible. Most of the heads of departments who attended the meeting learned about schistosomiasis, blister snails, and integrated control of schistosomiasis for the first time. Participants put forward questions and suggestions on the knowledge of schistosomiasis prevention and control, the effect of drug control on snails, influencing factors, project background and the realization of specific goals, and supported the project work on Pemba Island and were willing to eliminate Pemba Island. Schistosomiasis worked together and decided to establish a regular communication and reporting system for schistosomiasis prevention work between departments (JIPD 2018e).

On 4 October 2019, the sixth group of experts from the China-aided Zanzibar Schistosomiasis Control Technology Project went to Zanzibar Island to hold a project promotion and exchange meeting. About 30 leaders and professionals from the Ministry of Health, Ministry of Agriculture, Ministry of Education, Ministry of Environment, NTD Office, Health Promotion Office and other departments attended the event. At the exchange meeting the Director of the Bureau of Disease Control and Prevention of the Ministry of Health delivered an opening speech. The project team leader introduced the achievements of the China-aided mulberry schistosomiasis control project in Pemba Island for 3 years. Schistosomiasis is also endemic in some areas of Zanzibar. I hope that this exchange will promote everyone's understanding of the project, understand Chinese schistosomiasis control, and help promote Schistosomiasis prevention and control work on Zanzibar Island. Salum, the head of the schistosomiasis department of the NTD office, introduced the current neglected tropical disease epidemic in Zanzibar. The project team introduced Chinese schistosomiasis control work and achievements, focusing on the experience of cooperation between Chinese agriculture, water conservancy, and forestry departments and schistosomiasis control work (JIPD 2019a). On 1 November, the sixth batch of project teams carried out a multi-departmental cooperation and exchange promotion meeting for schistosomiasis prevention and control work at the project site, the Chakechake District Office of Pemba Island, the Department of Health, the Department of Water Resources, the Environment Department, the Department of Education, Agriculture, Health Promotion, etc. 12 department heads from multiple departments attended the meeting. The director of the Pemba Island Health Department stated that after a multi-departmental exchange meeting on schistosomiasis control was held in Zanzibar, in order to strengthen the Pemba Island's multi-sector cooperation and jointly fight against schistosomiasis, the Pemba Island Health Department invited the project team to be held on Pemba Island. This exchange promotion meeting. The project team leader introduced that one of Chinese valuable experiences in schistosomiasis control is multi-departmental cooperation. He thanked all departments for their attention and support to schistosomiasis control and hoped to promote the establishment of a sound multi-departmental cooperation mechanism in Pemba Island through the meeting. The project team introduced schistosomiasis-related knowledge, focusing on sharing the "Chinese experience" of multi-sector cooperation in agriculture, water conservancy, and environmental protection, as well as the cooperation between the project team and the Pemba Island Ministry of Education and the Water Resources Department. Participants had in-depth exchanges and discussions. The Department of Agriculture expressed to the project team a strong willingness to cooperate in schistosomiasis control and invited the project to conduct schistosomiasis investigations on rice farmers and cattle farmers (JIPD 2019c).

12.6 Cooperating with the Health Department for Health Education

As one of the preventions and control measures for schistosomiasis, health education is an economical and effective working method due to its low investment, good social benefits, and high input-output ratio. It is also one of the basic components of many schistosomiasis control strategies. One. However, due to the lack of funds, teachers, teaching materials and work experience in Pemba Island, schistosomiasis prevention health education has not been carried out effectively. In order to comprehensively raise awareness of schistosomiasis prevention among local people, spread schistosomiasis prevention knowledge, and promote healthy behaviors, the China-aided mulberry schistosomiasis prevention project has actively planned and carried out health education work based on local conditions: Mtangani, kiyuyu and wingwi are the three pilot projects and Uwandani demonstration area. Key work areas. The project team first carried out educational work in the above-mentioned areas. The third batch of project teams focused on schools and communities, and carried out 40 health education publicity lectures in schools, community primary health centers (PHC) and communities in the above-mentioned areas, with an audience of 5000 people. In addition, the project team also set up a demonstration zone sign in Uwandani demonstration zone, 13 wall slogans, and 22 snail pond warning signs. The printing of warning signs and slogans gave the people in the demonstration area a more intuitive impression, and played a very good role in warning and publicity. In Pemba Island, the proportion of young people is relatively high, and the students at school are women and farmers. The main people who are exposed to water are also the key people infected with schistosomiasis, and it is of great significance to do their health education work well. The project team carried out systemic schistosomiasis prevention health education classes in three pilots and 12 primary and secondary schools in Uwandani Demonstration Zone, disseminating schistosomiasis knowledge to them through videos, examples, exhibition samples, and on-site questions and answers. At the same time, the "Schistosomiasis Cup" essay competition was held. A total of 195 articles in English and 315 in Swahili were received in the competition. Six award-winning essays were selected and three schools were excellently organized, which effectively deepened students' understanding of schistosomiasis knowledge; While doing a good job in the health education of the pilot areas and students, the project team also meticulously designed and printed 5000 Swahili propaganda posters and 10,000 Swahili schistosomiasis propaganda brochures. These propaganda posters and brochures Posting and distributing to local schools, community primary health care centers, government agencies and other departments and villagers' homes to increase the project's influence and increase the coverage of health education; the team members and non-party employees go to each community in advance, communicate and coordinate with the village chief, and arrange health education place. Bring the equipment the next day, drove more than 10 km, and carried the equipment. When you arrive at your destination, you will set up your own cable, set up the table, set up the stereo, etc., and then give a few hours of lectures. In order to conform to the living habits of the people, the project team sometimes had to go to the village to carry out health education activities after get off work in the evening. After several months of hard work, the evaluation of the effectiveness of health education showed that the awareness rate of schistosomiasis control among the masses increased from 61.36% to 86.97%, the average score of residents' knowledge about schistosomiasis control increased from 6.99 to 7.94, and the average score of students' knowledge about schistosomiasis control increased from 6.66. As of 7.93, the population's awareness of schistosomiasis prevention and control has been continuously improved (JIPD 2018c). On 21 August 2018, the award ceremony of the "Schistosomiasis Cup" composition competition for the Chinese aided schistosomiasis control project was successfully held on Pemba, Zanzibar. Zanzibar's Pemba Disease Control Bureau, Education Bureau, NTD Office and China's aid for schistosomiasis control A total of more than 50 people from the third and fourth project groups participated in the award ceremony. The award ceremony also marked the end of the two-month "Schistosomiasis Cup" essay competition. Starting in mid-May, the China-aided mulberry schistosomiasis control project has carried out systematic health education courses in the three pilots of Wingwi, Kiyuyu, Mtangani and all primary and secondary schools in the Uwandani demonstration area. In order to further deepen local students' knowledge and understanding of schistosomiasis, the group organized and carried out the first Pemba "Schistosomiasis Cup" composition competition in these schools, encouraging and calling on students to actively participate in schistosomiasis control work. This move was also warmly welcomed by local schools and teachers and students. Students actively participated in writing and submitting essays. In the end, a total of 510 contest essays were received, including 195 in English and 315 in Slovak. There were six award-winning essays and three excellent schools. The Pemba Island's Disease Control and Prevention Bureau of Zanzibar expressed its gratitude to the careful organization of the Chinese aid schistosomiasis prevention project team and the serious cooperation of various schools. He hoped that this move can be expanded and implemented in other areas of Pemba, so that more Students participate in schistosomiasis control work. The Pemba Education Bureau of Zanzibar spoke highly of the innovative work of this essay contest, and hoped that the education and health departments could keep in touch and cooperate closely, and strive to work together for the health of all Pemba people (JIPD 2018b).

The Uwandani Demonstration Area is a comprehensive demonstration area for the Zanzibar schistosomiasis prevention and control project. After more than a year, with the joint efforts of Chinese and African staff, it has taken measures to strengthen disease inspection and treatment, consolidate snail detection, and carry out health education. With comprehensive prevention and control measures, the population infection rate in the Uwandani Demonstration Area has dropped from 8.92% at the beginning of the project to the current 1.36%, and the prevention and control work has achieved remarkable results. In order to further consolidate the results of schistosomiasis control in the Uwandani Demonstration Zone, on 3 November 2018, the fourth batch of members of the Zanzibar Schistosomiasis Control Project went to the Uwandani Demonstration Zone to carry out schistosomiasis health education activities. School teachers and students from the demonstration zone More than 500 people, including community health doctors and local people participated in the event. The location of this health education activity was determined to be a school in Uwandani demonstration area. The members of the project team explained schistosomiasis prevention and control knowledge to students and local people, broadcasted videos related to schistosomiasis infection prevention on the spot, distributed schistosomiasis prevention knowledge brochures and health education materials such as buckets, flashlights, cloth bags, etc., for the public in Uwandani demonstration area Learn more vividly about the hazards of schistosomiasis and related knowledge about the prevention and control of schistosomiasis in daily life (JIPD 2018a).

On October 12, 2019, the sixth batch of project teams held a handwriting competition for primary schools across the island on Pemba Island. A total of 42 schools in 35 areas with severe schistosomiasis epidemics participated in the competition. The Minister of Health of Zanzibar, the Director of Pemba Island Health Department, and the head of the SMZ President's Office attended the kick-off meeting. The school is responsible More than 40 people or school representatives attended the meeting. The head of SMZ President's Office, who is in charge of all primary schools in Pemba Island, thanked the project team for organizing and supporting this activity. Through this multi-sectoral schistosomiasis prevention activity, it can help primary school students to improve their health awareness. The project team thanks the various departments of Sangfang for their continuous support to the project. The project has reached the goal of eliminating schistosomiasis in the pilot and demonstration areas, effectively improving the health of residents. The project will further promote comprehensive measures such as snail detection, disease detection and treatment, and health education to consolidate the results of schistosomiasis elimination. The primary school students' hand-written report competition was held to increase the awareness of schistosomiasis prevention among primary school students as a high-risk population and reduce the risk of infection. The Minister of the Ministry of Health of Zanzibar believes that this activity is a continuous deepening of the project work. Based on the huge achievements of the project, some successful schistosomiasis control experience has been extended to areas outside the project. The sixth batch of expert groups of the project team has carried out the project work for more than 2 months, in addition to completing routine snail detection and disease detection and treatment, various trainings have been carried out for professionals, community health workers, and other schistosomiasis control departments. Four times, schistosomiasis health promotion work was extended to different regions, different departments and other fields (JIPD 2019b).

12.7 Cooperating with the Water Resources Department

In most schistosomiasis endemic countries, the natural water source containing cercariae of Schistosoma is still the only source of domestic water for residents. Even if the residents in high-risk communities get effective treatment, they cannot avoid reinfection. At the same time, for economic development in many African countries, in the absence of environmental hygiene assessment, large-scale water conservancy projects such as dams and irrigation facilities were built, which expanded the epidemic water, increased the risk of residents contacting the epidemic water, and led to the increase of infection rate. However, in African countries, because water resource management and health facilities construction are usually not under the jurisdiction of the health sector, and the cost of providing safe water is high, so in the past, schistosomiasis control agencies in the health sector of various countries have not focused much on such high-cost interventions, and epidemic areas have not actively pursued cross sector cooperation to improve the awareness of safe water.

On 27 October 2019, at the invitation of Pemba Island Water Conservancy Department, the Zanzibar Schistosomiasis Control Technical Assistance Project (hereinafter referred to as the project team) and the Mtangani Community Villagers Committee jointly held a seminar of water supply project for schistosomiasis control. The purpose of this seminar is to publicize and implement the cooperation between the project team and the water conservancy department in the water supply project of schistosomiasis control, and further explore the plan of the project. Leader of the project team introduced that the construction of the water conservancy project not only can improve the water intake and use of residents, but also has a very important significance for maintaining the results of Schistosomiasis elimination. The history of the cooperative water conservancy project was introduced in detail. The two sides cooperated closely on the preliminary demonstration, regional selection, material preparation and other aspects of the project, and jointly contributed to the development of the project. Later, the village head of Mtangani community made a speech, expressing his gratitude to the project team and the water conservancy department, and introducing the situation that residents have been facing the shortage of domestic water supply for a long time. This water supply project is a great blessing for the residents, which can greatly improve the living standards of them. Subsequently, the engineers of the water conservancy department and the village committee discussed the details of the construction. The village committee said it would actively mobilize the labor force in the village, and the construction could be completed in a week.

12.8 Agricultural Control Intermediate Host Bulinus of Schistosomiasis

In Chinese schistosomiasis endemic areas, the practice of reducing the density of oncomelania and even killing the snails is beneficial to aquatic products and poultry breeding, and the effect is good, and it has produced considerable economic benefits (Gong and Liu 1996; He et al. 2019; Lei et al. 2010; Shi 2018). The sixth group of experts in the Zanzibar region through muscovy duck breeding for the prevention and control of schistosomiasis in Egypt is still in the preliminary exploration stage. Experiments have proved that the breeding of muscovy ducks for the ecological snail eradication of water snails has a significant effect, especially for the suppression of water snail reproduction. The snail-killing organisms mainly reduce the number of harmful snails in three aspects: first, directly reduce the number of harmful snails by preying on harmful snails; second, reducing the number of harmful snails from the root source by preying on snail eggs; third, destroying harmful snails by preying on plants The living environment of harmful snails affects the reproduction of snails and indirectly reduces the number of harmful snails (Bi 2007). Judging from the fact that Muscovy ducks prey on blister snails, Muscovy ducks can effectively control the population of blister snails. Despite the existence of optional foods, Muscovy ducks still use their beaks to crush and swallow during the process of filtering water and foraging. Experiments have shown that each muscovy duck can prey 0.4 adult blister snails per day. Although 0.4 is a small number, long-term predation by group-bred Muscovy ducks will greatly reduce the density of blister snails. In addition, the offspring of blister snails and Due to their extremely small size, the snail eggs are more likely to be swallowed by ducks during filter feeding, so Muscovy ducks can be considered as a biological choice for snail killing. The effect of breeding Muscovy ducks for ecological snail eradication may be related to the fact that the habitat and activities of Muscovy ducks affect the growth process of aquatic plants, destroy the food and living environment of the water snails, and have a smaller relationship with the direct reduction of the density of the water snails by the Muscovy ducks. . In the experiment, it was found that the aquatic plants in the experimental group's pots were foraging and mechanically damaged by the Muscovy ducks. The water body was fluctuating and more turbid, and the water level dropped significantly compared with the control group. The aquatic plants in the control pots grew normally and were used as food by blisters Or shelter, calm water. Muscovy's damage to aquatic plants and the water fluctuations caused by it will highly affect the attachment of blister snails and their eggs on aquatic plants, which will negatively affect the growth and reproduction of blister snails and indirectly lead to the decrease of snail density. In the experiment, the blister snails were not only tenacious in vitality, a total of 120 adult blister snails were cultivated in a simulated natural environment for 40 days, only two died naturally, and the blister snails reproduced in a single body with strong reproductive ability. The offspring snails were cultured in a simulated natural environment for 20 days. The reproduction rate is as high as 15.15 per snail, and it is very meaningful to cultivate Muscovy ducks to produce strong reproduction inhibition on the water snails. Although the experiment simulates the natural environment, the water surface area is small, which has certain limitations. If the breeding scale of Muscovy ducks is not reached, it may have a small impact on the control of large ponds, especially water snails in the middle waters. However, the distribution of blister snails in the pond decreases with the decrease of aquatic plants from the side of the pond to the center of the pond. At the same time, human beings are the main source of infection of schistosomiasis, the activity habits of Muscovy ducks and the law of human activities at the side of the pond. It is consistent, so the disadvantages of muscovy ducks' semi-loving water have little effect on their use in ecological snail control and schistosomiasis control. Muscovy ducks have a wide range of breeding in Zanzibar, with strong adaptability to life and rich nutritional value. Especially for ponds that are not suitable for large-scale repeated drug snail eradication, biological snail eradication has broad prospects. If the mulberry schistosomiasis control project is adopted, the local cooperation of schistosomiasis control, agriculture, water conservancy, and environmental protection will be adopted to breed muscovy ducks in enclosures in the blister snail breeding grounds, which will not only effectively reduce the local blister snail population density, eliminate the harm of schistosomiasis, and protect water Ponds, streams and other natural environments are not affected or affected by drugs to eliminate snails as little as possible, and good economic benefits of muscovy duck breeding will be achieved, thereby changing the local industrial structure and promoting local agricultural and economic development. At the same time, Muscovy ducks can prey on mosquitoes and other insect larvae in the water body, reducing the incidence of local mosquito vectors and other insect-related infectious diseases such as malaria, lymphatic filariasis, and dengue fever. It is good for purifying the environment and protecting the health of local people.

12.9 Conclusion

Under the leadership of the Chinese government, after nearly 40 years of efforts, starting from the reality of our country, we adhere to the principle of putting prevention first, combining prevention with scientific research, and under the guidance of the principles of highlighting key points, classified guidance, comprehensive planning, and step-by-step implementation, actively adopt comprehensive treatment measures such as disease detection, snail elimination, publicity and education, and personal protective treatment, and gradually formed a schistosomiasis prevention and control mechanism with Chinese characteristics, and established and perfected a sustainable development of schistosomiasis suitable for Chinese national conditions. The prevention and control model has played a key role in effectively controlling the prevalence of schistosomiasis and has also provided experience for schistosomiasis in the world. Most of the schistosomiasis prevention and control work uses a combination of (MDA) and health education in abroad, and less snail control and other work, which is different from the schistosomiasis prevention and control work in China. The schistosomiasis prevention and control work mostly used the combination of mass drug administration (MDA) and health education in other countries, and rarely does snail control, which is different from the schistosomiasis prevention and control work in China.

From the previous related content, we can see that for the prevention and control of schistosomiasis in Africa, MDA alone cannot completely eliminate trematodiasis. At this stage, there are no reports on comprehensive management of schistosomiasis in Africa, but comprehensive management is adopted through Chinese aid to the African schistosomiasis project. The measures have achieved great results, and the infection rate of schistosomiasis in some areas has been lower than 1%. On the surface, Chinese experience in schistosomiasis control is applicable to the prevention and control of schistosomiasis in Africa and is worthy of promotion and application. However, there are also certain difficulties in the control process, such as how to adopt different control strategies in areas with different infection rates, which can reduce unnecessary manpower and financial resources, so that resources can be invested in more places for schistosomiasis control. The problem needs to be further resolved in future work.

References

Bi YY (2007) A study on the influencing factors of procambarus claki's predation of ncomelania hupensi. Capital Normal University

Boisson S, et al (2016) Water, sanitation and hygiene for accelerating and sustaining progress on neglected tropical diseases: a new Global Strategy 2015–20 (1876–3405 (Electronic))

Cao L, Hu GL, JG G (2003) The impact of the development of new drugs and new technologies on the strategy of schistosomiasis control in my country. Chin J Parasit Dis Con 16(6):380–382

Carlton EJ, Bates MN, Zhong B et al (2011) Evaluation of mammalian and intermediate host surveillance methods for detecting schistosomiasis reemergence in southwest China. PLoS Negl Trop Dis 5(3):e987

Committee OotLGfSPaCotCC (1986) Thirty years of schistosomiasis control. Shanghai Science and Technology Press, Shanghai

Doenhoff MJ, et al (2009) Praziquantel: its use in control of schistosomiasis in sub-Saharan Africa and current research needs (1469–8161 (Electronic))

Fenwick A, et al (2009) The Schistosomiasis control initiative (SCI): rationale, development and implementation from 2002–2008 (1469–8161 (Electronic))

Garcia HH, et al (2014) Efficacy of combined antiparasitic therapy with praziquantel and albendazole for neurocysticercosis: a double-blind, randomised controlled trial (1474–4457 (Electronic))

Gong XF, Liu JL (1996) Observation of duck and goose eating Oncomelania hupensis. Hubei J An Vet Sci (2):40–40

Gryseels B, Polman K, Clerinx J, Kestens L (2006) Human schistosomiasis (1474–547X (Electronic))

Guo JG (2006a) The history and current situation of comprehensive management of schistosomiasis in China. Chin J Prev Med 4:225–228

Guo JG (2006b) Status and role of strategy for control of sources of infection of schistosomiasis in China. Chin J Schistosomiasis Control 03:231–233

Guo J, Guo JG (2005) Research progress in early immunodiagnosis of schistosomiasis. Foreign Med Sci Parasit Dis 032(006):248–253

Guo J, Booth M, Jenkins J, Wang H, Tanner M (1998) Field activity cost estimates for the first 3 years of the World Bank loan project for schistosomiasis control in China. Southeast Asian J Trop Med Public Health 29(4):706–713

Guo JG, Zheng J (1999) Research progress on the prevalence and control of schistosomiasis in my country. Chin J Parasitol Parasit Diseases 17(005):2–263

Guo JG, Zheng J (2000) Schistosomiasis endemic and cintrol in China. Chin J Dis Control Prev 04:289–293

Guo JG, et al (2001) A baseline study on the importance of bovines for human Schistosoma japonicum infection around Poyang Lake, China (0002–9637 (Print))

Guo JG et al (2005) The role of 'passive chemotherapy' plus health education for schistosomiasis control in China during maintenance and consolidation phase. Acta Trop 96(2–3):177–183

Haggerty CAO, et al (2020) Aquatic macrophytes and macroinvertebrate predators affect densities of snail hosts and local production of schistosome cercariae that cause human schistosomiasis (1935–2735 (Electronic))

He R, Liu MB, Dou XR, Cui YF, L W (2019) Evaluation on effect of ecological duck farming on Oncomelania Hupensis elimination in hilly schistosomiasis endemic area. Parasitoses Infect Diseases 17(4):233–237

Hiv/Aids JUNPO (2010) Global report: UNAIDS report on the global AIDS epidemic. Joint United Nations Programme on HIV/AIDS (UNAIDS)

Hotez PJ, et al (2014) The global burden of disease study 2010: interpretation and implications for the neglected tropical diseases (1935–2735 (Electronic))

Hotez PJ, Engels D, Fenwick A, Savioli L (2010) Africa is desperate for praziquantel (1474-547X (Electronic))

JIPD (2018a) Aid Zanzibar's Schistosomiasis control project to carry out schistosomiasis health education activities in Uwandani demonstration Zone

JIPD (2018b) The awarding ceremony of the "Schistosomiasis Cup" composition contest was successfully held for the China-aided mulberry schistosomiasis control project

JIPD (2018c) China's aided mulberry schistosomiasis control project actively carried out health education work on Pemba

JIPD (2018d) The mulberry schistosomiasis control project carried out a multi-departmental cooperation and exchange promotion meeting for schistosomiasis control in Pemba Island

JIPD (2018e) The district-level promotion of the mulberry schistosomiasis control project in Pemba Island is carried out

JIPD (2018f) Pemba Island aided mulberry schistosomiasis work department coordination meeting held smoothly

JIPD (2018g) Strengthen project advancement and expand project impact-the district-level government promotion meeting for the mulberry schistosomiasis prevention project was held in Pemba

JIPD (2019a) China-aided Zanzibar schistosomiasis control technology project held promotion and exchange conference on Zanzibar Island

JIPD (2019b) China-aided Zanzibar schistosomiasis prevention and control technical assistance project held the kick-off meeting of Pemba Island schistosomiasis handwritten report

JIPD (2019c) The mulberry schistosomiasis control project carried out a multi-departmental cooperation and exchange promotion meeting for schistosomiasis control in Pemba Island

King CH, Dickman K, Tisch DJ (2005) Reassessment of the cost of chronic helmintic infection: a meta-analysis of disability-related outcomes in endemic schistosomiasis. (1474-547X (Electronic))

Kjetland EF, et al (2006) Association between genital schistosomiasis and HIV in rural Zimbabwean women (0269–9370 (Print))

Knopp S, et al (2013) Elimination of schistosomiasis transmission in Zanzibar: baseline findings before the onset of a randomized intervention trial (1935–2735 (Electronic))

Lei JS et al (2010) Application of technology for ecological snail control in large-area dike breeding. J Trop Diseases Parasitol 3:148–150
Li W et al (2019) Progress of interruption of schistosomiasis transmission in Jiangsu Province. Chin J Schisto Control 31(06):583–590
Mazigo H (2019) Participatory integrated control strategies and elimination of schistosomiasis in sub-Saharan Africa (2214-109X (Electronic))
Michuzi MI (2014) Ministry of health Zanzibar signs mou with China to implement a programme to elimanate schistosomiasis in Geneva
Mohammed AZ, Edino ST, Samaila A (2007) Surgical pathology of schistosomiasis. (0027–9684 (Print))
Rollinson D, et al (2013) Time to set the agenda for schistosomiasis elimination (1873–6254 (Electronic))
Savioli L, Gabrielli AF, Montresor A, Chitsulo L, Engels D (2009) Schistosomiasis control in Africa: 8 years after World Health Assembly Resolution 54.19 (1469–8161 (Electronic))
Shan LV (2017) China-Africa collaboration in schistosomiasis control
Shi XD (2018) Research progress on ecological snail control technology. J Trop Diseases Parasitol 016(002):117–12166
Steinmann P, Keiser J, Bos R, Tanner M, Utzinger J (2006) Schistosomiasis and water resources development: systematic review, meta-analysis, and estimates of people at risk (1473–3099 (Print))
Stelma FF, Talla I, Verle P, Niang M, Gryseels B (1994) Morbidity due to heavy *Schistosoma mansoni* infections in a recently established focus in northern Senegal (0002–9637 (Print))
Stothard JR, et al (2011) Closing the praziquantel treatment gap: new steps in epidemiological monitoring and control of schistosomiasis in African infants and preschool-aged children (1469–8161 (Electronic))
Stothard JR, Sousa-Figueiredo JC, Navaratnam AM (2013) Advocacy, policies and practicalities of preventive chemotherapy campaigns for African children with schistosomiasis (1744–8336 (Electronic))
Stothard JR, Sousa-Figueiredo JC, Betson M, Bustinduy A, Reinhard-Rupp J (2013) Schistosomiasis in African infants and preschool children: let them now be treated! (1471–5007 (Electronic))
Tchuem TL (2012) Control of schistosomiasis and soil-transmitted helminthiasis in sub-Saharan Africa: challenges and prospects. Curr Topics Trop Med. edn., ARodrigues-Morales AJ
Tchuem Tchuenté LA, Rollinson D, Stothard JR, Molyneux D (2017) Moving from control to elimination of schistosomiasis in sub-Saharan Africa: time to change and adapt strategies (2049–9957 (Electronic))
WHO (1984) World Health Organization new strategy on schistosomiasis (0125–1562 (Print))
WHO (1985) The control of schistosomiasis: report of a WHO expert committee [meeting held in Geneva from 8 to 13 November 1984]. World Health Organization, Geneva
WHO (2001) World Health Assembly Resolution WHA 54.19 Elimination of schistosomiasis. World Health Organization, Geneva
WHO (2006) Preventive chemotherapy in human helminthiasis: coordinated use of anthelminthic drugs in control interventions: a manual for health professionals and programme managers. World Health Organization Press, Geneva
WHO (2010) World malaria report 2010. Geneva Switzerland WHO 30(1):189–206
WHO (2012a) The London Declaration on Neglected Tropical Diseases. 2012. World Health Organization, Geneva
WHO (2012b) Roadmap for implementation: accelerating work to overcome the global impact of neglected tropical diseases. World Health Organization, Geneva
WHO (2012c) World health assembly resolution WHA 65.21 elimination of schistosomiasis. World Health Organization, Geneva
WHO (2013) World health assembly resolution WHA 66.12 neglected tropical diseases. World Health Organization, Geneva
WHO (2014) WHO signs Memorandum of Understanding with China and Zanzibar for collaboration on schistosomiasis elimination in Zanzibar

WHO (2015) Schistosomiasis: number of people treated worldwide in 2013 (0049–8114 (Print))
WHO (2016) Schistosomiasis: number of people treated worldwide in 2014 (0049–8114 (Print))
World Health Assembly Resolution WHA 28.53 Schistosomiasis (1975) World Health Organization. http://www.who.int/neglected_diseases/mediacentre/WHA_28.53_Eng.pdf?ua=1
World Health Assembly Resolution WHA 29.58 Schistosomiasis (1976) World Health Organization. http://www.who.int/entity/neglected_diseases/mediacentre/WHA_29.58_Eng.pdf?ua=1
World Health Assembly Resolution WHA 54.19 Elimination of schistosomiasis (2001). http://www.who.int/entity/neglecteddiseases/mediacentre/WHA_54.19_Eng.pdf?ua=1
Xu J, et al (2016) China-Africa and China-Asia collaboration on schistosomiasis control: a SWOT analysis (2163–6079 (Electronic))
Yuan HC, Guo JG, Robert B, Marcel T, Chen XY, HZ W (2000) The 1992–1999 World Bank schistosomiasis research initiative in China: outcome and perspectives. Parasitol Int 49 (3):195–207
Zanzibar (2018) Revolutionary Government of Zanzibar
Zhao WX (1996) Practical schistosomiasis. Beijing Medical Publishing House, Beijing
Zhou XN et al (2004) The current status of schistosomiasis epidemics in China. Chin J Epidemiol 25(7):555–558

Chapter 13
Construction of an Epidemic Management Information System Leading to Schistosomiasis Elimination in Zanzibar

Liang Shi and Yakub Shoka

Abstract The information system is an important part in schistosomiasis control work. The rapid development of information technology for epidemic data has promoted the establishment of China's schistosomiasis elimination monitoring system, and has accelerated the process of schistosomiasis elimination. The schistosomiasis information system has not been established at the national level in Zanzibar's and information management capabilities are very weak. There is a lack of data that can fully and dynamically reflect the population's disease and risk factors epidemic trends and their influencing factors. This situation cannot meet the goal of eliminating schistosomiasis. The China expert team assisted Zanzibar to build an African schistosomiasis control information system suitable for local epidemics, which improve data management application level and service capabilities of schistosomiasis control work.

Keywords Information System · Schistosomiasis control

13.1 Background

Zanzibar is an integral part of the United Republic of Tanzania, with an area of 2654 square kilometers. It consists of Unguja Island (also known as Zanzibar Island), Pemba Island and more than 20 small islands. Kilometers. At the end of 2019, the total population was about 1.625 million, and 98% believed in Islam. The administrative area of Zanzibar is divided into five provinces, namely Zanzibar City and

L. Shi (✉)
Jiangsu Institute of Parasitic Diseases, Wuxi, Jiangsu, China
e-mail: shiliang@jipd.com

Y. Shoka
Ministry of Health of Zanzibar, Zanzibar, Tanzania

K. Yang, H. Mehlhorn (eds.), *Sino-African Cooperation for Schistosomiasis Control in Zanzibar*, Parasitology Research Monographs 15,
https://doi.org/10.1007/978-3-030-72165-7_13

Southern Province, Zanzibar Northern Province, Zanzibar Western Province, Pemba Northern Province and Pemba Southern Province.

At present, schistosomiasis surveillance and information management capabilities in Zanzibar is weak, and no monitoring information system has been established at the national level. There is a lack of data that can fully reflect the population's prevalence of disease and risk factors and their influencing factors. This situation can no longer meet the goal of schistosomiasis elimination. To this end, it is necessary to carry out the construction of an information management system, establish a nationwide surveillance information system, collect and manage public health surveillance data, and focus on schistosomiasis incidence information in important local hospitals and case management information in primary medical and health institutions.

The information system is an important basis and support for schistosomiasis control. In China, the rapid development of epidemic situation information technology has promoted the improvement of the schistosomiasis surveillance system to a certain extent, and accelerated the process of schistosomiasis elimination. Expert Group on China Aid to Zanzibar Schistosomiasis control Project builds an African schistosomiasis prevention and control information system which suitable for the local epidemic. The system uses computer management software and network technology, combined with China's experience in schistosomiasis control, improves the current data management and application efficiency, comprehensively improves the quality, management and service capabilities, and promotes the standardization of schistosomiasis control, Standardized, scientific.

13.2 Demand Analysis

13.2.1 Government Analysis

As an important basis and support for schistosomiasis control, the information system should focus on the global health development goals. First, the information system should protect and improve people's livelihood, benefit the families of each affected area, especially children's health, and use information technology as the focus Means to promote the information application and interconnection of disease control and vector control; Second, the information system should realize cross-departmental information sharing, link up and down, dynamically track and monitor the occurrence and development of diseases, minimize their harm, and solve the strongest response from local people in Africa Issues such as public health and medical services. At the same time, it should optimizes resource allocation, changes the disease investigation and monitoring model, replaces manual paper management with computerized information management, and replaces the intermittent and solidified method with a continuous and dynamic model, which improves the efficiency and effectiveness of schistosomiasis.

13.2.2 Resource Analysis

The basis and core of schistosomiasis control is to collect, analyze the information, that is, the acquisition and use of information. According to different control methods, objects and purposes, control work is mainly case-based, event-based and laboratory-based, the entire population and sentinel hospitals, etc., including basic information (general situation of the case, morbidity, diagnosis), progression information (Retrospective information and case follow-up, laboratory testing, treatment, prognosis), as well as host, media, environment and other information. Multiple dimensions can be obtained by collecting raw information variables and performing statistical analysis, such as schistosomiasis infection rate and vesicular snail density. The main content is shown in the Table 13.1.

13.2.3 Process Analysis

The basic business process can be summarized as information collection, information management and information analysis and utilization. The information collection stage mainly realizes the acquisition of monitoring data. Collection methods can

Table 13.1 Schistosomiasis control information source

Source classification	Content	Investigation agency
Basic information	Information on village residents in endemic areas	Disease control institutions that carry out disease surveillance
	Resident information of village residents in endemic areas	
Survey information	Questionnaire survey of residents' schistosomiasis	
	Student schistosomiasis questionnaire survey	
	Investigation information of confirmed cases	
Media information	Environmental information on snail breeding	Disease control institutions that carry out disease surveillance
	Snail investigation information	
	Drug control blister conch information	
Case information	Urine sample information of schistosomiasis cases	Disease control institutions or medical institutions that carry out disease surveillance
	Schistosomiasis diagnostic information	
	Urine test quality control information for schistosomiasis	

be paper report card and questionnaire, electronic data exchange, electronic form import, etc. The procedure for information collection can be that medical institutions routinely report information to the disease control business management department in accordance with certain rules, that is, passive monitoring. Information collection can also be the disease control business management department to actively collect information as needed, that is, active monitoring. The process of information management is relatively complicated. In addition to the common operations of information query, deletion, correction, verification, update, etc., it is also necessary to use pipeline and account type monitoring. The former is based on the database during the year, and manages the information by year to realize the monitoring of the disease incidence trend; the latter is to establish case files, record each detection, treatment, follow-up information based on the case management of current cases, which achieve the patient dynamic information management of the entire process of suffering from disease. In the use of information analysis, in addition to the population health evaluation indicators such as infection rate, prevalence rate, mortality rate, summary report and analysis report, data space statistical analysis, data mining analysis and other activities can be used depending on the monitoring purpose and information type.

13.2.4 Business Analysis

The system is used for monitoring based on cases and water environment, to realize the collection, management and analysis of case information. The system is characterized by "online, real-time, case", that is, all county and above hospitals and township health centers report information to the state. At the same time, the system is also used to collect, manage, analyze, and use information on the breeding environment of snails and their prevention. The main contents of the construction include:

The network covers primary-level medical institutions and disease prevention and control institutions. The information collection function can be extend to primary medical institutions which improve disease surveillance coverage, accuracy and effectiveness through the Internet resources.

Achieve direct reporting of the epidemic situation, so that the frontline personnel of medical institutions and grass-roots disease prevention and control institutions can directly submit data to the monitoring system to maximize the timeliness of monitoring; realize monitoring network acquisition and management; information application for centralized management.

13.2.5 Coding Analysis

In order to achieve precise and long-term management of schistosomiasis control data, a unified population code and water environment code must be established. For coding rules, refer to Secs. 13.1 and 13.2 of this chapter.

13.2.6 Technical Analysis

13.2.6.1 System Performance Requirements

- The system has good responsiveness.
- The system computer webpage and mobile terminal support disease control and medical staff to conduct business online in real time. At the same time, considering the weak network information in remote areas of Africa, the system mobile terminal supports offline disease investigation and other services.
- The system should have massive data storage and management capabilities, stable data set optimization processing capabilities, and transnational stable data retrieval capabilities. During the project, the data center roughly estimated the data capacity to be around 5T. The system should have the optimal management of hierarchical, hierarchical, and partitioning of multinational data and parallel processing capabilities. It requires stable and fast access.
- System ease of use requirements: operation supports Chinese and English interfaces, easy to learn, and simple training can master system operation. The key steps need to set reminders and use the menu method, according to the business classification menu.
- The system should be able to provide 7 * 24 hours of continuous operation, with an average annual failure time of less than 24 hours and an average failure repair time of less than 60 minutes.

13.2.6.2 System Security Requirements

System security is the foundation of any business development. System construction needs to consider physical level, network level, system level, application level and security management requirements. It is also necessary to carry out procedures such as acceptance assessment, operation and maintenance on the system. Implement the requirement of not less than three levels of system level protection for the system. In addition, for data security, special attention should be paid to the privacy protection of residents' files during the construction of the system. Disease control, doctors or individuals need to pass identity authentication or security protection strategies to prevent the disclosure of patient personal factors before viewing the residents' health records.

13.3 Overall Design

13.3.1 Overall Architecture

Overall idea of system design: Taking the construction of schistosomiasis control information management system as the core, it covers the control information systems of the two island provinces, cities and districts of Wenguja and Pemba in Zanzibar. The system architecture is based on the construction of network infrastructure, building a data center for schistosomiasis prevention and control, and building a management application module for prevention and treatment. Eventually, a management system with the main functions of disease investigation and treatment, and snail control as the main body will be formed, and a safety management system and a standard system will be established and improved (Fig. 13.1).

13.3.2 Functional Model

The system should have three basic functions of information collection, information management and information analysis. The system has online and offline input functions, while providing interfaces to support data exchange with other

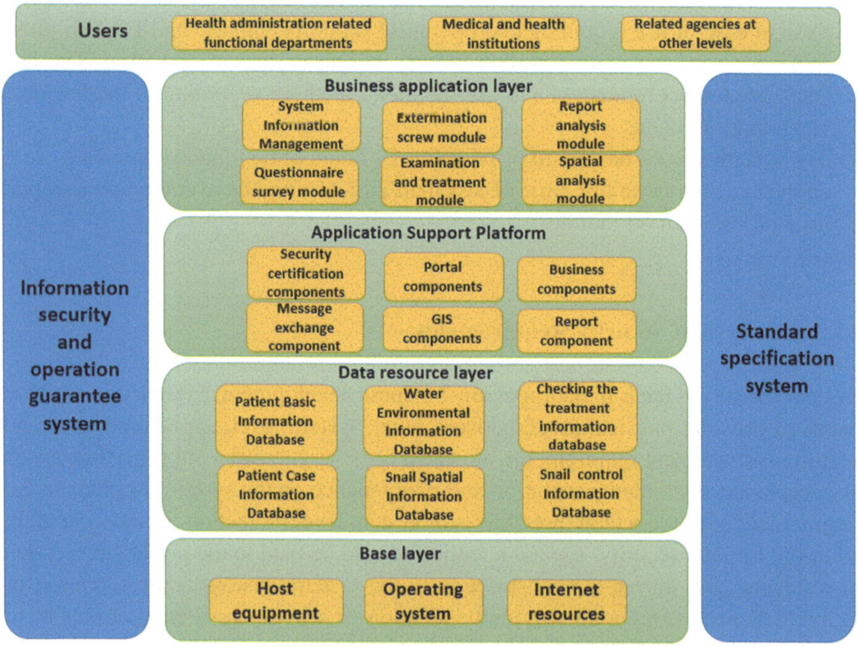

Fig. 13.1 Overall framework of information system

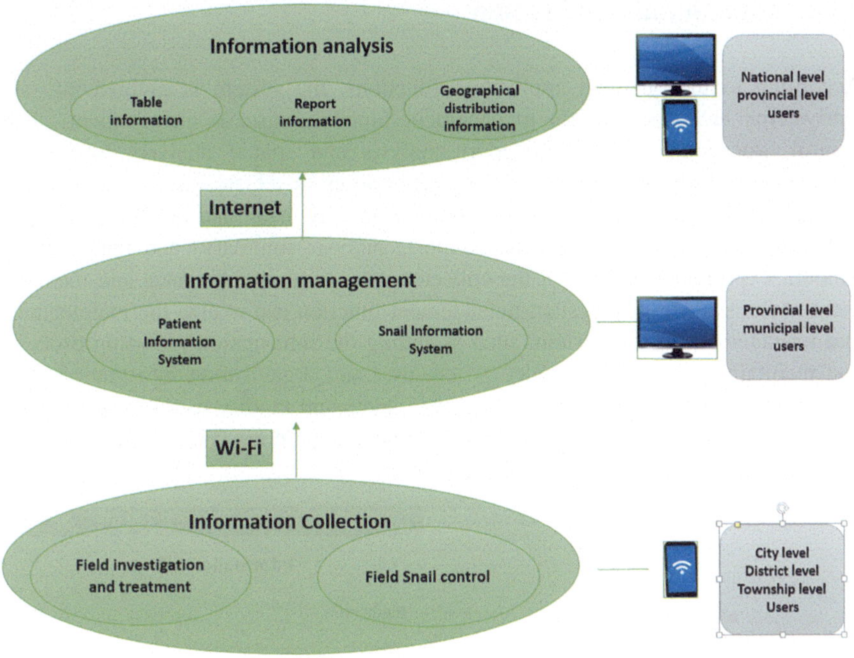

Fig. 13.2 System operating modes and relationships

information systems. The collected information will be aggregated to form a data set after operations such as review, correction, duplication check, and deletion. Reports, distribution maps, and disease measurement indicators will be produced after statistical analysis (Fig. 13.2).

13.3.3 Application Architecture

Responsible reporting units such as disease control institutions and medical institutions at all levels can access the system portal through a browser, log in to specific information systems according to their authority, and realize information entry. Each agency is responsible for managing and analyzing monitoring data, forming analysis reports, and responding to abnormal situations in a timely manner. The health administrative department uses the analysis results of monitoring data to make decisions and release information to the public.

13.4 Functional Application

According to the business application functions, the system is divided into six subsystems including system management function, disease management, snail control, questionnaire survey, report analysis and spatial analysis subsystem. Through these sic subsystem modules, information on schistosomiasis control can be realized: First, collection (entry, report, review, revision and deletion), Second, statistical analysis (query, statistics, report, analysis summary) and management (import, export and backup) of the collected data, third, maintenance and management of the system. The system management function can realize the maintenance and management of the system, such as setting the information reporting process, and maintaining the village code, system user and organization information. The system management function can also add, delete and modify user or organization information, and can output query results to generate reports (Fig. 13.3).

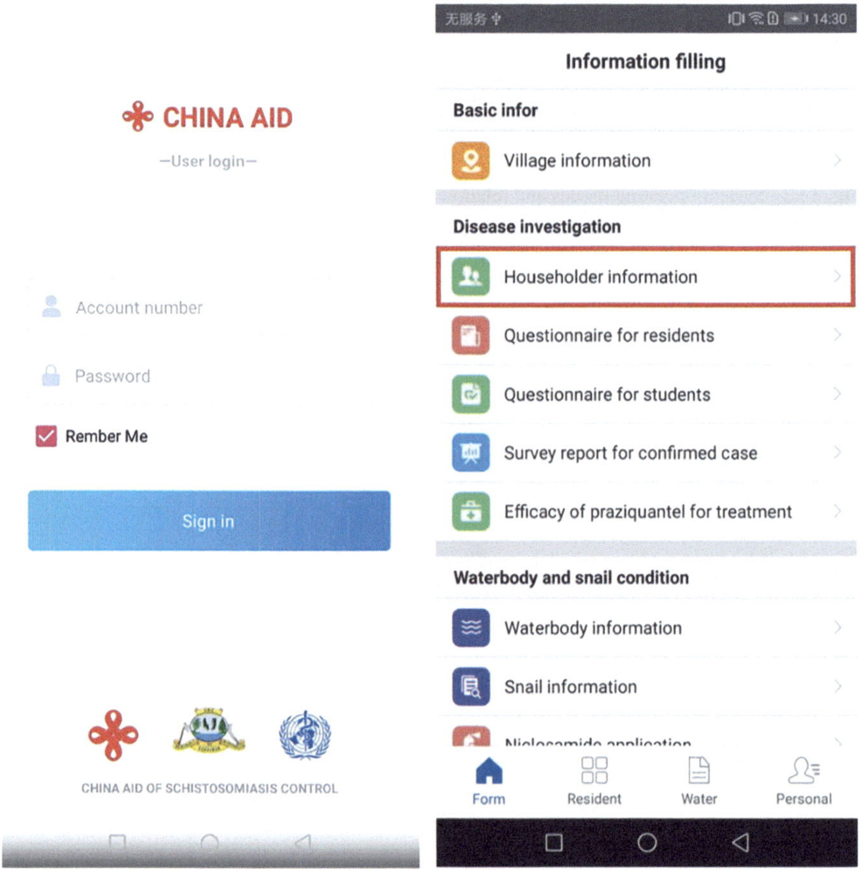

Fig. 13.3 Mobile terminal APP operation interface for medical and health institutions

13.4.1 General Functions

13.4.1.1 Information Collection

In the process of carrying out field work such as resident information survey and questionnaire survey, collection work can be done by manually or through the corresponding system of the mobile terminal APP.

13.4.1.2 Entering Information

There are generally two ways, one is to directly fill in the entry after entering through the control information management system; the other is to automatically synchronize the information collected by the mobile terminal APP system when the network is connected to control information management system. There are two ways to enter the field information. The laboratory test results are directly entered into the system. If the comprehensive information cannot be collected at one time, the information collected by the line should be entered in time, and the additional information will be added after the full information is collected.

13.4.1.3 Information Review

The information is entered by the reviewer for verification and confirmation. If any information is found to be incomplete or in error during the review, it should be verified with the information person and re-filled.

13.4.1.4 Information Quality Control

The data management personnel performs quality control work such as omissions and errors. After verifying and correcting the duplicate items, wrong items and missing items, re-fill. At the same time, quality control personnel organize data auditors in various regions to conduct regular inspections and intensively train similar problems to reduce errors at the data collection end.

13.4.1.5 Report Generation

For the information that has been verified to be correct, the information system regularly generates relevant summary tables.

13.4.1.6 Information Inquiry

Information inquiry implements authority management, and the information officer can inquire relevant information within the authorized scope.

13.4.2 Main Module Function Description

13.4.2.1 Inspection and Treatment Management Module

The system collects schistosomiasis cases through a combination of qualitative and quantitative working methods such as field investigations and expert interviews. The information system can describe the current status and trends of the disease, dynamically track and follow up the diagnosis and treatment of patients. The main functions include resident user information, questionnaire survey, urine test results, confirmed case reports, and praziquantel efficacy observation (Figs. 13.4 and 13.5).

- Resident information collection: mainly used to collect basic information of villages and residents.
- Questionnaire survey of residents and students: mainly used for questionnaire survey of residents and students that have been saved and record the survey information in real time. Mainly fill in two parts of information, such as basic information and questionnaire information.
- Entering urine test results: collecting sample test results through the web client input system.
- Confirmed case report: mainly used for questionnaire survey on residents with positive urine test results, and record survey information in real time.
- Praziquantel efficacy observation: mainly used for follow-up of confirmed cases and medication cases, and timely record the case. The main content includes three parts: treatment observation and diagnosis record (cannot be modified), treatment medication record, and response after treatment observation.

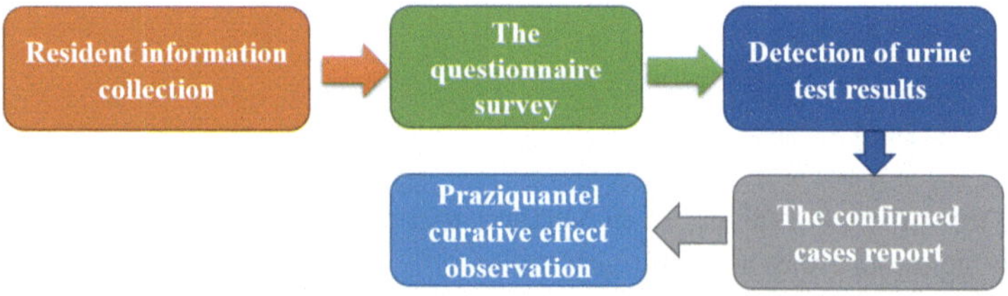

Fig. 13.4 Inspection and treatment information management process

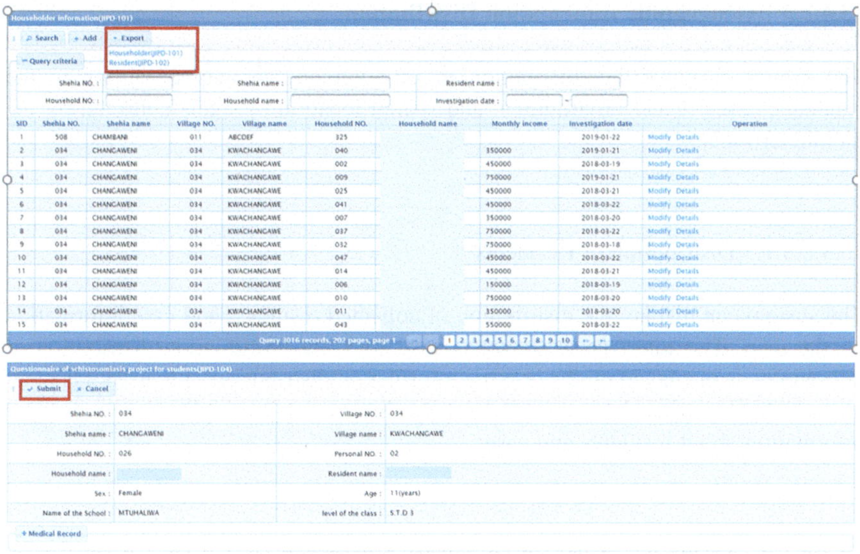

Fig. 13.5 Web interface portal interface diagram for medical and health institutions

13.4.2.2 Water Body Snail Situation Management Module

13.4.2.2.1 Basic Information of Water Body

It is mainly used by investigators to record the information of water bodies. The types of water bodies include permanent ponds and temporary ponds. Water body information is mainly divided into three parts: water body basic information, water body way village, water body picture (Fig. 13.6).

13.4.2.2.2 Snail Control Information

It is mainly used for checking and kill snails on submitted water bodies. Checking the snail is mainly divided into basic information and checking snail sub-information. Snail killing is mainly divided into two parts: basic information and Snail killing picture information (Fig. 13.6).

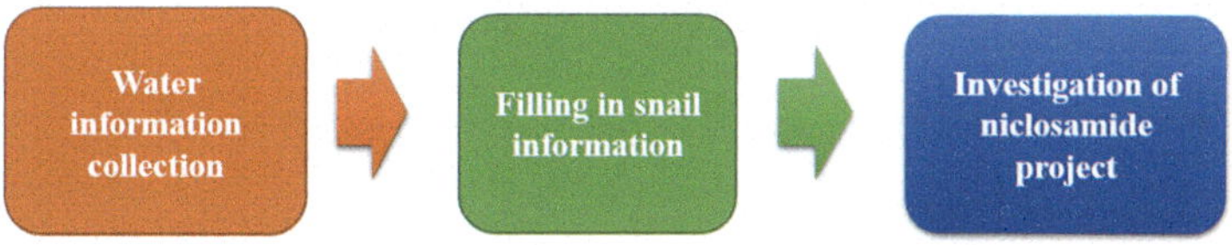

Fig. 13.6 Snail control information management process

13.4.2.3 Geographic Information Distribution

The system displays the distribution information on the map according to the residents, cases, water environment and spatial location of snails collected by the mobile terminal.

13.4.2.3.1 Distribution of Residents

The system can display the distribution of collected residents and case information on the map. The interface is shown in Fig. 13.7, the information distribution of households is displayed on the left; the information distribution of households in all areas is displayed on the right. Click on the list area on the right to view detailed information about residents and cases.

13.4.2.3.2 Display of Water Environment Distribution

The system can display the water environment and the distribution of snails on the map. The interface is shown in Fig. 13.8, the information distribution is displayed on the left; the area information is displayed on the right. Click the list area on the right to view the water body name, water body type, and view the water body and snail condition detailed information.

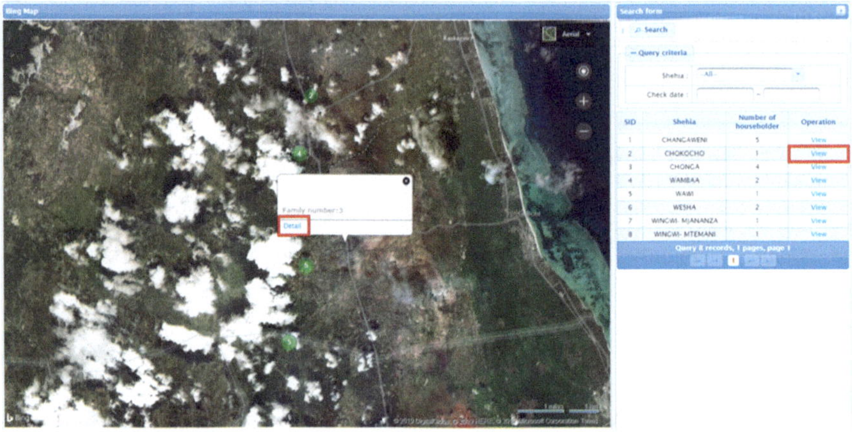

Fig. 13.7 Geographical distribution map of Pemba Island residents

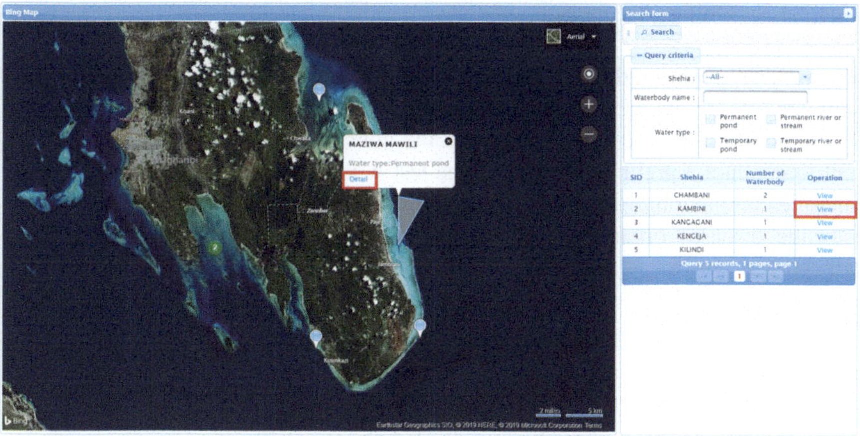

Fig. 13.8 Pemba Island water environment geographical distribution map

13.4.2.4 Deployment and Implementation

13.4.2.4.1 Implementation Plan

During the planning period of the project, a population case database and a water environment database focusing on the pilot area will be implemented, combined with local reality, to realize the main business applications of snail control and disease treatment, covering the pilot areas and demonstration areas. In the course of the project, the development tasks are reasonably arranged according to the priority. The project contractor and the construction party jointly guarantee the project construction and smooth implementation. The unit sends the project contact person and implementation plan to the Jiangsu Institute of Parasitic diseases, actively communicates on a regular basis, and provides timely feedback on the project construction.

13.4.2.4.2 System Debugging

After the system has been developed, deployed, and tested, the system starts joint debugging. The project site timely feedback the joint debugging situation to ensure that the system and the access unit can complete the data exchange through the platform standard interface requirements. Joint debugging mainly includes: internal joint debugging of the system to achieve stable operation of the platform; docking of the site, laboratory and data center, realization of docking between the mobile terminal and the computer network terminal; verification of collected data, and joint debugging test of system services.

13.4.2.4.3 Online Acceptance

The project team conducted a system self-test on the spot in Africa, and then submitted an acceptance test application to Jiangsu Institute of Parasitic diseases. The institute organized and coordinated the construction unit and the supervision unit to conduct a preliminary acceptance test with the project team. Three months after the completion of the initial inspection and trial operation, Jiangsu Institute of Parasitic diseases organized an expert team to conduct final acceptance of the project.

13.4.2.4.4 Operation and Maintenance Management

After the on-line acceptance, the system shifts to the daily management stage and ensures the normal operation of the system according to the development of special standard operating specifications. The project team and the Pemba Tropical Disease Control Office formulated the specification, including system support management, system use management, system operation management, and emergency response.

In order to create a "convenient, efficient and stable" information management system for control schistosomiasis, and provide assistance to elimination of schistosomiasis in Zanzibar. The design and development of the information platform began in September 2018, and the trial began in February 2019. It is mainly used for China's assistance to the Zanzibar schistosomiasis control project. There are 73 shehia in the control area of Zanzibar pemba Island in Tanzania. The application terminal is divided into two parts: computer WEB terminal and APP mobile terminal, which mainly covers the control of schistosomiasis and the control of snails. Jiangsu Institute of Parasitic diseases is responsible for the operation management and training of the platform, and Nanjing Zhongweixin Company is responsible for Technical Support. The user end is a user of a professional prevention organization and manages and maintains users at all levels in accordance with the principle of level-by-level management.

13.4.3 For the Community

13.4.3.1 Unified Crowd Coding

Break the repeated entry of each survey, increase workload and reduce errors in information review.

13.4.3.2 Innovative Application

The application of an information system based on mobile terminals has been realized, which provides important technical support for the registration and testing of subsequent grassroots medical and health institutions and provides more convenience for medical shrinking public services for local people.

13.4.4 For Medical Institutions

13.4.4.1 Disease Reference

I n township hospitals with network coverage on Pemba Island in Zanzibar, hospital doctors can access disease files based on system workstations. The doctor can understand the patient's past history, medical records and other information to the registered residents, which facilitates a more effective diagnosis, reduces unnecessary repeated examinations, and improves work efficiency. Through the sharing of disease file information, the system also effectively promotes the sharing of inspection and examination results between medical institutions and disease control centers.

13.4.4.2 Environmental Inspection

Disease control agencies can check the registered snail breeding environment, master the basic information of the snail breeding environment, the spatial distribution of the environment, the drug snail elimination and other information for the registered environment, provide important information support for effective snail control work, improve work efficiency (Figs. 13.9, 13.10).

13.4.5 For Government Supervision

Relying on the information system, it is possible to collect and analyze the population and environmental data of the whole island to form a comprehensive prevention and control information management system; the system develops and realizes remote login; In the pilot area, real-time monitoring and management of basic population information, patient case information, medication treatment follow-up information, snail breeding environment information have been realized. Based on these data, the management of the local health administrative department

Fig. 13.9 Snail control field application

can strengthen the prevention and management of schistosomiasis, improve the overall management level and provide decision support. Within 1 year after the system was launched, more than 20,000 pieces of crowd information, more than 600 water environment information, and more than 500 patient case information were entered. The Zanzibar Minister of Health and the Pemba Governor praised the intelligence and efficiency of the information platform when they visited the test of the information system. They said that the system can collect data offline and can remotely manage and statistics Real-time data update and epidemic situation dynamics which overcome many problems such as unstable network signals in Africa and delayed update of epidemic data. They believed that this information system will strengthen the disease monitoring network system and have great significance for the elimination of schistosomiasis in Africa. Many Chinese and foreign media such as Xinhua News Agency, People's Daily, CCTV, BBC and Daily News in Tanzania conducted in-depth interviews and publicity on informatization work (Figs. 13.9 and 13.10).

Fig. 13.10 Disease investigate application

References and Further Reading

Abdel-Rahman MS, El-Bahy MM, Malone JB, Thompson RA, El Bahy NM (2001) Geographic information systems as a tool for control program management for schistosomiasis in Egypt. Acta Trop 79:49–57

Barbosa VS, Loyo RM, Guimarães R, Barbosa CS (2017) The geographic information system applied to study schistosomiasis in Pernambuco. Rev Saude Publ 51:107

Lanjuan L (2012) Construction and utilization of regional health information platform. China Science Publishing, Beijing

Liang H, Xue Y (2004) Investigating public health emergency response information system initiatives in China. Int J Med Inform 73:675–685

Roberts M (2018) Successful public health information system database integration projects: a qualitative study. Online J Public Health Inform e207:10

Shen K (2017) Health information management, 4th edn. People's Medical Publishing House, Beijing

Wang Y, Zhuang D (2015) A rapid monitoring and evaluation method of schistosomiasis based on spatial information technology. Int J Environ Res Public Health 12:15843–15859

Wei H, Fei Y, Guo-Hua P (2017) Development and application of WEB-based information management system for chronic schistosomiasis patients. Chin J Schistosomiasis Control 29:363–365

Xiaonong Z (2009) Space epidemiology. China Science Publishing, Beijing

Xinzheng J (2014) Health information system. People's Medical Publishing House, Beijing

Yang GJ, Vounatsou P, Zhou XN Utzinger J, Tanner M (2005) A review of geographic information system and remote sensing with applications to the epidemiology and control of schistosomiasis in China. Acta Trop 96:117–129

Yang L, Hao W, Jia-Jing Z, Yu-Ting Z, Yue-Lin X, Hua-Tang L, Ye-Hua Z, Ming-Xing X (2019) Analysis of endemic situation of schistosomiasis based on geographic information system (GIS) in Wuhan City in 2017. Chin J Schistosomiasis Control 31:410–413

Yang Kun LY, Yang G, Huang Y, Hong Q, Zhu Y (2005) Research and design of internet geographic information system on schistosomiasis. Chin J Schistosomiasis Control 2005 (17):365–367

Zhang, Xuegao HJ (2019) National health informatization survey report- regional health informatization and hospital informatization (2019). People's Medical Publishing House, Beijing

Zhi-Hua C, Yi-Sheng Z, Zhi-Qiang X, Xue-Bing L, Yi-Min D, Li-Jun B, Kai-Min G, You Z (2017a) Application of electronic fence technology based on GIS in *Oncomelania hupensis* snail monitoring. Chin J Schistosomiasis Control 29:615–618

Zhi-Hua C, Yi-Sheng Z, Zhi-Qiang X, Xue-Bing L, Yi-Min D, Li-Jun B, Kai-Min G, You Z (2017b) Implementation of *Oncomelania hupensis* monitoring system based on Baidu map. Chin J Schistosomiasis Control 29:795–797

Zhou XN (2020) A high-quality driver to accelerate the progress towards schistosomiasis elimination by science and technology-led innovation in Jiangsu Province. Chin J Schistosomiasis Control 31:573–575

Zhou XN, Malone JB, Kristensen TK, Bergquist NR (2001) Application of geographic information systems and remote sensing to schistosomiasis control in China. Acta Trop 79:97–106

Chapter 14
Management of China-Zanzibar Cooperation Project: Lessons Learned

Yun Feng

Abstract China-Zanzibar cooperation project was the first international public health project that China sent batches of experts and worked in local field. This was the first exploration for such kind and worth looking backward and summarizing the lessons learned.

Keywords Implementation · Management · Coordination · Capacity building · Infrastructure

14.1 The Implementation and Management Principles

Faced with a completely different situations in Zanzibar as it was in China, the project was challenged in many aspects like languages, culture, politics, habits, etc. Based on the various problems solved, we summarized three principles the project was managed according to.

14.1.1 Comply with Local and International Rules to the Maximum Level in the Project Planning and Execution

For instance, the Chinese produced molluscicides and the praziquantel were all certified and registered by Zanzibar Food and Drugs Board. These provided legitimacy for their use in the local areas, as well as field data and application experiences for the WHO PQ certification in the future.

Y. Feng (✉)
Department of Schistosomiasis Control, Jiangsu Institute of Parasitic Diseases, Wuxi, Jiangsu, China

K. Yang, H. Mehlhorn (eds.), *Sino-African Cooperation for Schistosomiasis Control in Zanzibar*, Parasitology Research Monographs 15,
https://doi.org/10.1007/978-3-030-72165-7_14

14.1.2 Meet the Local Needs in all Aspects of the Project

As a field-based project, the first task of the expert team was to investigate the local needs. Many problems were discovered in the investigation, such as the lack of financial funds, the single control strategy of MDA, the dependence on foreign aid, the lack of cross-sectional cooperation, the lack of disease control facilities and schistosomiasis relevant knowledges in local community, the lack of disease control skills in the local health workers, etc. The project designed a comprehensive strategic package to control schistosomiasis in Zanzibar and explored ecological snail control strategies. For instance, raising ducks were found to have definite effects on inhibiting *Bulinus* particularly in the aspects of reproduction. Water supply project brought benefit to at least 2500 local residents, improved their health behaviors and would finally reduce the incidence of parasitic diseases including schistosomiasis (Figs. 14.1 and 14.2).

14.1.3 Fuse the Merits of Chinese Experiences with the Local Situations in the Execution of the Project

With 70-year schistosomiasis control experiences, the Chinese expert team created a brand-new model of schistosomiasis control strategy package in Zanzibar. For MDA, the Chinese made praziquantels were given to schistosomiasis patients and pills were taken immediately with the health workers stood by. Data were collected

Fig. 14.1 Ducks raised in Zanzibar for snail control

Fig. 14.2 Water supply project launching ceremony in Zanzibar

and all patients were followed up to evaluate the efficacy of MDA. These experiences were used to maximally prevent the patients avoiding taking the pills, which was quite common before. Snail control was given the same importance as MDA in the strategy package, which was proved to be effective in China. Molluscicides produced in China were put into use and all procedure were standardized to control the quality. All infected snails were killed in the 3 years of project execution. Innovated health education tools derived from successful Chinese experiences were created. Demonstration area board, wall slogans, warning signs for risk areas, automatic voice warning devices, most of which were in Swahili, were put into use. Schistosomiasis control themed lectures, essay contest, and poster contest were warmly welcomed in local schools.

14.2 Organization and Responsibilities

At the very beginning of the project, a tripartite project steering committee was set up to promote the coordination. Include:

14.2.1 China

14.2.1.1 National Health and Family Planning Commission

China's National Health Commission is the national coordination agency for Zanzibar schistosomiasis control projects. It is responsible for the consideration and approval of Zanzibar schistosomiasis control projects, raising funding for the project, supervision of the project implementation, while it also liaises with the relevant ministries, the Zanzibar Government, the World Health Organization and coordination between non-governmental organizations, and informs the progress of the project to relevant departments regularly.

14.2.1.2 Ministry of Commerce

China's Ministry of Commerce is responsible for the consideration and approval of Zanzibar schistosomiasis control projects with the National Health Commission, and it is also in charge of raising funds from the relevant departments of the Ministry of Commerce's foreign aid projects to provide financial support to the project.

14.2.1.3 Ministry of Foreign Affairs

According to China's foreign policies, Ministry of Foreign Affairs of China is responsible to guidance the project and deal with multilateral relationship.

14.2.1.4 Jiangsu Province Health Department

Jiangsu Province Health Department is the executing agency of the project and it is responsible for the relevant units of Jiangsu Province (Africa Aid Medical Team) and report the project progress to National Health Commission.

14.2.1.5 Jiangsu Institute of Parasitic Diseases (JIPD)

Jiangsu Institute of Parasitic Diseases is responsible to manage and organize the implementation of project, and develop the project plan and budget, and send experts and provide training, and report the project progress to the National Health Commission regularly.

14.2.1.6 National Institute of Parasitic Disease, China CDC

National Institute of Parasitic Disease, China CDC is responsible to provide the necessary technical support to each implementation units who participate in the project in Zanzibar and assist the implementation units to evaluate process and effectiveness of the project.

14.2.1.7 Chinese Medical Team in Zanzibar

Chinese medical team is responsible to suggest the appropriate local working mode and other work operating mode to aid agencies and Chinese experts, and support the aid agencies and expert's work for schistosomiasis control project.

14.2.2 Zanzibar

14.2.2.1 The Ministry of Foreign Affairs

The Ministry of Foreign Affairs is in charge of Zanzibar's international relationship and help coordinate with China and other relevant countries.

14.2.2.2 The Ministry of Finance

The Ministry of Finance is the focal point for the financial transactions and Memorandum of understanding (MOU) in collaboration with Ministry of Health Zanzibar.

14.2.2.3 The Ministry of Health

Zanzibar government shows strong political will for the multilateral cooperation to eliminate schistosomiasis under the support of China government and World Health Organization., and are willing to promise to mobilize all forces of government, non-government and community participation on schistosomiasis control project, and it is also responsible to coordinate multiple departments including the ministry of health, ministry of water resources, the environment ministry, the ministry of education to participate and ensure the implementation of the project.

The Zanzibar Ministry of health is responsible to organize the medical institutions at different levels in Zanzibar to participate in and implement schistosomiasis control project. Under the guidance of experts, the Ministry of Health carry out the work, such as health education, media surveys, sampling collection, investigation and treatment of patients. It is responsible to supervise the project and ensure the normal operation of project management. It maintains regular communication with the National Health and Family Planning Commission of China the World Health Organization and non-governmental organizations, and regularly informed of the progress of the project. In addition, reconcile NTD Zanzibar help aid agencies guidance and training the local public heath team and provide the field control services for the aid agencies and personnel of project in Zanzibar.

14.2.3 WHO

Department of Neglected Tropical Diseases WHO-HQ is responsible of project guidance and organization, coordination with relevant international organizations and funds involved in the project, provide the technical support, policy guidance and financial support, monitor and evaluate the overall project, and coordinate the National Health and Family Planning Commission of China, Zanzibar governments, NGO coordination among relevant authorities, and regularly informed of the progress of the project, and WHO Zanzibar is the liaison office to assist coordination mechanism, and coordinate the SCORE project for data sharing.

14.2.4 Project Organization Structure

With the supervision of the national schistosomiasis committee (NSC), and the full support from the national institute of parasitic diseases, China CDC, Jiangsu institute of parasitic diseases as well as the NTD office of Zanzibar, the project management organization (PMO) office executed the project. The PMO office consisted of two parts. One was located in and supported by Jiangsu institute of parasitic diseases in China, the other was in Pemba Island and supported by NTD office in Zanzibar. The project organization structure was shown in the following figure (Fig. 14.3).

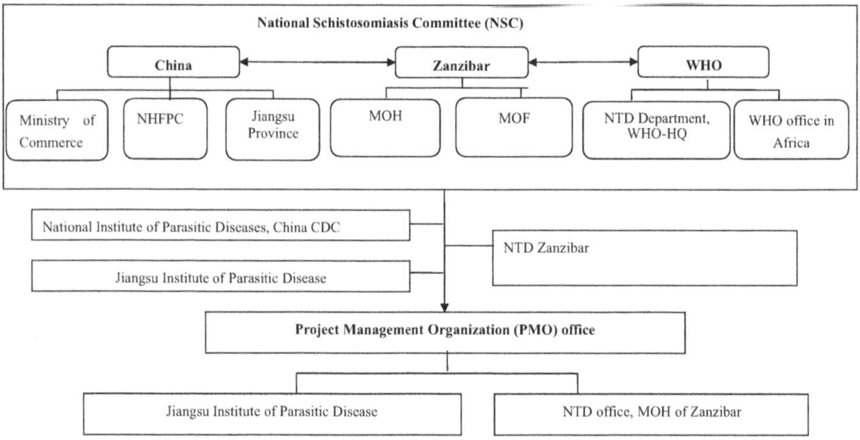

NHFPC: National Health and Family Planning Commission
MOF: Ministry of Finance.
NTD: Negative Tropical Diseases

Fig. 14.3 Project organization chart

14.2.5 Coordination

The success of any international project is closely related to the support of local stakeholders, particularly local government or leaders. The execution of the project was influenced by local politics, economics, culture, religion, resources and so on. In the past 50 years, China's Africa aid activities had been well received by Zanzibar people and government, especially medical aid. Zanzibar schistosomiasis project was a novel Chinese public health aid to Africa. It focused on a disease of poverty and benefit mostly rural and poor people. Thus, it had a potential to receive support from most of the population. It also had the internal motivation to seek support from local and international sectors and establish a coordination and partnership system.

14.2.5.1 Coordination among the Project, Local Government and Sectors

Schistosomiasis control is a multi-sectoral action. It is not only the task of health sector but also need the collaboration among agricultural, water conservancy, educational and other sectors. In Africa, lack of coordination among different sectors was an important reason that hindered the development of schistosomiasis control. In China, we had a mature system to coordinate among different departments. It was a great challenge to duplicate a similar system to a country with a different language, culture and political system. First, the project needed to let the government know about its targets, strategies, the progress and people who worked for it. After the connection was made, the project could find a way to solve the problems it met with the help from the government. Second, the project needed to learn the local policy and culture so as to adjust its strategies and avoid potential conflicts. Third, a close partnership and working on the same page would greatly accelerate the progress of the project, particularly the execution of essential strategies. After 3 years of exploration and practice, we summarized the project-government coordination system in Table 14.1. It was a two-direction and mutual communication system that helped establish mutual relations and trusts. A list of sectors that collaborated in different activities had also be tableted here (see Table 14.2).

14.2.5.2 Partnership on an International Level

The international partnership on schistosomiasis were limited between China and other countries. Situations has improved in the past 10 years, for example, the RNAS + in the southeast Asia. These international collaborations helped exchange information and knowledge, strengthen capacity building, and help hold stakeholder meeting at the national and international level. In Zanzibar project, we kept close contacts with the Chinese Embassy and the WHO Zanzibar office. Thanks to their

Table 14.1 The project coordination system

Participants	Communication ways	Communication contents
Project→government	Initial visit	Each batch of Chinese expert group will pay an initial visit to local government after settling down in Zanzibar. They will explain the project targets, strategy, and general plans; and introduce the new batch of experts and the work plans in the following months.
	Regular report	Project progress, significant changes, coordination requests and problems that need government help
Government→project	Visit the project field	Get an in-depth knowledge about the project through project demonstration and field survey
	Policy making coordination	Consult and coordinate with the project leaders before and after relevant policies are made and published, particularly those policies that will make great impact to the project;
	Regular information release system	The government will inform the project about local government information through government bulletin, media, internet or telephone, etc., particularly about the solutions to the problems the project faced.
Project←→government	Monthly informal discussion/themed discussion	The project will discuss with relevant government sectors each month to discuss about the problems and relevant solutions; the project will discuss with relevant government sectors at the following timing: Project launch, periodical progress, major study issue, project conclusion.
	Telephone, email, internet communication	A specific project member is in charge of the communication, consulting about the local laws and policies and coordinate about the operation.

help, the project could successfully coordinate multi-sector collaboration, run complex programs, and adjust health policies, etc.

14.3 Human Resource Management: Capacity Building Oriented Approach

14.3.1 Chinese Team Recruitment and Training

The Chinese employees of the project mainly worked in the field office and the home office in China. The home office members included the project leader, the project

Table 14.2 The sectors and relevant coordination activities

Sectors	Coordination activities
Zanzibar Health Ministry	Seek support for major project operation and activities; invite leaders to important project activities
Zanzibar Health Study Center	Collaborate multi-sectors on schistosomiasis control
NTD office	Coordinate as a belt between local organizations and the project
Zanzibar Ministry of Water Resources	Cooperation on clean water strategies and other water resource protection actions in snail control
Zanzibar Agriculture Ministry	Cooperation on the agricultural strategies for snail control
Zanzibar Education Ministry	Cooperation on health education of schistosomiasis control in schools
Zanzibar Sports Department and other departments	Cooperation on schistosomiasis control related sports activities and other activities

secretaries, the accountants. They work full-time or part-time for the project. The field office was the working base for each batch of Chinese expert team. The project sent a total of six batches of Chinese experts to Zanzibar. Each batch was composed of five personnel, who were mainly experts from JIPD and other disease control centers in Jiangsu Province, beside of one cooker helping with food preparation.

Before leaving for Zanzibar, the Chinese expert team was trained mainly on English language as well as Zanzibar customs and religious habits. These were aimed to prepare them for the language and culture barrier in the future work and life in a different country.

14.3.2 Local Personnel Recruitment and Training

The project had an idea to teach local people the skills of schistosomiasis elimination. To build the local capacity of Schistosomiasis control, it was necessary to establish a well-trained local team. Meanwhile, the Chinese team alone were faced with problems in many aspects: First of all was the language barrier. The official languages of Zanzibar were English and Swahili. It was not difficult to communicate with government workers in English only. However, in field work, most of the local residents particularly rural people did not understand English. It would be a great barrier for field work like disease survey and health education. The second problem was the religious customs. In Pemba Island where the project located, most of the residents were Muslims. Misunderstandings or even conflicts might occur if the field workers were not familiar with local customs. Thus, for both training and working purposes, it was an urgent must to establish a local team who spoke local languages and was familiar with local customs and public health works.

14.3.2.1 The Recruitment and Labor Division of Local Team

In the beginning, most of the local team were recruited from the NTD office in Pemba Island. These team members used to distribute donated praziquantels to local people before, thus they had relevant treatment experiences. However, they were lack of other schistosomiasis control skills in the field. Owing to the uneven level of education and work experiences of the local team members, the project communicated with each of them about their control experiences, language skills, their insight into the job and personal wills. Based on this information, the local team were divided into the following groups (see Table 14.3).

Due to limited group size in the beginning, all members in a group category did all the jobs in the category. For example, all field work members participated in the household survey and sample collections in the baseline study. This kind of arrangement helped finish the job more efficiently and also expanded the job skills for the group members, preparing for the following enrollment for group expansion.

However, the limited group size impeded the project progress as time went by. Moreover, the generally low education level and the average 46.7-year of the first 13 group members hindered them from learning new disease control technologies. Most of them did not have even high school education. Some of them were nearly at retiring age. With these considerations, the project preferred to recruit new personnel at the age of 25–35, with relevant health work experiences.

Younger personnel could ensure the implementation of field work in complicated environment. As the field work was mainly in rural area with complex traffic situation, walking to targeted areas bearing heavy devices was required particularly in raining seasons when the muddy road blocked the cars. Manpower handling of the

Table 14.3 Work groups of the local team

Group category	Group name	Main responsibility
Group leader	Group leader	Assist with the work plan and execution, manage with the local group member and communicate with local government, etc.
Field work	Population survey group	Information and questionnaire survey, urine collection, etc.
	Snail survey group	Local water body survey, snail survey, etc.
	Snail control group	Molluscicidal activities to water body, etc.
Laboratory work	Laboratory group	Laboratory snail farming, anatomy, urine tests and other pathological study, etc.
Data related work	Data group	Data check, input and management, etc.
Administrative work	Secretory	Reception, coordination and daily routine work
Cleaning work	Cleaning group	Daily clean the office and laboratory, sometimes assist with the test work of laboratory group
Transportation	Driving group	Material transportation and field work

heavy sprayer machines was sometimes required especially in the snail control activities. Younger people could share tough heavy jobs while the older and more experienced members could help teach them the skills. In this way, the efficiency of schistosomiasis control activities was improved.

The recruitment process was cooperated between the project and the government. The project proposed the requirement and enrolling criteria, and then the government coordinated and dispatched the personnel from local health organizations. To our experience, we mainly recruit laboratory workers from clinical laboratories and pathology departments of local hospitals, survey workers from local primary health centers, and other group members from local disease control organizations. In Zanzibar, there was no comprehensive disease control organizations such as CDC. Local public health disease control organizations were mostly composed of some specific disease control offices such as NTD offices. The NTD office had its own independent office but still belonged to local health department. Human resources from this kind of offices might have different skills for different disease control programs. Their work experience should be screened in the enrollment processes. Besides, these specific offices may also be sponsored by other project so it would be difficult to dispatch personnel from them.

Our project also preferred to hire personnel with higher education so that they might learn knowledge and skills faster in computer technology, spatial epidemiology, or pathogenic biology. Limited to the actual educational level in the local health organizations, the enrollment criteria did not include the educational level. But we still recruited at least one college graduator for each of the survey group, data group and lab group. After they learnt new technology such as informationalised data collection or spatial measurement, they could teach the others in the respect group. The age and education information of the local group was summarized in Table 14.4.

Table 14.4 Age and education summary of the local team

Category	Group name	No. of workers	Average age	Education (college/secondary/primary education no.)
Field work	Population survey group	2	47.5	1/1/0
	Snail survey group	3	33.0	1/2/0
	Snail control group	5	38.0	3/2/0
Laboratory work	Laboratory group	2	32.5	2/0/0
Data related work	Data group	3	31.7	1/2/0
Administrative work	Secretory	1	42.0	1/0/0
Cleaning work	Cleaning group	3	46.0	0/2/1
Transportation	Driving group	2	57.5	0/2/0
Total		21	40.7	9/11/1

14.3.2.2 Training of the Local Team

The training of the local team arranged at two phases. The first phase of training was arranged before the launch of the project. A group of Zanzibar NTD office members attended in JIPD a training course on prevention and treatment of schistosomiasis for developing countries. Some of them later became the backbone strength of Zanzibar team of the project. In the course, they were trained for the biology and vector of schistosome, molecular biology techniques of schistosomiasis, the Chinese experience of schistosomiasis control, molluscicide and anti-schistosomiasis medicine, etc. They learnt laboratory skills and visited local schistosomiasis control fields. They also learnt about Chinese culture. This kind of theories and practices combined learning helped trainees quickly control the new knowledge and skills.

The second phase of training occurred in Zanzibar after the project was launched. The Zanzibar employees were trained according to the actual needs of practical work. For instance, before conducting snail control activities, the Chinese team would give theoretical lectures to the Zanzibar team and taught the skills hand-to-hand. Exams would take place afterwards. Only qualified workers would be allowed to attend the field or laboratory activities. After 3 years of learning plus practicing training, the Zanzibar team was able to work independently and continue with the schistosomiasis control activities after the project ended. Moreover, all practices were trained according to standardized operation procedures (SOP). These SOPs covered the practices including the field investigation and control of *Bulinus* snails, resident tests and treatment of schistosomiasis haematobium, health education, sample preparation, information system management, questionnaire investigation, etc. They helped the project members control the quality of these procedures and provide sustainable working and training standard in the long run (Figs. 14.4, 14.5, 14.6).

Fig. 14.4 Skill training in office

Fig. 14.5 Skill training in the field

Fig. 14.6 Skill training in the field

14.3.3 Human Resource Management System Construction

It was our goal to construct a harmonious and sustainable human resource management system during the period of the project. To realize this, we summarized three principles in the complex work.

First, all employments should be in compliance with local laws. There was a long history of friendly relationship between China and Zanzibar. As many China-aid infrastructure construction projects and a fixed-stationed medical team were located in the island, the locals were quite amicable to Chinese experts. However, due to various differences in religions and cultures between two countries, carefully complying to local laws in all ways was still the best choice with least legal risks for an international project, particularly for the employment of local people.

Second, the Chinese employees should inspire the ownership of local employees from the top to the bottom. It was essential for the local employees and also the local residents to realize the elimination of schistosomiasis was for the benefit of their own people, their own nation and their own younger generation. They had to understand that the payment of the project was not only for the job but also for the hope of health they brought to the community. The Chinese leader and the local leader would communicate about these ideas and adapted them to local concepts. These localized concepts would then be delivered to all project members for them to understand the meaning of their works. They would discuss them and give feedbacks to the leaders for improvement. After several rounds of perfection, these ideas would be put in a health education campaign to local residents.

Third, "more work and more gains" principle should be insisted in salary management. In designing the salary package, we set the goals in the very beginning: we must establish a stable project team, which paid a competitive salary that satisfied various living needs of the members; it also inspired the work enthusiasm of the team members and created higher work value; it would also coordinate with the personal career goals of the team members.

The salary package was designed by the Chinese side and jointly negotiated by the two sides. It was composed of the basic salary, the performance salary, the bonus, the welfare and subsidy. The basic salary was the payment fixed to the job position and the required skills. The performance salary was a normative systematic remuneration paid according to the performance evaluation. The bonus was an incentive remuneration for the extra works or outstanding performances. The welfare and subsidy were payments based on the employee's organization membership. The salary package was designed according to the "more work, more gains" principle, which is in contrast with the local government's fixed payment. With strict work discipline implementation, this principle was proved to be effective.

14.3.4 Cultural Integration

In Zanzibar, 98% of the population believe in Islam while most of the Chinese people are not familiar with this religious belief. Before arriving in Zanzibar, the Chinese team had been trained about relevant rituals and taboos, such as the prohibition of alcohol and pork. Therefore, the work plan was made with fully consideration of local habits. In the work agenda, for example, the team leaders arranged two breaks for local employees to do their routine prays. Moreover, weekly opinion exchange meetings provided a platform for team members to communicate with each other, particularly on cultural stuffs like cooking. Several cultural exchange events were arranged. Many Chinese team members learnt some Zanzibar cuisines and it was the same for Zanzibar members. In this way, the project deepened the mutual understanding of the two sides (Fig. 14.7).

14.4 Infrastructure Construction and Supply Management

Infrastructures and supplies were the base for the success of a project. They not only provide the working environment and materials, but also work as a sustainable milestone for the project. In the Zanzibar schistosomiasis control project, we built infrastructures including a laboratory and office building for the local community. Besides, a transnational supply system of materials and devices was run smoothly.

Fig. 14.7 China-Africa cultural exchange event

14.4.1 Laboratory & Offices

The laboratory and offices are essential for the efficient executing and effect of project tasks. It was designed and built before the launch of the project. We followed three design principles: functional, elegant and cost-effective.

Offices are also designed differently according to the people who use them. For example, health specialists' offices were relatively spacious, and arranged close to the conference room for convenience. All offices were supposed to be simple and elegant, embracing the aesthetic appreciation of both China and Zanzibar. Reception rooms and reference rooms are common in a Chinese office building but we did not build them in Zanzibar. Instead, we put reception chairs, printers and file cabinets in the health specialist's offices.

The establishment of a laboratory was a complex systematic project consisting of the overall plan, graphic design and construction of the infrastructures, including the building structure, the electricity and water supply, illuminating equipment, air ventilation and purification, security and environmental protection measures, etc. With the idea to build a "safe, functional, environmental-friendly, durable, elegant, cost-effective, and advanced" laboratory, we illustrated the lab-design in the following aspects:

1. Layout graphic design

   Security, lab purpose and demand, ergonomics, lab accommodating capacity, device layout, design flow, etc. were the main consideration in the graphic design for our Zanzibar project base.

   Security was the first priority among all considerations as laboratory was the most vulnerable place for the risks like explosion, fire, hazardous gas leak, etc. Therefore, we should design as good escape passage and ventilation system as we can. For example, aisle between the experimental stations should all connect directly to the corridors for better evacuation in case of any accident.

   In Zanzibar project, the laboratory was composed of chemistry lab, instrumental analysis lab, bio-clean lab, auxiliary lab, and supply room. Different labs and devices had their own purposes and demands for ventilation, water and electricity supply and cleanness. These all brought challenges to the graphic design.

   In the first step of our design flow, Party A (the user) proposed the basic function requirement and rough layout plan; Party B (design institute), Party C (design company) discussed with Party A and confirmed together about the plan. Traditional laboratory design focused on the appearance and indoor structure rather than lab function. That could result in the disfunction of the labs or essential devices. Therefore, the lab function should be focused in the lab design and professional experts should be consulted with in the very beginning.

   The second step was to design the single lab stations. Each device, each lab station, each lab and each floor should be confirmed one by one until the whole building was totally confirmed.

The third step was to invite the users to participate in the demo meeting of the lab to make a final decision.

2. Device-based structural functional system

   Device-based structural functional systems consisted of experimental stations, device stations, functional cabinets, devices and export systems.

   Worksurfaces were essential for experimental stations, which accounted for 1/3 of the price. In Zanzibar project, we chose veneer like Trespa board and solid core board like epoxy resin board as worksurfaces. They were both durable and cost-effective. Device stations were mainly for the instrumental analysis lab. A single device station was supposed to bear at least 500 kg of weight.

3. Water supply and drainage system

   Water supply and drainage system provided water system for experimental stations. In Africa, as the water supply system was not stable, we had to install water storage devices. At least a 10-ton water tank was required for office and laboratory each.

4. Electric control system

   Laboratory electricity included lighting power consumption and motive power consumption. The former was the prerequisite of a laboratory. In Pemba Island where our Zanzibar project located, the power supply was extremely unstable. It was quite common to be cut power every day. Therefore, we needed to have our own electric generator. Our experience was that at least an 80 KW power generator could support a 400 m^2 laboratory with devices. Meanwhile, each device including computers should be connected to uninterruptible power supply to protect the device as well as data stored.

5. Special gas distribution system

   In a common laboratory, only nonflammable gases (N_2, CO_2) and inert gas (argon, helium) were allowed to connect in by pipes for use. Combustion-supporting gas like O_2 and inflammable gas like H_2 could be used in GC-MS room. All pipes had to be tested for air-tightness and remove oil before use.

6. Hazardous gas export system

   Sometimes there would be chemical pollutions in laboratory, particularly hazardous gas. It was important to remove these gases with the least energy. The latest idea was to treat the laboratory as a smoke exhaust cabinet. In Africa, the most commonly used ventilation systems were fume hood, universal hood and ceiling hood. And the mainly used fans were axial fan (diagonal flow fan, duct fan) and centrifugal fan (Figs. 14.8 and 14.9).

14.4.2 Material & Device Supply

In order to tightly administrate the material and device supply of the project, we established the following system and appointed a specific team member for the management work in each batch.

Fig. 14.8 Before and after the re-construction of the office building

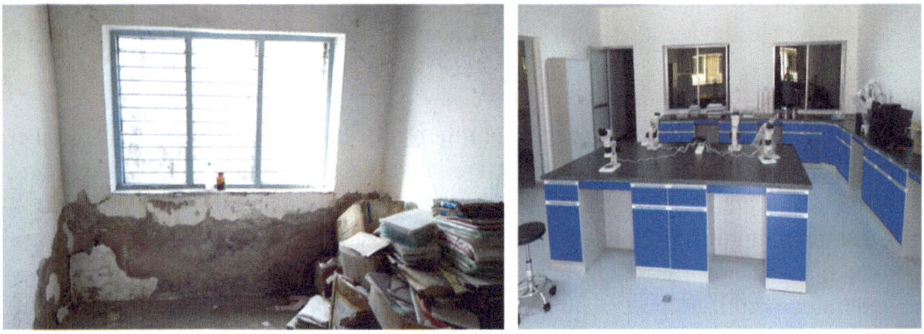

Fig. 14.9 Before and after the renovation of the laboratory

14.4.2.1 Material Procurement, Commodity Inspection and Shipping

All materials were purchased from enterprises in the supply list of the Ministry of Commerce of China through public bidding. Good quality, suitability in practice and cost-effectiveness were the major criteria for procurement. Besides, considering the requirement of the Ministry of Commerce, details like the specification and model of the products chosen, transporter, delivery time had to be fixed in the execution. Therefore, during the process of procurement, products index, performance, goods preparation cycle, applicability of products to recipient countries and after-sales *service* support capability were also in the concern. All the efforts were to maximize the social benefits of the project.

According to the Chinese law "Administrative measures for inspection of foreign aid materials", all equipment and supplies procured by the financial aid from China should be qualified by commodity inspection before shipping abroad. In the commodity inspection process, JIPD took the responsibility of timely monitoring and supervision. Technical experts were invited to the factories to communicate about the situation of recipient countries to make sure of the future smooth operation of the devices produced.

Considering the complex of the custom declaration, JIPD out sourced the process to the transportation company of the recipient country as they were more familiar.

In preparing commodity inspection documents, export documents, arranging inland transportation, sea transportation and customs declaration, JIPD kept close contact with Administration of Quality Supervision, Inspection and Quarantine (AQSIQ), suppliers, commodity inspection bureau in product origin, port commodity inspection station, customs and logistics subcontractors, to keep up with the latest status of document process, commodity inspection transportation and carrier ships. All these helped ensure that all links were closely connected, and the goods were shipped out on time.

Sufficient communication between the Chinese sides and the Zanzibar sides was a must particularly for the local custom declaration documents. Handover point should be confirmed clearly in the contrast and handover status should be signed off in time.

14.4.2.2 Technical Material Management

Technical materials refer to the technical equipment, production supplies, etc. They were procured, used and managed by JIPD in the period of the project. But the property right of technical materials belonged to the recipient. The Ministry of Commerce and JIPD confirmed with the list of technical materials and then informed the list to the commerce organization of the recipient country. After the project execution concluded, the recipient commerce organization would conduct a comprehensive inventory and arrange transfer confirmation procedures. All relevant documents were kept on record in the Ministry of Commerce of China.

14.4.2.3 Life Material Management

Life materials included the transportation vehicles, office equipment, durable goods, consumable goods and other materials that supported the work and life needs of the project team in Zanzibar.

Life materials were also procured, used and managed by JIPD during the project execution. After the project finished, the property of fixed assets like vehicles, office equipment and durable goods belonged to the recipient. The procedures of inventory, transfer and documentation were the same as technical materials. The remaining consumable goods would be transferred to the recipient by JIPD itself after the project ended.

14.5 Financial Management

14.5.1 Financial Personnel Management

The project office in China was allocated with full-time financial personnel, including financial supervisor and cashier, who had accounting qualification certificate and more than 3 years of relevant experiences and intermediate or above accounting titles. A part-time financial worker was assigned to each batch of the expert group to be responsible for the use and management of field funds.

14.5.2 Financial Management System

According to the relevant provisions of "the accounting standards for business enterprises", the project accounts were recorded according to the actual income and expenditure; at the same time, the management system of assets was established to strengthen the management of bulk materials. According to the relevant national laws, regulations and financial rules, the Ministry of Commerce supervised the use of project funds and inspect the completion of performance objectives and the effectiveness, etc. in each budget year. For projects crossing many years, a whole process evaluation would be conducted after completion. JIPD executed the performance evaluation suggestions raised by the Ministry of Commerce.

Further Reading

Wang XL, Zhu JM, Zhang DL (2013) China's foreign aid on human resources for health: experience and implications. J Africa Stud Dev 5(5):90–98. https://doi.org/10.5897/JASD2011.063

Yang K (2019) A new exploration into the model of China's foreign aid to Africa. China Invest 8:68–69

Yang K, Yang HT, Liang YS, Dai JR, Li W, Zhang JF, He J (2019) A path analysis on China's participation in global health governance: a case study of China aid of schistosomiasis control in Zanzibar. Chin J Schistosomiasis Control 1:14–18. https://doi.org/10.16250/j.32.1374.2019041

Chapter 15
Evaluation of China–Zanzibar Cooperation Project of Schistosomiasis: No-Stakeholder

Yu-Zheng Huang and Dirk Engels

Abstract To evaluate the effect of China-aid project of schistosomiasis control in Zanzibar from 2017, a non-stakeholder team visited the base in Zanzibar in May 2019. Evaluation activities include field and laboratory investigation, health education and information system, training and working ability of staff.

The evaluation team considered that the comprehensive control strategy based on integrative interventions has reduced the rate of human and intermediate snail infection, the project showed great effective in Pemba Zanzibar. These interventions provided a value option to the African countries as national strategy to control schistosomiasis haematobium.

Keywords Schistosomiasis control · Evaluation · Pemba Zanzibar · No-stakeholder

15.1 Background for Evaluation

Zanzibar is part of the United Republic of Tanzania, including Zanzibar and Pemba and nearby more than 20 small islands, covering an area of 2657 square kilometers. It is located on the xiangani Peninsula in the Central West Bank of Zanzibar Island. It used to be the economic and trade center of Zanzibar empire. With a population of about 1.3 million, 98% believe in Islam. The administrative region of Zanzibar is divided into five provinces, namely Zanzibar urban area and southern province, Zanzibar northern province, Zanzibar western province, Pemba northern province, and Pemba southern province. The climate is humid and hot, the annual average temperature is 25 °C, the annual average temperature difference is only 4 °C, the annual precipitation is 1500–2000 mm, and the seasonal distribution of precipitation

Y.-Z. Huang (✉)
Center of Global Health, Jiangsu Institute of Parasitic Diseases, Wuxi, Jiangsu, China

D. Engels
Uniting to Combat Neglected Tropical Diseases, Geneva, Switzerland

K. Yang, H. Mehlhorn (eds.), *Sino-African Cooperation for Schistosomiasis Control in Zanzibar*, Parasitology Research Monographs 15,
https://doi.org/10.1007/978-3-030-72165-7_15

is relatively uniform. From April to May and from November to December every year, the rainfall is too concentrated and the flood disaster occurs. Sometimes, hurricanes also hit Zanzibar, leaving large numbers of people homeless and destroying large areas of clove and coconut trees. Pemba Island is an agricultural Island famous for its cloves. The map of Pemba Island is like a full-term baby. The largest port is mkoani in the south, chakchak in the central city, and weti wete in the north.

Schistosomiasis is a global public health problem. As one of the local major public health problems, it seriously endangered the health of local residents in Africa. Zanzibar is located in east sea of East Africa. It is part of Tanzania and includes two islands, Unguja and Pemba islands. Zanzibar is highly endemic for schistosomiasis haematobium. In 2011, a survey of 24 schools showed that the infection rates were 8% (0–38%) and 15% (1–43%) in Unguja and Pemba islands. Despite several interventions to control the disease that have been conducted by Ministry of Health Zanzibar in collaboration with other international partners like SCI, SCORE, and others, more effort is needed to control and eliminate the disease due to high transmission.

In 2012, during the fifth Forum on China-Africa Cooperation (FOCAC), the Chinese government promised to increase its investment to African countries, expand cooperation areas including public health and disease control, and promote the sustainable development of Africa. On 21 May 2014, a tripartite memorandum of understanding was signed between China, WHO, and Zanzibar which was followed by the launch of the collaboration of schistosomiasis control in Zanzibar, a three-year (2017–2019) schistosomiasis elimination pilot project on the island of Pemba. It was expected to explore the strategy of schistosomiasis elimination in Africa through extending the Chinese experience in prevention and control of schistosomiasis within multilateral collaboration and to support China–Africa cooperation. The project is financed and technically supported by Chinese government, to explore localized strategy of schistosomiasis control. In August, 2016, Jiangsu Institute of Parasitic Diseases was formally committed to Zanzibar schistosomiasis control project ("the project").

The project lasted from January 2017 to January 2020. Its main tasks include: understanding the transmission pattern of schistosomiasis in Pemba Island; establishing comprehensive control systems of schistosomiasis in Pemba Island with assessment to pick up the optimized one; health education; capacity training for Zanzibar personnel to understand basic disease control knowledge; drafting up optimized interventions for schistosomiasis control and standard operation procedures in Pemba Island. China planned to send six batches of project teams, each with four Chinese health specialists to conduct work for 6 months. In December 2016, infrastructures including offices and laboratory started construction; on 15 February 2017, the first batch arrived in Pemba; currently, the third batch of team members is in work.

Till now, the project has been nearly half done. It is necessary for China–Zanzibar–WHO to conduct a comprehensive evaluation on the operational progress, expected effects of schistosomiasis control and research outcomes, as well as potential social impacts. Therefore, recommendations of improvements for the next stage could be made. In May 2019, a review of the project was requested by China Aid, to assess progress of the project, understand how the project would further need to evolve, assess the feasibility of an expansion to the whole of Zanzibar, and whether the project can serve as a model for schistosomiasis elimination in other countries in Africa.

15.2 Objectives and Methods

This evaluation will mainly focus on the following objectives:

- To review the establishment of a sustainable schistosomiasis control system in Zanzibar;
- To review the progress of the extension of Chinese experience in Africa;
- To review the construction of well-going health partnership.

It is proposed that the review adopts a strengths based approach (SBA). SBA will promote and facilitate self-reflection among project stakeholders about: their own advantages; what has worked well and what have been the factors for success; the shared objectives; the priorities for the next phase of collaboration; and steps which will help achieve agreed priorities. This approach is also culturally respectful and more likely to contribute to sustainable benefits over time.

15.3 Preparation Information for Evaluation

15.3.1 Review Team

The evaluation will be conducted by a team representing the participating/third-party countries, led by an independent evaluation specialist (team leader). The team leader shall have strong understanding of health sector of parasitic disease control, particularly in African context. He/She will provide leadership and oversight to the review team in the planning and completion of this evaluation. Team members will be expected to bring a mix of relevant knowledge and expertise. The team leader will be supported by a review team comprised of representation from various fields and each of the project partners/third-party countries.

The team leader will be responsible for developing the detailed evaluation plan, applying the review's principles, ensuring the review process is effective and efficient, and oversighting data collection, analysis, and report writing. The team members will be responsible for contributing to the development of the evaluation plan, data collection, analysis, and report writing.

15.3.2 Working Group

A working group for preparing the evaluation will be formed with representatives from each participating country, to inform the review, support its implementation and take responsibility for ensuring the review process is relevant, efficient, effective and inclusive.

The Project Management Unit (Project Manager in Zanzibar and China Coordinator) will be responsible for supporting the evaluation in various ways, having incorporated comments from partners into the evaluation. The director will provide administrative and operational support to the evaluation team, including for travel and accommodation; provision of documents; organization of meetings and events; and liaison with stakeholders where required.

15.3.3 Key Review Stakeholders

Project partners within Zanzibar, China and WHO are the primary stakeholders for the evaluation. The key project partners comprise the following stakeholders, e.g. Local people in Pemba Island; Local media; Local District Department of Health, Finance, Education, Environment, Agriculture, Water Conservancy, Security, etc.; Executive Officer of Southern and Northern Province in Pemba Island; Health Ministry of Zanzibar; Agriculture Ministry of Zanzibar; Local NGO partners; African Office of WHO; Tanzania and Zanzibar Office of WHO; Consulate General of China in Zanzibar; Project team members of first three batches; relevant Chinese media; relevant China pharmaceutical companies; China aid of Zanzibar Medical Team; etc.

15.3.4 Data Collection

A mix of methods is proposed, in order to achieve the review purposes and answer the key questions. Each evaluation method has strengths and limitations, so a mix of methods is useful for minimizing limitations and thus confirming key findings.

A range of suggested methods for consideration includes: review of project implementation documents and available reports (including data on treatment of schistosomiasis, snail control, training courses, participants assessments and other events); listen to the presentation of project team; Interviews with representatives from local government, WHO offices, etc.; Focus group discussions with project team members of the first three batches, local people, local media, relevant Chinese media, relevant China pharmaceutical companies, China Aid of Zanzibar medical team, etc.; Videos/photo-voice for gathering information from participants in remote locations; Collective sense-making workshop for stakeholders to analyze initial findings and contribute to a shared understanding of project progress and achievements to date. The interviews/focus group discussion could include but not limited to the following questions, as shown in Table 15.1.

15.3.5 Data Analysis

The data analysis will be an ongoing process throughout the review. The review team will consider findings on a daily basis, assessing the information gathered and how this contributes to the next day interview strategy and content. Considering the various nationality and professional background of review team members, it is expected that some different views and perspectives will emerge from data. The analysis will aim to examine why such differences have arisen and how they reflect the different interests and perspectives of partners. This analysis is also expected to lead to clearer identification of lessons learned and provide a detailed basis for recommendations about how the project should move forward. The timeline of evaluation in Zanzibar showed in Table 15.2.

15.3.6 Key Deliverable

The Review Team will need to take into account cultural and translation differences when drafting related documents. This may involve paying careful consideration to language which is culturally respectful and relatively easy to translate. The following documents will be submitted by the Review Team:

- Evaluation Plan (final)—approx. five to ten pages.
- Indicative Schedule (to enable meetings/travel/accommodation bookings)—approx. one to two pages.
- Review Report (draft and final)—limited to 30 pages.

15.4 Evaluation Approach and Findings

The review took a strengths-based approach (SBA), meaning that it facilitates self-reflection among project stakeholders about:

1. What has worked well and what have been the factors for success;
2. The priorities for the next phase of collaboration; and
3. Steps that will help achieve agreed priorities.

The following mix of methods was used, to achieve the evaluation purposes and answer key questions:

Table 15.1 The questions for interview by the evaluation experts team

	The establishment of a sustainable schistosomiasis control system in Zanzibar	The progress of the extension of Chinese experience in Africa	The construction of well-going health partnership
Project implementation	(1) What kind of changes have the project brought to local infrastructure and device on schistosomiasis control? How long will the change last? (2) What kind of skills have local specialist acquired? What kind of skills need to be further trained? (3) How are the system of working management and finance management? Which ones are reasonable and why? Which ones need to be improved and how? What kind of difficulties have you met in implementing the systems? (4) Which medicines ad devices have been put into use? How are the effects? How are the prospect of their use in the future? Which medicine or devices need to be further introduced into the project?	(1) How is Chinese management system and experience used in the project? (2) How is Chinese disease control technologies used in Zanzibar project? What kind of adaptions need to be made? (3) How is the Chinese training pattern with Zanzibar health specialists? How do this pattern adapt with their educational level and local culture? (4) How are Chinese health technologies, medicine and device received by local people?	(1) What are the essential factors that contribute to the health partnership? (2) What experiences about partnership could be gained from this project? (3) What improvement in capacity was made for participants? (4) How could the cooperation of this project be further enhanced?
Health education	(1) What is the need of local people for health education as reflected by KAP investigation? (2) What kind of health education has been conducted and what is their respective advantage and limitation? How to improve in the future? (3) What has changed of the local people after receiving health education?	(1) What language barriers exist in health education? How do they affect health education? How are they solved? (2) What cultural difference exist in health education? How do they affect health education? How are they solved?	(1) How does China-Zanzibar partnership cooperate in health education? (2) What challenges is such cooperation pattern faced with? How to improve it in the future?

(continued)

Table 15.1 (continued)

	The establishment of a sustainable schistosomiasis control system in Zanzibar	The progress of the extension of Chinese experience in Africa	The construction of well-going health partnership
Science investigation	(1) How are schistosomiasis epidemics in Pemba Island? What are its spatial and temporal distribution patterns? (2) What are the spatial and temporal distribution patterns of the water bodies in Pemba Island? What are the spatial and temporal distribution patterns of snails in Pemba Island? (3) What is the effect of molluscacide in laboratory? What is the effect of molluscacide in the field? If the effect is not satisfactory, what is the reason and how could it be improved? If the effect is satisfactory, how is the molluscacidal standard operation procedure established? (4) How is Bulinus reproduced in laboratory? How is the biological study of Bulinus going on in laboratory?	(1) What are the similarities and differences between the schistosomiasis epidemics in Pemba Island and in China? How will they affect control strategies? (2) What are the similarities and differences between the snail situation in Pemba Island and in China? How will they affect control strategies? (3) What are the similarities and differences between the molluscacide application in Pemba Island and in China? (4) What are the advantages and limitations between Chinese praziquantel and other praziquantel in Zanzibar application?	(1) How do China/Zanzibar/WHO cooperate in research in the project? (2) How could research capacity be enhanced through cooperation?
Social impact	(1) What is the impression of local government/people/partner international organization and NGO about China? (2) What is the impression of local government/people/partner international organization and NGO about Zanzibar schistosomiasis control project? (3) What do Chinese people know about the project? How do they think of it?	(1) What have China/Zanzibar/WHO learned from the project?	(1) What impact has the project had on the cooperation between China and Africa?

(continued)

Table 15.1 (continued)

	The establishment of a sustainable schistosomiasis control system in Zanzibar	The progress of the extension of Chinese experience in Africa	The construction of well-going health partnership
	(4) What does international organization/ NGO know about the project? How do they think of it?		

Table 15.2 The timeline of evaluation in Zanzibar

Date	Activity	Location	Milestone
5 March–1 May	Draft the mid-term evaluation plan of Zanzibar schistosomiasis control project	China and Zanzibar	Plan draft
20 April–2 May	Confirm the member list of review team and working group	China/Zanzibar/WHO	Name list
25 April–4 May	Teleconference for evaluation team members to discuss indicative evaluation schedule; team leader to draft schedule and plan to working group	China/Zanzibar/WHO	Evaluation plan skampskamp; schedule
3–5 May	Confirms travel/accommodation/meetings/ bookings based on evaluation group feedback on schedule	China/Zanzibar/WHO	N/a
5 May–8 May	Evaluation team conduct field visit, data collection, stakeholder's interview, and other material/ reports as identified in the evaluation plan.	Zanzibar	N/a
8 May–10 May	Data analysis, clarification, report drafting and submission.	China/Zanzibar/WHO	
10 May–12 May	Finalise report incorporating comments from all working group members	China	

- Review of project implementation documents and available reports;
- Interviews with policy and technical partner representatives, programme management and coordination personnel, facilitators (trainers and researchers) and participants in project activities in Zanzibar NDT and China team;
- A collective workshop for stakeholders to analyze initial findings and contribute to a shared understanding of project progress and achievements to date;
- Field visits and discussions with managers and staff of participating institutions (in selected locations);

15.4.1 Project Documents and Reports Review

The Evaluation Team reviewed the progress reports prepared by the China-Zanzibar Project Team. Most of the data was on epidemiological surveys done at baseline and subsequent follow up surveys. The project has been carried out in seven shehias, four intervention shehias (Mtangani, Wingwi, Kiyuyu, and Uwandani—the latter also serving as demonstration site), and three control shehias (Wambaa, Vitongoji and Shengejuu). Intervention and control shehias were stratified according to prevalence of infection at base line, showed in Fig. 15.1.

The populations in all seven shehias—intervention and control—as in all the endemic areas of Pemba, received praziquantel through mass drug administration twice a year, in April and November, as part of the NTD Control programme. In addition to this, in the four intervention shehias, project activities consisted of:

- Baseline survey of infection rates in humans, snail survey, and knowledge, attitudes and practice (KAP) survey, with regular surveillance thereafter.
- Snail control with niclosamide-metaldehyde in all water bodies where snails were found at baseline and/or during surveillance checks.
- Treatment with praziquantel of all those found infected during epidemiological surveys (test and treat).
- Health education and interventions to solicit behavioural change.

In the intervention shehias, the use of snail control, the test and treat approach and health education resulted in a greater overall reduction in prevalence of schistosomiasis infection at 18 months than in the control shehias where only 6-monthly mass drug administration was used. The beneficial effect was more marked in the high transmission shehias as compared to those at lower baseline transmission levels where only a slight difference in reduction of prevalence was observed. Additional data reviewed concerned epidemiological data for baseline and 6 months post-treatment, as well as water contact sites or transmission zones.

It is tempting to think that the baseline prevalence of schistosomiasis infection in the intervention shehias is related to the number of water bodies in each community, but probably there is need to know more about water-contact behavior of the population around these water-contact sites as well as he degree of human contamination and its interaction with snail density to produce cercarial risk in these ponds. In comparing the follow up data Tables 15.1 and 15.2, it is also tempting to conclude that the impact in the intervention shehias achieved 6 months after base line could be maintained throughout 18 months.

Because there were so many ($n = 30$) ponds in Uwandani, it was also considered a demonstration site for test and treat and snail control.

With a population of 2502 people in Uwandani, population compliance with project activities, especially urine surveys are particularly high, varying from 79% to 99.9%. It is of interest to have more details on the people diagnosed with schistosomiasis after each survey. It is not clear whether these are from clusters or families, and which age-groups are most infected. It may be entirely possible that they are

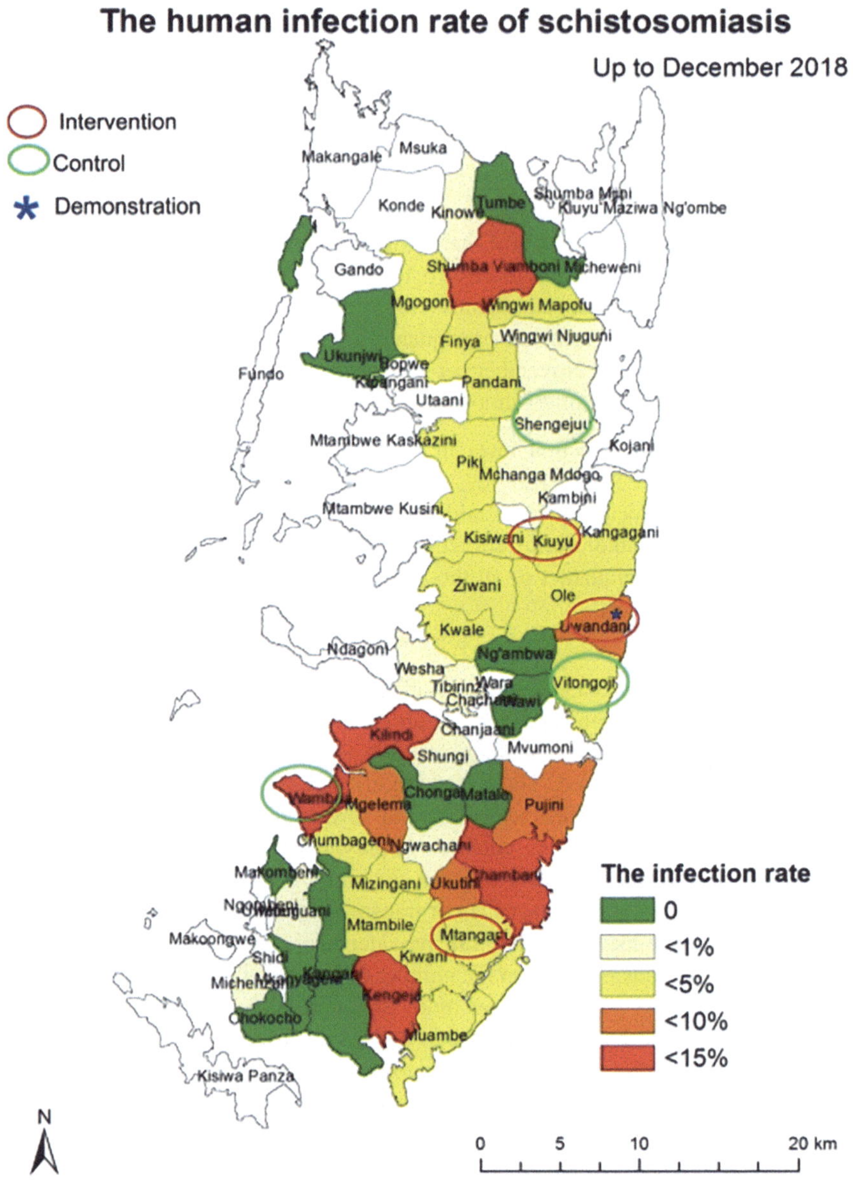

Fig. 15.1 Map of Pemba, indicating intervention and control shehias. Red cycles: Intervention shehias; Green cycles: Control shehias

Fig. 15.2 The snail endemic in Pemba Zanzibar

completely different people at each survey. More detailed "micro-epidemiological" analysis could possibly elucidate water-contact behavior, hot-spots for transmission, and risk factors. Presumably, the screening and treatment coverages attained in Uwandani are achieved in the other intervention shehias as well.

The impact of interventions, reducing prevalence of infection below 1%, in the intervention shehias over the 2-year period of the project are the best that have ever been achieved on the island.

While during this evaluation the evaluation team did not review data on the snails, it was informed that following application of molluscicides to ponds and streams, there are no snails found for at least 6 months. The project has two dedicated teams, for snail surveys and snail control respectively. The evaluation team visited the snail laboratory where various studies are on-going. The use of loop-mediated isothermal amplification (LAMP) to detect parasite DNA in pooled snails (Fig. 15.2) could facilitate detection of infected snails and differentiation of *S. haematobium* and *S. bovis* cercariae.

The diagnostic laboratory is well maintained as shown in Fig. 15.3, and the staff are well trained to detect *S. haematobium* eggs. Given the likely low intensity infections, it may be useful to use Lugol's Iodine to stain the parasite so that they are more visible. In such a low transmission area, it may be that pooled urine samples, with an appropriate diagnostic technique such as LAMP could help detect hotspots of transmission.

The evaluation team had the opportunity to see health education materials. Of interest, were messages painted on houses in intervention communities. These had been agreed with the homeowners and community leaders. It was clear from speaking with secondary school students and community leaders that there was

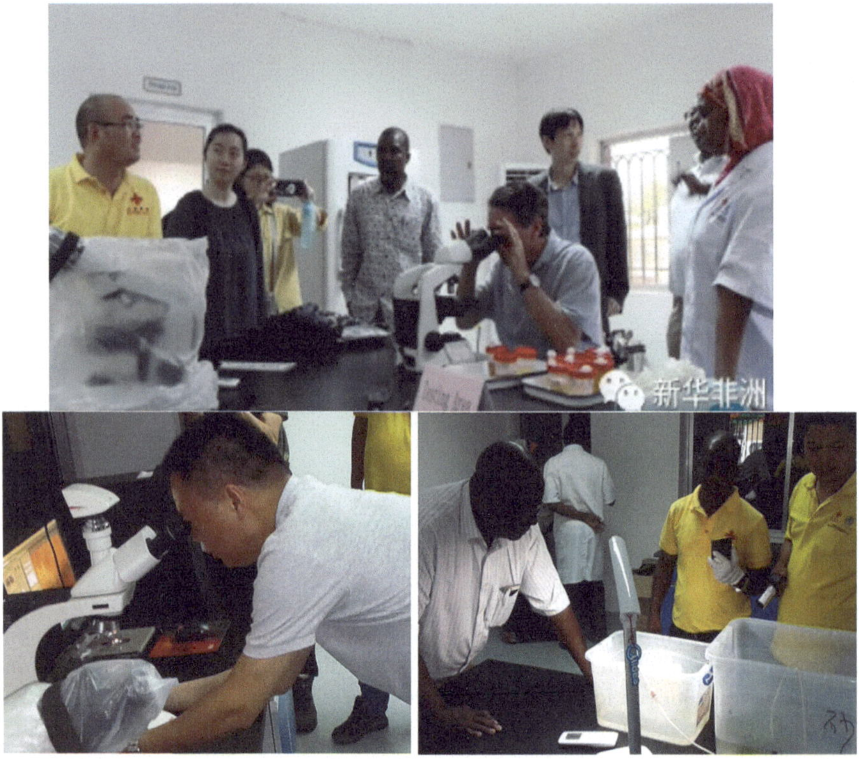

Fig. 15.3 Evaluation of laboratory capacity in Pemba Zanzibar

Fig. 15.4 Health education for local students in Pemba Zanzibar

high awareness of schistosomiasis and of the control activities taking place. There were also notices and warning messages near transmission sites (Fig. 15.4).

This evaluation took place during the long "rains", which made it possible to appreciate the amount of water bodies on Pemba Island. Data on the number of water bodies in study villages has been given above in Table 15.2. Access to water was

Fig. 15.5 Field investigation in Pemba Zanzibar

discussed with the project team and the evaluation team was able to see that piped water is present if not accessible to all in the visited areas. Clearly people continue to have at-risk contact with potentially infested waterbodies because the available potable water may not meet all the domestic requirements, especially not washing and bathing.

Access to water was discussed in the context of multisectoral collaboration with agricultural and water authorities. There was no documentation as to the strength of such collaboration. Involvement of other sectors could result in different crops that could use more water or result in the drainage of standing water, and thus reducing the transmission potential of water bodies (Fig. 15.5).

15.4.2 Field Visits and Local Facility Investigation

The second full day of the evaluation team on Pemba began with a visit to Uwandani shehia. During the visit, the molluscicide application procedure was demonstrated at a large pond. A liquid formulation of niclosamide was used as contrasted with the powder form used recently by the SCORE project in Pemba. In addition to the snail surveys, some water samples are collected for analysis to establish the niclosamide concentrations (Fig. 15.6).

After the molluscicide demonstration the team visited the project base station in the Mkoroshoni-Chake district which provides renovated office space for both the Chinese team and local staff. This is also the site of the laboratories for urine analysis of human infections and for the culture of Bulinus snails. These facilities, including computer facilities for data storage and analysis, appear to be well-suited for the needs of the project to date.

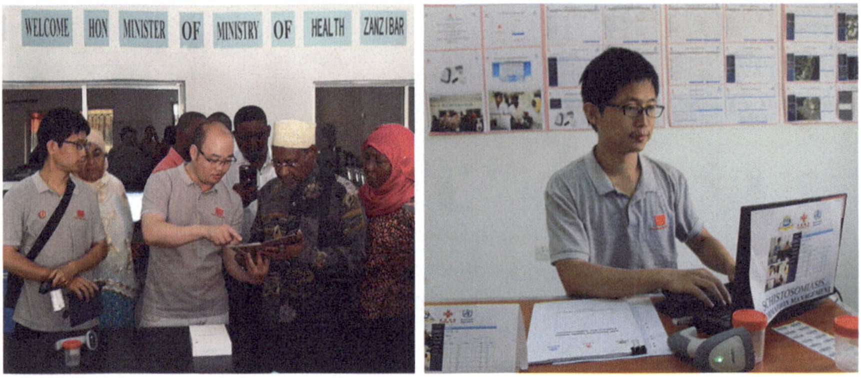

Fig. 15.6 Information system of schistosomiasis control

Fig. 15.7 The workshop for the schistosomiasis control in Pemba Zanzibar

The following morning the team visited the Kiuyu shehia and met with community leaders to gain their perspective on the project. The meeting was well-attended. After the meeting with community leaders the evaluation team had the opportunity to assist in a health education session in a secondary school class (Fig. 15.7).

15.4.3 Workshop for Schistosomiasis Elimination

A collaborative workshop took place on the day 3 of the mission, on Pemba island. Dr. Yang Kun from the Jiangsu Institute of Parasitic Diseases, P.R. China, gave a summary of the pilot project achievements on behalf of China, while Dr. Fatma Kabole, Head of the Zanzibar NTD Programme and Mr. Saleh Juma gave an account of the achievements on behalf of Zanzibar.

The two presentations were followed by a vivid discussion with constructive suggestions for a potential extension of the project.

15.4.4 Interviews with Representatives from China and Zanzibar NDT

To assess the effectiveness of the project through interviews with these representatives from China and Zanzibar NDT to get recommendations and feedback.

15.4.4.1 Minister of Health

The Minister of Health highlighted the long-term relationship between China and Zanzibar that has started as early as 1965 and thanked the People's Republic of China for its continued engagement. He also thanked the World Health Organization for having initiated the current tripartite collaboration project which is the first of its kind and may provide a model for similar collaborations elsewhere.

He expressed satisfaction with the contribution of the current schistosomiasis elimination pilot project towards the renovation of infrastructure, the building of Zanzibari capacity and the practical approach of pushing towards an impact on transmission. He expressed appreciation for the way the Chinese teams have integrated with their national counterparts and have adjusted to the local situation and culture.

Zanzibar has a long-standing commitment towards controlling and eliminating Neglected Tropical Diseases or NTDs, namely lymphatic filariasis, schistosomiasis and soil-transmitted helminthiasis, and more recently trachoma and snakebite envenoming. Control of morbidity due to schistosomiasis started in the early 1980s and has yielded good results. However, to have a more lasting impact, a more comprehensive approach aimed at reducing transmission was needed. In this respect, the contribution by China with expertise gathered at home in eliminating schistosomiasis was well appreciated.

The results of the pilot project speak for themselves. A reduction in infection rates of affected populations from as high as 8% to less than 1% in a short period of less than 2 years is impressive. The Minister expressed hope and expectation that based on such results the project would be expanded to the whole of Zanzibar–Pemba and several remaining "hot spots" of schistosomiasis transmission in Unguja—and become a model of what could be achieved also in other countries in sub-Saharan Africa.

He nevertheless expressed the wish that during such an extension of the project due attention be given to sustainability to allow the Zanzibar team to take over and ensure maintenance of the achievements beyond the end of the project. Zanzibar is economically doing well with an annual growth of around 7%, increasing the likelihood that it will be able to take charge of such maintenance.

He also expressed the wish that during a potential extension due attention is given to collaboration with other sectors such as water, education and local administration with a view of decentralizing activities into primary health care and local governance. He also asked China for specific advice on how to limit the risk of schistosomiasis while intensifying rice culture, as Zanzibar wishes to become as self-sufficient as possible in terms of rice production in the coming years.

He finally made the case for a practical research component to accompany the implementation programme. Zanzibar has recently created a National Research Institute that is to coordinate and oversee all research institutes and initiatives in the country.

15.4.4.2 Deputy Minister of Health

The Deputy Minister echoed the general message articulated earlier by the Minister that there was considerable satisfaction about what had been achieved by the project to date. She indicated that the model of the project, with interaction, training and participation of local staff was very unusual for other externally funded projects focused on disease control. Moreover, she was clear about the need for the second phase of the project, if funded and approved, to develop a clear plan for the transition from reliance on substantial Chinese assistance in the field, the laboratory and in data management to a sustainable operation dependent on local personnel and resources.

Although that had gone very well to date, she expressed some concern that developing an action plan for Zanzibar from the perspective of the very much larger Chinese schistosomiasis experience presented a continuing challenge to a successful transition. Given the low prevalence levels now being achieved, she also stressed the importance of education of the various constituencies in Zanzibar to the importance of sustained schistosomiasis control programs. Continued education was likely to be needed from the level of those at risk to those at higher levels whose fiscal and institutional support would be necessary in the long term.

The Deputy Principal Secretary expressed gratitude that the Chinese government chose to collaborate with the Zanzibar government on the control and elimination of schistosomiasis. This follows on the long-term collaboration of the Chinese Medical Missions to Zanzibar to provide clinical care as well as to control malaria. The Chinese government had also built a hospital at Mkoani.

The project has resulted in the building of offices and laboratories for diagnosis of schistosomiasis infection and work on several aspects of snails. New Chinese medicines (praziquantel) and molluscicides are being imported. Their importation and use in Zanzibar were subject to approval by the regulatory authority. Approval included visits to factories in China to examine manufacture of the active ingredients of the chemicals and medicines to be used in Zanzibar.

The Zanzibar government has supported the project financially and by seconding staff. The WHO has also supported the Zanzibar staff with in-country and regional training.

There have been five teams of Chinese experts seconded to the project from Jiangsu Province. The Zanzibar government seconded public health staff members to work with the Chinese experts. Several Zanzibari staff members have been trained in China and on the job on use of molluscicides, diagnosis of schistosomiasis, and on preventive chemotherapy. The Chinese and Zanzibari teams are working and collaborating well.

The fifth Chinese team has very young experts, but they seem willing to learn about things in Zanzibar.

With respect to health education and community participation, things are now working well. There had been some confusion with respect to messages in Kiswahili on every other house. The communities had first objected but understood the need for the messages once they were explained. Similarly, alarms warning people from transmission sites were removed by community members. There is more engagement with local leaders. They are supportive of interventions and have expressed appreciation of the project.

The DPS appreciated the progress being made in the project areas. Epidemiological and malacological results are very encouraging. There has been a significant impact on prevalence of infection, as well as on community behaviour. She expressed interest in the expansion of the project to the rest of Pemba Island as well as to Unguja.

She also thanked the WHO for encouraging the tripartite agreement. The success of the project is appreciated at the highest level of the Zanzibar government.

Mr. Yacoub Shoka, Operation Officer, Department of Preventive Services.

Mr. Shoka expressed appreciation for the Chinese intervention that has made a big difference. A 5-year extension would be most welcome, for both Pemba and Unguja islands, with the understanding that most of the work and challenges will reside in Pemba. In such an extension provision must be made for further provision of equipment and renovation of infrastructure, even though the needs may be less than during the pilot phase. The need for training and capacity building, on the other hand, will remain, if not increased.

During a potential extension, more attention needs to be given to collaboration with other sectors to permanently lower the risk of infection, and from the start of an extension one needs to think of sustainability. A model for this can be the Zanzibar malaria programme that currently is already in its maintenance phase.

Mr. Shoka also mentioned that soon a research section of the NTD programme will be created.

15.4.4.3 Coordinator

Some challenges relating to language remain even though Chinese Team members help each other communicate with the local citizens. The 6-monthly rotation of Chinese experts means that they return to china just when they are beginning to master the situation and work.

Saleh Juma has been to China several times, for training on schistosomiasis and laboratory techniques. He has also participated in international meetings to advocate for and present on the China-Zanzibar control and elimination project.

He believes the conditions of service are fine, and the Government of Zanzibar is meeting its obligations by providing staff—22 personnel—and making sure that their needs are met. The government supervision of the project through the Directorate of Preventive Health Services under Drs Fadil Abdallah and Fatma Kabole in Unguja and Yacoub Shoka on Pemba, is good. Expansion of the project to the whole of Pemba would require an additional 15–20 staff, and the government is supportive of the extension and would avail the requirements for them. It would also require the registration of Chinese made praziquantel and molluscicides, which are only used in the project shehias.

The Coordinator and Team Leaders can access data and generate reports. They also have access to satellite maps.

In terms of dealing with communities, this is through leaders at shehia or ward (diwani) level. Shehias select Community Drug Distributors. Other health problems encountered in the communities are reported directly to the Directorate of Preventive Health Services through Primary Health care structures.

15.4.4.4 Salim Ali Khatib, Team Leader, Snail Survey Team

Salim leads a four-person team and joined the project in 2014 from the PHC department of the MOH. The team is responsible for snail surveys all over Pemba and in the four project shehias. The team also participates in the health education interventions as required. Another team deals with snail control through mollusciciding.

He was trained on Environmental Health in Tanga and at the Tropical Pesticides Research institute in Arusha. Since joining the project, he has undertaken training in the use of molluscicides at the Jiangsu Institute of Parasitic Diseases in China.

Duties of the snail survey team are to conduct surveys across Pemba. Snail surveys take place at any site for 15 minutes. If snails are found, further searches are done at least 100 metres in either direction from the site. Surveys are done every month, and if molluscicides are used, repeat surveys are done 2 weeks after.

Reporting of results is done through the Coordinator, Saleh Juma, even though Team leaders can also report to the Chinese Team Leaders.

The working relationship with the Chinese team is good, and there are no problems with respect to logistics for work. However, some challenges include lack of transport to and from work. The respondent lives at least 7 km from work. Pay could be improved as those working directly in the MOH receive higher pay than those seconded to the China-Zanzibar project. More office space is also needed, and teams want to be able to access data on snail surveys.

For expansion to the project, between 15 and 20 additional staff would be needed for the snail survey team, along with requisite training, transport and pay.

15.4.4.5 Local Staff Responsible for Data Management

The issue here concerned the diversity of data being generated by different segments of the team and its accessibility for subsequent analysis and to management for

Fig. 15.8 The team of local staff and China experts

further analysis and integration. The responsibility of the two data analysts interviewed is to input field and laboratory data into the computer system which is then accessible to both those who generated the data and to management as required. These data range from field reports of snail density by location to laboratory data of various sorts. There appears to be at least an elementary level of ability to map the spatial characteristics of the data. The systems and software now in place seem well suited to the needs of the project and the staff seemed clear about their responsibilities. Mr. Saleh evinced satisfaction with data related issues from a management perspective. Mr. Saleh further clarified some detailed issues of the project to date. His broad perspective combined with detailed knowledge of the project and enthusiasm for its continuation was impressive. He presented a useful summary of the organization of local government but left some questions as to the linkage of control activities with the existing local infrastructure of Ministry of Health. This should be clarified in the context of the eventual transition of the project to long term local operation and management (Fig. 15.8).

15.5 Results of Evaluation

15.5.1 Shehias Level

The impact of interventions, reducing prevalence of infection below 1% in the intervention shehias over the 2-year period of the project is the best that has ever been achieved on the island. Continued intervention may result in the interruption of transmission.

The comparative SCORE study was conducted twice a year MDA, although behavioral interventions and local mollusc killing were alike. But ChinaZanzibar project could remove snails from the environment for at least 6 months, and across transmission seasons. it is likely that the difference with the China-Zanzibar project results is the robust molluscplay regime.

15.5.2 *Africa and International Level*

The work carried out during this pilot project has contributed to the establishment of WHO guidelines for laboratory and field testing of molluscicides for the control of schistosomiasis.

(Dr Rajpal Singh Yadav (2019), Guidelines for laboratory and field testing of molluscicides for control of schistosomiasis. https://www.who.int/schistosomiasis/resources/9789241515405/en/)

The tripartite collaboration between the Zanzibari Government, the Chinese Government and WHO appears to have worked well. The Chinese teams have collaborated well with their Zanzibari counterparts. The evaluation team has been able to see that Chinese expertise has been adapted to the African context with a different form of schistosomiasis and intermediate snail host. The Chinese teams have also been respectful of the Zanzibar culture and traditions. As such, the successful tripartite collaboration can indeed become a model for other schistosomiasis-endemic countries in Africa to benefit from the long-standing Chinese expertise with this disease in the future.

Nowadays, poor infrastructure, too single economies and high unemployment are the status quo in many African countries. As China's "Belt and Road" initiative has been responded to and welcomed in African countries along the route, more and more Chinese enterprises have come here with projects and funds to help the local economy develop and improve people's livelihood. The improvement of the people's level can also help the local people to solve the water problem of tap water and improve public health, and help to eliminate schistosomiasis. At the same time, China's "Belt and Road" initiative to strengthen cooperation and between China and Africa, will further contributes to the promotion and implementation of the project.

15.6 Discussion and Recommendation

15.6.1 *Extension of Project*

15.6.1.1 Material Point of Extention

The results attained in the intervention shehias should be consolidated by a stepwise extension to rest of Pemba Island, and later to the remaining schistosomiasis endemic areas on Unguja.

Regular, large scale mollusciciding during an expansion of the project must, in its early stages, be accompanied by an environmental impact assessment to detect and mitigate potential negative consequences.

Start preparing relatively early for a maintenance phase that could be run by a Zanzibari team to maintain schistosomiasis elimination after the tripartite project has finished. This entails the keeping of detailed records on interventions providing a means for rigorous follow up and analysis of data, the establishment of standard operating procedures for surveillance and response to potential resurgence, as well as thorough planning of resources, staff and capacity.

15.6.1.2 Extension of Cooperation

When moving forward with an expansion, there needs to be close collaboration with other sectors and three sectors are particularly important in this respect:

- There is first the agricultural sector, that could be consulted to explore if improved water storage, use and channeling for agricultural practices could improve crop production, such as rice, while limiting the risk for schistosomiasis.
- Then there is the sector of water resources, that could be solicited to make available larger quantities of safe water—potentially including collection and storage of rain water—for activities such as washing and bathing beyond what is currently available for drinking and cooking.
- Thirdly there is local governance and community involvement. Village health committees are keen to closely collaborate with the project, and close links with local governance has been a factor of success in schistosomiasis elimination in China.
- Lastly, livestock and fisheries should be consulted to find out about presence of animal schistosomes such as *S. bovis* in Zanzibar.

15.6.2 Further Development of the Project

As the project will gradually move towards low transmission of schistosomiasis, more innovative and sensitive diagnostic techniques should be used to detect schistosomiasis infection in humans and snails. With respect to the detection of parasite eggs following urine filtration, staining the slide preparations with Lugol's iodine could improve the diagnostic sensitivity in the short term. This was the SOP under the SCORE project. Pooling of urine samples from several individuals or pooling of snails from several sites on a waterbody and detection of parasite DNA, offer means to more efficiently detect hotspots of transmission. Use of similar techniques for parasite detection in environmental water samples should be explored.

The manufacturer's authorised representative in the EU is Springer Nature Customer Service Centre GmbH, Europaplatz 3, 69115 Heidelberg, Germany. If you have any concerns regarding our products, please contact ProductSafety@springernature.com

Printed and bound by CPI Group (UK) Ltd, Croydon, CR0 4YY
30/09/2026
02225692-0008